The Arboviruses: Epidemiology and Ecology

Volume II

Editor

Thomas P. Monath, M.D.
Director
Division of Vector-Borne Viral Diseases
Centers for Disease Control
Public Health Service
U.S. Department of Health and Human Services
Fort Collins, Colorado

CRC Press
Taylor & Francis Group
Boca Raton London New York

CRC Press is an imprint of the
Taylor & Francis Group, an informa business

CRC Press
Taylor & Francis Group
6000 Broken Sound Parkway NW, Suite 300
Boca Raton, FL 33487-2742

Reissued 2019 by CRC Press

A Library of Congress record exists under LC control number:

Publisher's Note
The publisher has gone to great lengths to ensure the quality of this reprint but points out that some imperfections in the original copies may be apparent.

Disclaimer
The publisher has made every effort to trace copyright holders and welcomes correspondence from those they have been unable to contact.

ISBN 13: 978-0-367-23536-9 (hbk)
ISBN 13: 978-0-367-23537-6 (pbk)
ISBN 13: 978-0-429-28024-5 (ebk)

Visit the Taylor & Francis Web site at http://www.taylorandfrancis.com and the CRC Press Web site at http://www.crcpress.com

FOREWORD

The term "arbovirus" is used to describe a diverse array of viruses which share a common feature, namely transmission by arthropod vectors. This ecological grouping now includes over 500 viruses, most belonging to five families — the Togaviridae, Flaviviridae, Bunyaviridae, Reoviridae, and Rhabdoviridae. Over 100 of these agents have been associated with naturally acquired disease in humans and/or domestic animals, and among these approximately 50 of the most important pathogenic viruses have been selected for detailed review under this title.

The complexity of arbovirus ecology requires a fundamental understanding of the influence of each of the multiple components (virus, vector, viremic host, clinical host, and environment) on infection and transmission cycles. The first volume, devoted to the variables which affect arbovirus transmission, provides this background and also contains guidelines for the design of future epidemiological investigations. Armed with these general principles, the student, teacher, or research worker will be able to structure specific knowledge about individual arbovirus infections found in Volumes II through V.

Recent textbooks are available which provide comprehensive coverage of the clinical aspects, pathogenesis, virological characteristics, and molecular biology of arboviruses. The intent of this book is different, for it focuses on the epidemiology and ecology of the arboviruses, the risk factors underlying the appearance of disease in the community, and the roles of arthropod vector and vertebrate hosts in virus transmission. Emphasis is placed on the field and laboratory evidence for involvement of vector and host species and on the ecological dynamics which determine their ability to spread infection. Elements of transmission cycles which are susceptible to surveillance, field investigation, prevention, and control are elucidated.

A number of arboviruses which have caused human disease only on rare occasions are not included in the book or are mentioned in passing within chapters on related diseases. Although these viruses (for example, Spondweni, Ilheus, Rio Bravo, Usutu, Orungo, Wanowrie) are inherently interesting and may, with changing ecologic conditions, turn out to be medically important, little or no information about their epidemiology (insofar as it relates to clinical hosts) is available. Finally, the scope of this book has been limited strictly to arthropod-borne infections and other viruses sometimes considered under the aegis of arbovirology (e.g., rodent-borne viral hemorrhagic fevers) are not included.

The compilation of a book of this scope required sacrifices in time and energy by a large number of contributors, all of whom faced multiple other commitments. This sacrifice will, I expect, be partially compensated by the availability of a useful compendium of collective knowledge.

Thomas P. Monath
October 1986

THE EDITOR

Thomas P. Monath is Director of the Division of Vector-Borne Viral Diseases, Centers for Disease Control, and is an affiliate faculty member of the Department of Microbiology, College of Veterinary Medicine, Colorado State University.

He received his undergraduate and M.D. degrees from Harvard University and his clinical training in Internal Medicine at the Peter Bent Brigham Hospital, Boston. In 1968 he joined the U.S. Public Health Service, serving as Medical Officer in the Arbovirology Unit, Centers for Disease Control, Atlanta, and later as Chief of the Arbovirus Section. Between 1970 and 1972, he was assigned to the Virus Research Laboratory of the Rockefeller Foundation, University of Ibadan, Nigeria, where he conducted field research on the epidemiology of yellow fever and Lassa fever. Since 1974, Dr. Monath has been Director of the Division of Vector-Borne Viral Diseases, Fort Collins, Colorado. In 1984 — 1985 he spent a sabbatical year in the Gastroenterology Unit of the Massachusetts General Hospital.

Dr. Monath is a Fellow of the American College of Physicians, the Infectious Disease Society, and the Royal Society of Tropical Medicine and Hygiene. He is a member of the American Society of Virologists, the American Society of Tropical Medicine and Hygiene, and the Association of Military Surgeons. He serves on the Editorial Boards of the *American Journal of Tropical Medicine and Hygiene*, *Acta Tropica*, and the *Journal of Virological Methods*. Dr. Monath is a member of the Committee on Research Grants, Board of Science and Technology for International Development, National Research Council, and is currently Chairman of the AIBS Infectious Diseases and Immunology Peer Review Panel to the U. S. Army Medical Research and Development Command. He is a member of the World Health Organization Expert Committee on Virus Diseases and the Pan American Health Organization Scientific Advisory Committee on Dengue, Yellow Fever, and *Aedes aegypti*. He has served as Chairman of the Executive Council of the American Committee on Arthropod-Borne Viruses, as a Councilor of the American Society of Tropical Medicine and Hygiene, and as a member of the Directory Board, International Comparative Virology Organization and the U. S.-Japan Cooperative Medical Research Program Panel on Virus Diseases.

Dr. Monath has authored or coauthored over 140 scientific publications in the field of virology and is editor of the book, *St. Louis Encephalitis*, published by the American Public Health Association. His main research interests are the ecology, epidemiology, and pathogenesis of arbovirus infections.

TO THE MEMORY OF MY PARENTS

CONTRIBUTORS

G. Stephen Bowen, M.D., M.P.H.
Deputy Director, AIDS
Center for Prevention Services
Centers for Disease Control
Atlanta, Georgia

Alain-Jean Georges, M.D.
Director
Department of Microbiology
Institut Pasteur
Bangui, Central African Republic

E. Paul J. Gibbs, B.V.Sc., Ph.D.
Professor of Virology
Department of Infectious Diseases
College of Veterinary Medicine
University of Florida
Gainesville, Florida

Jean Paul Gonzalez, M.D., Ph.D.
Laboratory Chief
Department of Virology
Institut Pasteur
Bangui, Central African Republic

Ellis C. Greiner, Ph.D.
Associate Professor
Department of Infectious Diseases
College of Veterinary Medicine
University of Florida
Gainesville, Florida

Paul R. Grimstad, Ph.D.
Associate Professor and Director
Laboratory for Arbovirus Research and Surveillance
Department of Biological Sciences
University of Notre Dame
Notre Dame, Indiana

Duane J. Gubler, Sc.D.
Chief
Dengue Branch
Director
San Juan Laboratories
Division of Vector-Borne Viral Diseases
Centers for Disease Control
San Juan, Puerto Rico

William R. Hess, Ph.D.
Microbiologist
USDA, Agricultural Research Service
Plum Island Animal Disease Center
Greenport, New York

Harry Hoogstraal, Ph.D. (Deceased)
Senior Scientist
U.S. Naval Medical Research Unit
Cairo, Egypt

Peter G. Jupp, M.Sc., Ph.D.
Arbovirus Research Unit
National Institute for Virology
Johannesburg, South Africa

Thomas G. Ksiazek, Ph.D., D.V.M.
Veterinary Microbiologist
Department of Epidemiology
USAMRIID, Fort Detrick
Frederick, Maryland

Kenneth J. Linthicum, Ph.D.
Chief
Department of Arboviral Entomology
USAMRIID, Fort Detrick
Frederick, Maryland

Bruce M. McIntosh, D.V.Sc., D.Sc.
Arbovirus Research Unit
National Institute for Virology
Johannesburg, South Africa

Harry Standfast, D.Sc.
Senior Principal Research Scientist
Division of Tropical Animal Production
Commonwealth Scientific and Industrial Research Organisation
Brisbane, Australia

Toby D. St.George, D.V.Sc.
Chief Research Scientist
Department of Tropical Animal Science
Commonwealth Scientific and Industrial Research Organization
Brisbane, Australia

Douglas M. Watts, Ph.D., M.S.
Microbiologist
Department of Pathogenesis and Immunology
USAMRIID, Fort Detrick
Frederick, Maryland

TABLE OF CONTENTS

Volume II

Chapter 14

AFRICAN HORSE SICKNESS

William R. Hess

TABLE OF CONTENTS

I. HISTORICAL BACKGROUND

A. Discovery of Agent and Vectors

In a report published in 1921, Theiler[1] stated that at least seven enzootics of African horse sickness (AHS) had occurred in the Cape Province during the period from 1780 to 1918. Prior to that time, there were no horses in South Africa, and the disease was not known there or elsewhere. The disease emerged only after horses from Europe and the East Indies were imported to South Africa. It was not immediately recognized as a separate disease entity. Because it had some clinical features similar to anthrax and piroplasmosis, it was often mistaken for those diseases. Various bacteria, parasites, and even a fungus were suggested as possible causative agents of AHS. However, in 1900, M'Fadyean[2] succeeded in transmitting the disease with a bacteria-free filtrate of blood from an infected horse. These findings were confirmed by others 1 year later,[3-5] and it was generally accepted that the causative agent was a virus.

When it was realized that the disease was not spread by mere contact, it was correctly assumed that the virus (AHSV) was probably transmitted by a blood-sucking arthropod.[6] The list of suspected vectors was enormous. Mechanical transmission by *Stomoxys calcitrans* was demonstrated.[7] However, its importance as a transmitter was discounted because it was a daytime feeder, and it was found that horses confined to mosquito-proof stables at night were protected against infection.[6] Because of their prevalence during an outbreak of AHS in the Sudan, *Lyperosia minuta* was regarded as a possible transmitter.[8] Other investigators suggested that certain ticks and several members of the order Diptera were likely vectors of the virus (AHSV).[9,10] Most of the claims were conjectural, and early efforts to test the vector capabilities of some of the suggested insects were negative. Nieschultz and co-workers[11,12] studied the mosquito fauna at Onderstepoort and, after numerous transmission attempts, concluded that mosquitoes were not the vectors of horse sickness.

Du Toit[13] recovered AHSV from field-trapped *Culicoides spp.* by injecting a horse with a suspension of the triturated flies. Later, he succeeded in transmitting ASHV by feeding *Culicoides* on an infected horse and 12 days later refeeding them on a susceptible horse.

Ozawa and Nakata[14,15] succeeded in experimentally transmitting AHS by feeding artificially infected mosquitoes on susceptible horses. Three species — *Anopheles stephensi, Culex pipiens,* and *Aedes aegypti* — proved to be potentially capable of acting as biological vectors of AHSV.

Later studies at Onderstepoort once again cast doubts upon the capability of mosquitoes and also *Culicoides* to act as biological vectors of AHSV.[16] Finally, Boorman et al.[17] demonstrated that *C. variipennis* could be infected by feeding on AHSV-infected embryonated hen eggs.[17] After 7 days, the infected midges were able to transmit the virus by bite to uninfected eggs, thus proving that *C. variipennis* could act as a true biological vector of AHSV in the laboratory. The fact that closely related bluetongue virus (BTV) was found to be transmitted by *Culicoides spp.* has apparently been taken as further evidence that AHSV is similarly transmitted. At any rate, AHSV is generally assumed to be a *Culicoides*-borne arbovirus.[18] This assumption and studies on the distribution and possible windborne dissemination of *Culicoides* have led to the suggestion that the spread of AHŞ over long distances has been the result of windborne transport of AHSV-infected midges.[19]

Although *Culicoides* transmission of AHSV in nature is not to be denied, the experimental demonstration of infection, replication, and transmission of AHSV by at least three species of mosquitoes should not be ignored.[14,15] Also, the possible role of ticks in the transmission of AHSV in certain areas is worthy of serious consideration. The isolation of AHSV from street dogs in the Aswan Province in Egypt,[20] and the subsequent demonstration that the dog tick, *Rhipicephalus sanguineus sanguineus*, is experimentally capable of transmitting the virus, are indicative of the possible involvement of a tick-dog cycle in the year-round maintenance of the virus in an area where the disease is enzootic.[21] In the same area, AHSV has been isolated from camels and the tick species that is regularly associated with them, *Hyalomma dromedarii*.[22] The virus replicates and can pass transstadially through the developmental stages of the tick. Transovarial transmission was not demonstrated, but starved infected adults readily fed on and transmitted the disease to horses.

B. History of Epizootics

A report of a devastating disease of horses which, in retrospect, was considered to be AHS appeared in records of the Cape Colony of South Africa in 1719. About 1700 horses imported by the Dutch East Indies Company were destroyed by the epizootic.[23] Thereafter, the disease recurred annually after the spring rains and continued until the first frost. Epizootics producing extensive losses occurred about every 20 years in the Cape Colony. The most damaging ones were in 1780, 1801, 1819, 1839, and 1854—1855. In the epizootic of 1854—1855, nearly 70,000 or more than 40% of the total horse population of the Cape of Good Hope died of the disease. In addition to the Cape, AHS proved to be enzootic in most of Africa south of the Sahara. Wherever horses, mules, or donkeys were introduced, AHS was sure to appear.[23]

Aside from its annual or seasonal occurrences in the enzootic regions of Africa, AHS has spread a number of times during this century far beyond its usual confines. In 1928, the disease spread slowly along the Nile Valley from Sudan into Egypt.[24] Although the number of animals involved was not large, about 89% of the horses and 70% of the mules that contracted the disease died. There were also some deaths among donkeys. The disease died out during the winter, and no outbreaks were reported in the region during the next 15 years. In the summer of 1943, AHS again appeared in Sudan and spread north to the Nile delta. It subsided during the winter, but unexpectedly recurred in Egypt in 1944 and then spread to Palestine, Syria, and Jordan. The losses were relatively small but portentous.[18] It was clearly evident that AHS was capable of spreading rapidly and extensively.

A major epizootic occurred in 1959—1960 with the disease appearing in Iran, West Pakistan, and Afghanistan in the summer of 1959. It reappeared in the spring of 1960 and rapidly spread to India, Turkey, Cyprus, Iraq, Syria, Lebanon, and Jordan. It was estimated that 300,000 *Equidae* were destroyed by the epizootic.[25]

The last widespread epizootic of AHS occurred during 1965—1966.[26] Algeria, Morocco, Tunisia, and Spain were affected. By this time, vaccines were available and control methods were fairly well established. Animal losses were undoubtedly reduced substantially by the forceful application of control measures. Nevertheless, the epizootic was costly.

C. Social and Economic Impact

The horse has always occupied a somewhat unique position among domesticated animals. It may have tremendous sentimental as well as economic value, and in either case, its loss may be deeply felt. The histories of African exploration, conquest, and agricultural development were profoundly affected by AHS. Because the mortality rate among horses south of the Sahara was close to 90%, the early explorers rode oxen, followed navigable rivers, or walked; military expeditions were without mounted cavalry; and the early settlers often were obliged to use animals other than horses to till their fields. The disease is still a real threat to the livelihood of the small farmer who depends on the horse as a beast of burden. However, in the more developed countries, the horse is now kept almost entirely for recreational purposes. In some countries, it has actually increased in numbers, quality, and monetary value. With its high mortality rate and capability of rapid spread, AHS could still cause considerable losses before being brought under control.

II. THE VIRUS

A. Antigenic Relationships

AHSV is classified along with the viruses of BTV, Ibaraki disease, epizootic hemorrhagic disease of deer, and a number of other related viruses that have been isolated from insect vectors as an orbivirus, a genus within the family Reoviridae.[27] The orbiviruses are arboviruses that consist of ten segments of double-stranded RNA surrounded by a double-layer protein shell. They have large doughnut-shaped capsomeres and are ether resistant and acid labile. Although these viruses are structurally and biochemically very similar and have a common replication strategy, it is believed that the present nomenclature does not adequately point up the differences and the relationships among the members of the group.[28,29] For example, BTV and AHSV are almost indistinguishable by physicochemical means. Some hybridization between their nucleic acids has been demonstrated.[30] Both are transmitted mainly by *Culicoides;* however, their host ranges are different, and they are serologically distinct.

Over the years, marked differences in the efficacy of vaccines produced from different AHSV isolates were noted.[31] As a group, the AHSV isolates were identifiable by complement fixation (CF), but within the group, nine distinct serological types were identified by neutralization (N) tests. Recognition of the serotypes was particularly important for the fabrication of effective vaccines.[32,33]

B. Host Range

Natural infections with AHSV have been found in equines, dogs, and camels.[22,34] Among the equines, susceptibility is highest in horses, somewhat lower in mules, and lowest in donkeys. Mortality rates follow the same order, with rates as high as 95% for horses in some outbreaks and about 80% for mules. African donkeys are quite resistant to AHS, and few deaths have occurred among them. Mortality rates have varied considerably from one epizootic to another, but they have always followed the same general pattern. However, in the 1960 epizootic in the Middle East, substantial losses occurred among donkeys. The donkeys in that region are apparently more vulnerable than the African donkey.[18] Zebras are normally highly resistant, but some deaths from AHS have been reported.

The susceptibility of the dog to AHS was observed over the years by a number of investigators,[35-38] and it was generally believed that the disease could be acquired by the dog only by eating infected meat or by experimental injection of AHSV. In a search for animals other than equines that might harbor the virus over the winter in Egypt, Salama and co-workers[20] examined 111 blood samples collected during the winter from street dogs in an enzootic area in Aswan Province. Isolations of AHSV were made from three of the samples, thus proving that some dogs, at least in that area, were naturally infected and carried the virus during the winter.[20] Two AHSV isolations were also made from camels in the Aswan Province of Egypt.

Also in the search of reservoir hosts in Africa, some sera of both wild and domesticated animals have been tested for the presence of AHSV-specific antibodies. Although the surveys were quite modest, significant antibody levels were found in elephants and zebra.[39-42] Virus was isolated from zebra, but not from elephants. In a survey of domesticated animals other than equines in Egypt, antibodies to AHSV were found in sheep (23.5%), goats (14%), dogs (7%), camels (5%), and buffalo (4%). Antibodies were not found in cattle. As stated previously, virus isolations have been made from dogs and camels in Egypt.[44] Judging from the antibody levels found in sheep and goats, they too may be naturally infected.

Experimental infections have been produced in a wide variety of animals. Some of the early workers claimed to have infected cattle and goats. Some of the observations must be regarded as suspect because they were apparently unable to distinguish between AHS and heartwater in cattle and goats.[34] It was found that white mice could be infected by intracerebral (i.c.) inoculation, and that virus serially passaged became neurotropic. With the exception of rabbits, most of the common laboratory animals (guinea pigs, hamsters, and rats) may be infected with the mouse-adapted neurotropic strains of AHSV.[44]

C. Strain Variation

The antigenic plurality of strains of AHSV was recognized by Theiler[45] as early as 1908. The degree of virulence, the nature of the disease produced, and the levels of immunity elicited were among the various strain differences noted. From a study of 84 virus strains on hand at the Onderstepoort laboratory, McIntosh[32] established the existence of 7 distinct immunological types of AHSV. A few years later, Howell[33] added two more serotypes to the array and speculated on the origin and stability of the various types. There are numerous strains in each serotype. Because some give rise to better immunizing agents than others, great care must be used in selecting strains for vaccine production. The task of selecting strains to be used as live vaccines was especially difficult. The so-called attenuation of virus by serial passage in mouse brains or cell cultures was laborious and time consuming. The task was lightened somewhat when it was found that there was a positive correlation between plaque size and pathogenicity in AHSV strains grown in cell cultures.[46] However, there is no correlation between plaque characteristics and immunogenicity. The expensive and time-consuming animal testing of clones of low virulence for immunogenicity is still required. So far, effective vaccine strains selected by these means have proved to be stable.

D. Methods for Assay

Isolation and titration of AHSV may be accomplished by i.c. inoculation of young mice.[44,47] In a typical virus isolation and titration, suckling mice are inoculated i.c. with 0.025 mℓ of a serial 10-fold dilution of the substance to be tested (blood, tissue suspension, triturated vectors, etc.). Five to ten mice are used per dilution. They are observed daily for 2 weeks for nervous signs and prostration. Deaths are recorded, and the virus LD_{50} dose is calculated. The brains of some mice *in extremis* are harvested for further passage and virus identification by CF or typing by N tests. Also, the virus may be typed by testing a precipitating antigen extracted from the infected mouse brains with reference sera of the nine AHSV serotypes in an Outcherlony gel diffusion test.

Cell culture methods are also being used in isolating, identifying, and titrating AHSV.[46,48,49] Roller tube cultures of baby hamster kidney cells (BHK21) are used for primary virus isolation and titration, and virus typing may be done by a routine plaque N test on African green monkey kidney cells (VERO).[47]

III. DISEASE ASSOCIATIONS

A. Humans

Infections with AHSV have not been reported in man.

B. Domestic Animals

Aside from the horse, mule, and donkey, the dog is the only other domestic animal to display clinical signs and sometimes die of infection with AHSV. The disease in equines is of greatest concern; therefore, the brief descriptions of clinical features and pathology presented here will refer to the disease as it occurs in horses. Four forms of the disease have been distinguished:[50,51]

1. **The peracute or pulmonary form.** This form of AHS is usually seen in severe epizootics where the mortality rates are high. The incubation period is 3 to 5 days. There is an acute febrile reaction that may last only 1 or 2 days and reach 104 to 105°F. This is followed by progressive respiratory involvement usually including severe dyspnea and spasmotic coughing. The animal stands with its legs spread and its head extended. Its nostrils are dilated, and at the time of death, a frothy fluid may flow from them. The animal actually drowns in its own fluids. Death usually follows within a few hours after the first clinical signs appear. The mortality rate is over 90%. The most characteristic changes seen at necropsy are edema of the lungs and an effusion of serous fluid in the pleural cavity. The lymph nodes, especially those in the thoracic and abdominal cavities, are enlarged and edematous. Periaortic and peritracheal edematous infiltration, hyperemia of the glandular fundus of the stomach, congestion of the renal cortex, hyperemia and petechial hemorrhages in the mucosa and serosa of the large and small intestines, and subcapsular hemorrhages in the spleen are other changes often seen. Petechial hemorrhages occur in the pericardium, and the pericardial sac may contain fluid. Epi- and endocardial petechial hemorrhages are occasionally seen, but cardiac lesions are usually not outstanding. The disease in dogs is usually the pulmonary form.
2. **Cardiac or subacute edematous form.** This form of AHS is caused by virus strains of lower virulence, or it may occur in immune animals infected with heterologous strains of the virus. The incubation period is about 7 to 14 days, and the first clinical sign is a febrile reaction that lasts 3 to 6 days. As the fever begins to subside, characteristic edematous swellings develop. They first appear in the supraorbital fossae and eyelids and then extend to the lips, cheeks, and tongue. This subcutaneous edema may extend down the neck and involve the shoulders, the brisket, and the thorax. Before death, petechial hemorrhages appear in the conjunctivae and under the ventral surface of the tongue. Colic may precede death from cardiac failure. Death usually occurs within 4 to 8 days after the onset of fever. The mortality rate is about 50%. In animals that recover, the edema subsides in 3 to 8 days. Hydropericardium is the most prominent and constant change seen at necropsy. The pericardial sac may contain more than 2 ℓ of fluid. Usually there are petechiae and ecchymoses on the epicardium and endocardium. These hemorrhages are often most prominent along the course of the coronary vessels and beneath the bi- and tricuspid valves. The lungs may be normal or only slightly engorged. There is rarely an excess of fluid in the thoracic cavity. The gastrointestinal tract usually has lesions similar to those that occur in the pulmonary form of the disease. However, submucosal edema is usually much greater.

3. **Mixed form.** This is essentially a combination of the cardiac and pulmonary forms of the disease. The majority of fatal cases of AHS may be classified as the mixed form with lesions of either the pulmonary or cardiac form predominating.
4. **African horse sickness fever.** In its mildest form, AHS may appear as no more than a thermal response of 1 to 5 days duration. The temperature may go as high as 105°F, but usually after 2 days, the fever subsides and the animal recovers. This is the form of AHS which almost always occurs in goats or donkeys that are experimentally infected. In nature, this form of the disease might escape detection.

The fatal disease in dogs has usually been the pulmonary form.

C. Wildlife

It was clearly evident that the causative agent of AHS was present in South Africa before horses were introduced. The disease did not occur outside the continent of Africa, and it appeared only when horses were taken into certain areas of Africa. Therefore, a reservoir for the disease agent was sought among the indigenous animals. The zebra was a prime suspect, and some observers reported deaths among them that they attributed to AHS.[6] However, it was found that the disease was also contracted in areas where zebra and other game animals did not exist, and the injection of blood from small mammals, birds, reptiles, or amphibians into susceptible horses failed to produce the disease.[6] Much later, AHSV N and CF antibodies were found to be quite common in the blood of zebra and elephants in Kenya and South Africa.[39,41,42] Clinically recognizable AHS has been seen in zebra, and virus has been isolated. The presence of antibodies in the elephant has not been explained. Clinical disease has never been observed in elephants, nor has virus ever been isolated from them. A wildlife reservoir for AHSV has not been definitely established.

D. Diagnostic Procedures

In an area where the disease is enzootic, it may often be diagnosed in the field with a fair degree of accuracy. The clinical signs and gross lesions are usually characteristic enough to enable a presumptive diagnosis to be made. For example, edema of the supraorbital fossae is pathognomonic for the cardiac form of the disease.

However, some of the clinical signs and post-mortem findings of AHS may appear in other equine diseases such as equine infectious anemia, equine piroplasmosis, purpura hemorrhagica, and rhinopneumonitis. A confirmed diagnosis requires isolation and identification of the virus. This has usually been achieved by i.c. inoculation of unweaned mice with defibrinated blood taken at the peak of fever. Spleen suspensions (10%) have also been used. A litter of eight to ten mice is used. Each mouse is given 0.025 mℓ of blood diluted 1/10 in sterile distilled water or phosphate-buffered saline. They are observed for 2 weeks, and the brains of those that show nervous signs and prostration are removed and subpassed i.c. as a 10% suspension to another litter of suckling mice. After three to five serial passages, mortality is usually 100%. Viruses isolated in this manner are neurotropic and may no longer produce clinical disease in horses, although they are still antigenic. They have been especially valuable in the diagnosis of AHS. Antigens for CF test,[31] Outcherlony agar gel precipitin test (AGD),[43] and virus neutralization (N) test[44] have been prepared from brains of moribund mice. Stocks of type-specific reference viruses have also been prepared in mice, and they have been used to hyperimmunize rabbits to obtain type-specific antisera. Using these procedures, reagents, and tests, it is possible to isolate, identify, and type AHS viruses. It is also possible to survey animal populations for AHS antibodies. The CF test has been used for rapid diagnosis. It is group specific, but CF antibodies are rather short lived. For typing, the N test must be used. N antibodies are present for a much longer time.

In efforts to improve upon and possibly supplant the slow and tedious procedures of propagating AHSV by i.c. inoculation of unweaned mice, the use of cell cultures was introduced.[48,49,52] The methods were first applied to propagate and attenuate virus strains that had been isolated and typed by the established mouse brain methods. It was soon found that the viruses could be typed by N in cell cultures.[49] Finally, it was found that cell cultures could be used to isolate virus directly from naturally infected animals.[46] Of the numerous cell cultures tested, the stable monkey kidney cell lines MS and VERO and the baby hamster kidney cell line BHK21 proved to be most useful. Coexistence of virus and antibody in the blood of the infected animal has apparently accounted for some of the difficulty in isolating the virus. However, most of the virus in the blood appears to be firmly associated with erythrocytes, and these may be washed comparatively free of antibody. Virus isolation is facilitated by using an inoculum of washed erythrocytes hemolyzed by sonication or the addition of distilled water. The use of roller tube cultures also appears to favor virus isolation.[47]

Fluorescent antibody techniques have also been applied to AHSV detection. Although the methods are only group specific, they are more convenient and rapid than CF. Indirect immunofluorescence has been used successfully in an antibody survey in wild zebra in Kenya.[40]

A number of new tests are being developed, and promising preliminary trials have been conducted with some of them, but much more testing is required before they are likely to be accepted as standard procedures. Among them are microadaptations of the CF and AGD tests, hemagglutination tests with erythrocytes coupled with type-specific AHS antibodies used to type virus isolates, and an indirect enzyme-linked immunosorbent assay (ELISA).[53]

E. Effects of Virus on Vectors

Although some of the transmission studies with mosquitoes, *Culicoides,* and ticks have been quite detailed and quantitative with respect to size of blood meal, multiplication, and persistence of the virus in the vector, no mention has been made of adverse effects of the virus on the vectors.

IV. EPIDEMIOLOGY

A. Geographic Distribution

As its name implies, AHS apparently originated in Africa and at the present time appears to be confined to that continent. As stated previously, there were substantial incursions of the disease in the Middle East in 1944, Afghanistan, West Pakistan, Turkey, Iraq, India, and Cyprus in 1959 to 1961, and Spain in 1966. It was last reported in Pakistan in 1974 and Saudi Arabia in 1975.[54] Clinical AHS was reported in Yemen in 1980 and 1981, but may have been present as early as 1976.

Until 1930, the enzootic areas in Africa were south of the Sahara, mainly in South, East, and Central Africa. In 1930, the disease extended up the Nile Valley into Egypt and has occurred periodically in Southern Egypt ever since.

B. Incidence

No major epizootics of AHS have occurred since 1966. However, cases occur sporadically in the enzootic areas of Africa. In fact, there are 10 to 12 countries that have reported cases almost every year for the past 10 years or more.[54] Although the number of cases reported during the past several years is relatively small, the potential for the occurrence of a major epizootic is still present. The virus need only reach a country where there is a large susceptible or unvaccinated horse population. A capable vector is almost certain to be present.

C. Seasonal Distribution

Some of the countries of Africa where AHS is enzootic may be listed from north to south as follows: Egypt, Suda, Ethiopia, Kenya, Tanzania, Zambia, Zimbabwe, Botswana, and South Africa. These countries form an unbroken chain spanning the entire length of the continent and present a wide range of climatic conditions. Because temperature and moisture are the main factors determining the incidence of the disease, there is no common time of season that applies to the occurrence of AHS throughout the enzootic areas of Africa. In the equatorial regions, the incidence of the disease and the rains are closely correlated. In temperate zones and altitudes where there is a winter season accompanied by frost, the rains continue to be of utmost importance, but the disease abruptly disappears after the first frost.

D. Risk Factors

A horse's susceptibility to AHS is unrelated to its age, sex, or nutritional conditions. Accounts from the very beginning of written history in Africa indicate that horses of all breeds brought to the continent were fully susceptible and were likely to succumb to AHS. However, there are horses in North Africa and just south of the Sahara in West Africa which are descendants of horses that were present there since at least 2000 B.C.[55] Over the years, these local breeds apparently progressively acquired a natural resistance to AHS. Unvaccinated local breeds did not display clinical signs of the disease, but a serological survey revealed that they had been exposed to the virus.[56] The disease was observed in animals sired by imported stallions out of native mares.[57] Apparently, natural resistance is not a dominant trait passed from the female to her progeny. The genetics of natural resistance to AHS requires further study.

E. Serologic Epidemiology

A country that has had outbreaks of AHS and wishes to make a serious effort to control the disease should start with a serological survey to determine the immunological status of its animal populations that are at risk. Group-specific tests such as CF may indicate that contact with the disease has occurred within the last 6 months. The N test detects the longer-lasting antibodies that are type specific and indicate the virus strains to which the animals may be immune. The information obtained may be of great value in locating points of vulnerability and form the basis for selecting appropriate preventive measures to be instituted.

A serological survey was conducted recently in Egypt in an effort to detect a reservoir in animals other than equines which might be the source of the sporadic cases of AHS which occur despite the policy of annual immunization of all equines in the country.[43] A total of 1000 sera were collected from sheep, goats, dogs, camels, cattle, and buffaloes. They were tested by CF and AGD against a type 9 reference antigen. The two test procedures yielded somewhat different results. The CF test produced the highest number of positive reactions: sheep (23.5%), goats (14%), dogs (7%), camels (5%), buffaloes (4%), and cattle (0%). Results were as follows with the AGD precipitin test: sheep (2.5%), goats (8%), dogs (6%), camels (1.5%), buffaloes (1%), and cattle (2.5%). With both tests, the incidence of positive reactions was lowest in sera collected in the winter. Although it is impossible to say which test produces the most significant results, it appears that many farm animals in Egypt are exposed to and produce antibodies against AHSV. This and other studies indicate that the dog, at least, may play a role in the maintenance of the virus in parts of Egypt.

Because there have been no major epizootics of AHS in Africa and no extensive incursions elsewhere in recent years, some complacency appears to have developed. Despite the fact that tests suitable for conducting extensive serological surveys have been developed, none of the countries that are especially at risk has sought to determine the immunological status of its equine population.

V. TRANSMISSION CYCLES

A. Evidence from Field Studies

1. Vectors

Although a wide variety of arthropods have been suggested as possible vectors of AHS, it is generally accepted that the disease is transmitted in nature by species of *Culicoides.* This belief had its origin in the work of Du Toit[13] in South Africa. He was able to produce AHS in a horse by injecting an emulsion of wild-caught *Culicoides,* but at the time, his efforts to transmit the virus from an infected to a susceptible horse by *Culicoides* bites were unsuccessful. Later, he stated in a personal communication in Wetzel et al.[16] that transmission by bite was subsequently achieved. Wild-caught *Culicoides* were fed on an infected horse. After 12 days, they were fed on a susceptible horse, and after another 12 days, the horse died of AHS.

The start of all major outbreaks and sporadic cases of AHS in Egypt have invariably occurred in the Aswan and Qena Provinces and the borderlands between Egypt and the Sudan. Some of the sporadic cases could perhaps be attributed to illegal entrance of infected equines from the Sudan and failure to vaccinate all susceptible equines. The disease recurs during periods of hot weather following heavy rains. Mosquitoes and *Culicoides* abound under these conditions, but they do not account for the survival of the virus during cool, dry seasons. As stated previously, the Egyptian workers have tried to identify an animal reservoir that might maintain the virus during the offseasons. Their attention focused especially on dogs and camels because the former were known to be susceptible to AHS and were numerous in the area, and the latter were often brought in from the Sudan where control measures were far less stringent, and the incidence of AHSV-specific antibodies in their blood by CF (23.2%) was significantly higher than found in the local camels (5%). Their search yielded six AHSV isolations from dogs and four from camels.[20,22] The camel isolations were particularly significant because two were obtained from engorged ticks, *Hyalomma dromedarii,* removed from camels that were brought in from the Sudan. The animals did not show clinical signs of AHS. Later it was shown that *H. dromedarii* fed as larvae or nymphs on an infected animal were able to transmit the disease when subsequently fed as nymphs and adults, respectively, on susceptible animals.

2. Vertebrate Hosts

Both virus isolation and antibody detection have been achieved with naturally infected horses, mules, donkeys, zebra, dogs, and camels. Antibodies have been found in sheep, goats, cattle, and elephants, but AHSV has not been isolated from these species.[41,43]

B. Evidence from Experimental Infection Studies

1. Vectors

C. variipennis proved capable of acting as a biological vector of AHSV in the laboratory.[17] Midges placed on the air sac membrane of embryonated hen eggs that had been infected by inoculation of AHSV by either the yolk sac or intravascular route fed through the membrane and became infected. After 7 days, they were fed in the same manner on uninfected eggs and transmitted the virus to the embryos. It was further shown that virus multiplication occurred in two species of *Culicoides (C. nubeculosus* and *C. variipennis)* artificially infected by either intrathoracic inoculation or by allowing them to ingest virus through a chick-skin membrane. However, neither of these species is likely to serve as a vector of AHSV in Africa. *C. variipennis* is an American species and is not found in Africa; *C. nubeculosus* is not normally susceptible to oral infection with virus.

The weather and climatic conditions that favor the development and reproduction of *Culicoides* are also favorable for mosquitoes, and some of the experimental demonstrations of AHSV transmission by mosquitoes have been at least as impressive as the experimental transmission studies with *Culicoides*. After numerous unsuccessful attempts to demonstrate transmission by mosquitoes, workers in South Africa concluded that mosquitoes were not involved in the natural transmission of AHSV.[11] About 30 years later, Ozawa and Nakata[14,15] in Iran were able to experimentally demonstrate transmission with *Culex pipiens, Anopheles stephensi,* and *Aedes aegypti.* The mosquitoes were fed on a mixture of hemolyzed normal horse erythrocytes, glucose, and cell culture fluid containing AHSV. They transmitted the virus when fed on uninfected horses 15 to 18 days later. Virus titers in mosquitoes taken at various times after engorging on the infected mixture were markedly decreased during the first few days after feeding, but by 7 to 9 days after feeding, the virus titers had increased to levels higher than the amounts originally ingested. The maximum level was reached in about 2 weeks and was maintained in some mosquitoes for 3 or more weeks thereafter. In an experiment to determine the percentage of mosquitoes that maintained the virus for extended periods, it was found that the probability of infection remaining longer than 24 days after feeding was less than 10%. Finding the virus in wild-caught mosquitoes is likely to be possible only during a full-blown epizootic.

It has long been known that dogs are susceptible to AHS,[35-37] and it was generally believed that they acquired the infection only through ingestion of virus-contaminated meat. When the Egyptian workers isolated AHSV from naturally infected dogs in Upper Egypt, they conceded that virus-contaminated meat was the most probable source of their infection.[20,21] However, because dogs in the area were usually heavily infested with the dog tick *Rhipicephalus sanguineus sanguineus,* they tested the tick as a possible vector of AHSV. They did not succeed in recovering virus from ticks collected from dogs in the field, but virus was readily isolated from ticks fed on experimentally infected dogs. The infected ticks transmitted the virus when fed on susceptible dogs or horses. They were also able to transmit AHSV from infected horses to susceptible dogs and horses. The virus used in the experiments was a dog isolate, and when transmitted by tick bite, it produced only mild disease in both dogs and horses.

2. *Vertebrate Hosts*

Aside from the infection of dogs by ingestion of infected meat, it appears that AHSV is transmitted only by blood-sucking arthropods. Specific AHSV antibodies have been found in equines, dogs, sheep, cattle, goats, buffaloes, camels, and elephants, but the virus has been isolated from only equines, dogs, and camels.[6,22,41]

C. Maintenance / Overwintering Mechanisms

Soon after the disease appeared in South Africa, it was noted that it reached its peak during hot weather following heavy rains and quickly disappeared after the first frost. It reappeared several months later after the rains and warm weather returned. It was apparent that an insect vector was involved in transmitting the virus, but it was believed that an animal reservoir had to be involved in maintaining the virus during the winter months. However, animal reservoirs that might account for the maintenance AHSV in the diverse and often isolated areas of South Africa where the disease has occurred have never been found.

Based on the findings reported by Du Toit[13] in 1944, *Culicoides* are regarded as the most likely vectors of AHSV. If this is true, it is quite conceivable that among the nearly 1000 known species of the genus *Culicoides,* there are a number that might be capable of harboring the virus during the off-seasons. In the vast majority of species, the egg and pupal stages are of short duration, and hibernation as adults is unknown. *Culicoides* spend most of their life in the larval stage and overwinter as such. If the virus is maintained over the winter or dry seasons in larvae, we must assume that it was passed transovarially from the adult to the larvae. This is yet to be demonstrated.

It is thought that the dog may be the host in which AHSV overwinters in Egypt, but the virus has also been isolated from camels and donkeys, and other farm animals have significant levels of AHS antibodies. Species of *Culicoides* and mosquitoes that have been found to be experimentally capable of transmitting the virus abound and, in at least two species of ticks, the virus may be transmitted transstadially.[53] Clearly, there are a number of overwintering possibilities for AHSV in Upper Egypt.

VI. ECOLOGICAL DYNAMICS

Despite the overwhelming evidence that AHS is an arthropodborne virus, no species or even genus of arthropods can be considered solely responsible for the maintenance and spread of the disease. There have been numerous attempts to identify the specific vectors involved, but thus far, only one species of *Culicoides* and one tick species have been incriminated with any degree of certainty, and neither the dynamics nor the importance of their involvement in the epizootiology of the disease has been established. Therefore, the vector aspects of this section are limited and somewhat speculative.

A. Macro- and Microenvironments

Areas that are the most favorable for the occurrence and spread of AHS are warm coastal regions; warm, moist inland areas such as swamps, valleys, and wetlands along river courses.[25,59] These are usually low-lying areas with abundant flora and fauna. Most *Culicoides*, mosquitoes, and a number of tick species thrive in such areas. In South Africa, it was observed that AHS seldom occurred at altitudes above 500 m, but under unusual conditions of temperature and rainfall, it has been seen up to 1200 m. In a mountainous region of Kenya, close to the equator, the disease has occurred at 2400 m altitude.[55] At that altitude, perhaps a different vector species is involved. However, there are species of mosquitoes and *Culicoides* which manage to survive for months during unfavorable ecological conditions. Some mosquitoes survive as dormant eggs or are able to hibernate as inseminated females. Most species of *Culicoides* survive as larvae in the soil.

B. Climate and Weather

All who have dealt with AHS in the field have noted the positive effects of heavy rains followed by heat and high humidity on the occurrence and spread of the disease. They have also noted the negative effects of cold and drought. As stated previously, these factors govern the activities of the vectors. They may also govern the movement and distribution of the vertebrate hosts.

C. Vector Oviposition

The generally accepted opinion that AHS is a *Culicoides*-borne arbovirus is based primarily on Du Toit's[13] isolation of the virus from *Culicoides* caught in the field, and his subsequent demonstration that these insects were capable of transmitting the virus by bite to susceptible horses. The species involved was *C. (Avaritia) imicola* (= *pallidipennis*). It is widely distributed in Africa and the Middle East and is also found in Spain.[58] The biology of the species has been worked out mainly in connection with its role as a vector of BTV. It is closely associated with cattle as evinced by the fact that in South Africa it often oviposits in fresh cow dung.[60] In Israel, it breeds in and around animal pens in soil moistened by the overflow of water troughs or along animal sewage and drainage canals.[61] Information on the oviposition of *C. imicola* is not complete, and one cannot assume that it is the same as that which has been determined for another species. For example, *Culicoides* spp. may have 3 to 4 gonotrophic cycles, and the eggs are laid in batches which may vary from 30 to 40 for some species and up to 450 for others.[62,63]

The camel tick *Hylomma dromedarii,* which has been found capable of harboring and transmitting AHSV in Upper Egypt, occurs wherever there are camels.[64] It is found in southern Russia and in the Near, Middle, and Far East. It occurs in North Africa, in the transitional belt south of the Sahara desert, and as far south as Somalia and northeastern Kenya. Its normal number of hosts has not been definitely established. Experimentally, they have been reared as single-, two-, and three-host ticks. Field observations indicate that they are normally a two-host tick. The host change usually occurs after the nymphal-adult molt.[64] Engorged, ovipositing females have been found in camel yards, under stones, under desert shrubs, and in rodent burrows. Egg batches varying from 2000 to 14,000 have been attributed to individual females.[65] Females that engorge late in autumn undergo a winter diapause and do not oviposit until spring.

D. Vector Density and Fecundity

Using suction-type light traps operated almost nightly in various locations at Onderstepoort in South Africa, daily catches were analyzed and recorded over a period of several years.[66] Traps were located in close proximity to the nightly holding facilities of several different species of farm animals. The catches consisted mainly of *C. imicola,* and at times it made up more than 97% of the catch. The adult densities were usually greatest from August to May with the peak incidence usually occurring during the period from January to March. During the winter, the period from May to August, *Culicoides* were seldom found in the traps. However, in years when there were unusually heavy rains in late summer and winter temperatures were mild, *Culicoides* were active throughout the year.

Similar studies conducted in Israel during 1981 to 1982 indicated that *C. imicola* was also the most prevalent species of *Culicoides* feeding on cattle, horses, and sheep in that area, and the population was highest during August to September.[61] During this period, the average time between blood meals was 3.3 to 4.6 days. The estimated survival rates between blood meals were highest between April and October, with two high peaks of 0.75 in August. The highest parous rates were during August to October. Data of this kind could be extremely valuable for predicting periods of greatest risk of disease outbreaks. However, variations in weather conditions can cause considerable differences in the data collected from one year to another. Continuous observations are required.

Although *H. dromedarii* is probably the most completely desert adapted of all ixodid ticks, large numbers probably die in the egg and larval stage along irregularly traveled trails across barren deserts.[64] Life cycles have been observed to vary from 93 days in warm weather to more than 280 days in cold weather. In nature, two generations a year may occur. Climatic conditions affect oviposition, hatching, and premolting periods, and thus govern the length of the life cycle.[65]

E. Host Preference

By far the largest numbers of *C. imicola* were caught in traps operated near cattle and horse holding pens. A trap near a sheep pen yielded the next highest numbers, while only a few were found in a trap near a hen house. The host preferences determined by precipitin tests on 657 blood meals from engorged *Culicoides* matched those indicated by the sizes of the catches at the various trap locations.[66,67]

H. dromedarii is closely associated with camels and is found only within the normal range of these animals.[64] The adults feed mainly on camels but also are found on cattle and horses and, to a lesser extent, on sheep, goats, and dogs. Larvae feed on small burrowing mammals and hares and occasionally on lizards. Nymphs will also feed on these animals, but are found in large numbers on the larger hosts when they are available. Except for the small mammals that may be available in different locations, the host preferences are essentially the same in all areas where the tick exists.[64]

F. Vertebrate Host Immunological Background

Equidae that have survived infection with AHSV develop solid immunity to that particular virus type, but remain susceptible to the other serological types.[32] Foals from immune dams acquire a natural immunity that protects them from AHS during the first 8 months or so of their lives. In some areas where the disease is enzootic, mass vaccination of *Equidae* is practiced, but other animals such as camels, sheep, goats, and dogs that may be infected but seldom display clinical signs are not vaccinated. They perhaps serve as reservoirs of the virus.

G. Vector Competence

There has been no experimental determination of vector competence in AHS, but considering the fact that AHS is entirely vector transmitted, it may be assumed from the rapid and extensive spread of the disease during major epizootics that the vectors are highly competent and/or overwhelming in numbers.

H. Migration of Vectors and Hosts

Under normal conditions, many *Culicoides* species range less than 100 m from their breeding sites, but winds may substantially increase their range. One species whose density diminishes to a tenth of its original value at 70 m from its breeding site may be carried by the wind more than 1000 m without a substantial reduction in density.[68] Evidence indicating that *Culicoides* have been carried 5 to 6 km on winds have been presented.[19] It has further been suggested that midges have been carried 40 to 700 km by winds.[69] The suggestion is supported by circumstantial evidence derived from an analysis of meteorological records, geographic factors, host-vector cycles, and histories of outbreaks.

It is unlikely that the host animals of *C. imicola* are involved in their dispersal over any more than relatively short distances.

As stated previously, there is evidence that horses were introduced into Africa south of the Sahara about 2000 B.C.[55] This line of horses apparently acquired natural resistance to AHS, and there is serological evidence that they have had contact with the virus.[56] Perhaps they were involved in spreading the disease along some of the traditional migratory routes of the nomads in northwestern Africa. The highly susceptible horses brought to Africa from Asia and Europe from the 17th century on have been the victims and not the spreaders of AHS.

The traditional patterns of migration of animals and nomadic tribes have no doubt been substantially altered in Africa by the droughts, famines, wars, and political turmoil of recent years; nevertheless, it is safe to assume that seasonal migrations of wildlife still occur in many areas without regard for national borders, and the movement of people has, in fact, been intensified by some of the recent events. The full impact of such movements on the spread of animal diseases in Africa remains to be seen.

Nomads migrate between southern Morocco and Algeria and the Saharan regions of Mali and Niger. They take with them camels, donkeys, sheep, goats, and dogs. The spread of AHS along their migratory routes is evidenced by the presence of N and CF antibodies in donkeys living in the oases through which they pass.[55] Species of *Culicoides* probably exist at these oases, and we may be quite certain that the ticks *H. dromedarii* and *R. sanguineus sanguineus* not only exist there, but are transported by their hosts from one oasis to another. Egyptian investigators recently have presented evidence involving these two tick species in the maintenance and transmission of AHS in Upper Egypt.[21,22] The role of the camel and *H. dromedarii* in the spread of the disease from Sudan into Upper Egypt is convincing.

VII. SURVEILLANCE

The last time AHS extended beyond its enzootic enclaves in Africa was in 1965—1966 when it went from Northwest Africa to Algeria, Morocco, and Tunisia and on to Spain. Perhaps the success in eradicating the disease on the Iberian Peninsula has lead to complacency concerning the danger of future incursions of AHS outside the continent of Africa. At any rate, there is no indication that any of the countries along the Mediterranean and in the Near and Middle East which were struck by the disease in 1944, 1959—1960, or 1965—1966 are conducting surveillance of any kind. Aside from the work conducted in Egypt, little research dealing specifically with AHS has been done in the past several years. Most recent studies have dealt primarily with bluetongue, but AHS is often mentioned because the vector aspects of the two disease are thought to be similar. In African countries where the disease is enzootic, apparently it is being adequately controlled for the time being with vaccines.

VIII. INVESTIGATION OF EPIZOOTICS

In 1948, Alexander[18] presented a detailed analysis of the epizootic of AHS that occurred in the Middle East in 1944. The spread of the disease was traced from the Sudan into Egypt, down the Nile Valley to Cairo and the Nile delta. Eventually it crossed over into Palestine, Syria, Lebanon, and Jordan. Climate, weather, animal movements, possible vectors, the nature of the disease, mortality rates, and the effectiveness of vaccines were among the factors discussed. The great epizootic of 1959—1960 which spread through the Middle East and reached as far as India was similarly analyzed.[25] The studies yielded little or no new information on the epizootiology of AHS. In both epizootics, the disease advanced as might have been expected from the conditions that prevailed. The studies did reveal the value of vaccination, but aside from that, they are now of only historical interest.

IX. PREVENTION AND CONTROL

A. Vector Control

Although systematic tests were not made, it was believed that the local application of repellents and insecticides during outbreaks substantially reduced the incidence of the disease. In some countries, the interiors of airplanes arriving from regions where AHS is known to exist are promptly sprayed with insecticide. This is done with an aerosol spray before the passengers disembark.

B. Control of Vertebrate Hosts

Soon after AHS was encountered in South Africa, it was found that horses placed in mosquito-proof stables at night usually escaped infection.[34] Horses entering the U.S. from Asia, Africa, and the Mediterranean countries are held in quarantine in insect-proof stables for at least 60 days.

C. Use of Vaccines

During the 1965/1966 epizootic in Spain, the northward advance of the disease was halted and eradication was achieved by vaccinating all *Equidae* in a broad zone around the affected areas.[26] It appears that a number of satisfactory vaccines are now available, but supplies may not be sufficient to quickly arrest a fast-spreading major epizootic. When the disease first occurs in an area, affected horses should be quickly eliminated, and the noninfected *Equidae* should be vaccinated with a polyvalent vaccine. When the diagnosis has been confirmed and the virus isolate has been typed, a homologous vaccine may be employed at less cost and with greater efficiency.

X. FUTURE RESEARCH

Despite the fact that there has not been a major epizootic of AHS in the past 20 years, it is too early to relegate the disease to history. In most parts of the world, the horse is no longer of importance as a beast of burden, nor is it any longer of military importance. The horse is now becoming important mainly for recreational purposes. The breeds have been upgraded and so has their value. In many instances, they are quickly transported by air from one country to another, and there is still much to be learned about some of the diseases that may travel with them.

The reservoirs and vectors of AHSV are not well established. There are nine known serotypes of the virus. Type 9 virus, which is continuously active in Egypt, appears to be of rather low virulence, but it is not known how it might behave in a fully susceptible population. There may be more than nine serotypes, or perhaps strains of the known types have evolved against which the present vaccines may be ineffective. These are but a few of the aspects of AHS which should be further investigated.

REFERENCES

1. **Theiler, A.,** African horsesickness *(Pestis equorum), S. Afr. Dep. Agric. Sci. Bull.*, 19, 1921.
2. **M'Fadyean, J.,** African horsesickness, *J. Comp. Pathol. Ther.*, 13, 1, 1900.
3. **Theiler, A.,** Die Sudafrikanische Pferdesterbe, *Dtsch. Tieraerztl. Wochenschr.*, 201, 221, 233, 241, 1901.
4. **Nocard, E.,** La horse sickness ou "Maladie des Chevaux" de l'Afrique du Sud, *Bull. Soc. Cent. Med. Vet.*, 55, 37, 1901.
5. **Sieber, H.,** Experimentelle Untersuchungen uber die Pferdesterbe, *Z. Infkr. Haust.*, 10, 81, 1911.
6. **Theiler, A.,** The problem of horsesickness, *Rep. 13th Annu. Meet. S. Afr. Assoc. Adv. Sci.*, 65, 1915.
7. **Schuberg, A. and Kuhn, P.,** Uber die Ubertragung von Krankheiten durch Einheimische Stechende Insekten, *Arb. Gesundb. Amt. (Berl.)*, 40, 209, 1912.
8. **Williams, A. J.,** Notes on an outbreak of horse-sickness connected with the presence of *Lyperosia* as possible transmitter, *Vet. J.*, 69, 382, 1913.
9. **Van Saceghem, R.,** La pest du cheval ou horse-sickness au Congo Belge, *Bull. Soc. Pathol. Exot.*, 11, 423, 1918.
10. **Carpano, M.,** African horse-sickness as observed particularly in Egypt and Eritrea, *Minist. Agric. Tech. Sci. Serv. Bull.*, 115, 1, 1931.
11. **Nieschultz, O., Bedford, G. A. H., and Du Toit, R. M.,** Results of a mosquito survey at Onderstepoort during the summer 1931-32 in connection with the transmission of horsesickness, *Onderstepoort J. Vet. Sci. Anim. Ind.*, 3, 43, 1934.
12. **Nieschultz, O. and Du Toit, R. M.,** Investigations into the transmission of horsesickness at Onderstepoort during the season 1932-33, *Onderstepoort J. Vet. Sci. Anim. Ind.*, 8, 213, 1937.
13. **Du Toit, R. M.,** The transmission of blue-tongue and horse-sickness by *Culicoides, Onderstepoort J. Vet. Sci. Anim. Ind.*, 19, 7, 1944.
14. **Ozawa, Y. and Nakata, G.,** Experimental transmission of African horsesickness by means of mosquitoes, *Am. J. Vet. Res.*, 26, 744, 1965.
15. **Ozawa, Y., Nakata, G., Shad-del, F., and Navai, S.,** Transmission of African horse-sickness by a species of mosquito, *Aedes aegypti Linnaeus, Am. J. Vet. Res.*, 27, 695, 1966.
16. **Wetzel, H., Nevill, E. M., and Erasmus, B. J.,** Studies on the transmission of African horsesickness, *Onderstepoort J. Vet. Res.*, 37, 165, 1970.
17. **Boorman, J., Mellor, P. A., Penn, M., and Jennings, M.,** The growth of African horse-sickness virus in embryonated hen eggs and the transmission of virus by *Culicoides variipennis* Coquillet (Diptera, Ceratopogonidae), *Arch. Virol.*, 47, 343, 1975.
18. **Alexander, R. A.,** The 1944 epizootic of horsesickness in the Middle East, *Onderstepoort J. Vet. Sci. Anim. Ind.*, 53, 471, 1948.
19. **Sellers, R. F., Pedgley, D. E., and Tucker, M. R.,** Possible spread of African horse sickness on the wind, *J. Hyg.*, 79, 279, 1977.

20. **Salama, S. A., Dardiri, A. H., Awad, F. I., Soliman, A. M., and Amin, M. M.,** Isolation and identification of African horse sickness virus from naturally infected dogs in Upper Egypt, *Can. J. Comp. Med.*, 45, 392, 1981.
21. **Salama, S. A., El-Husseini, M. M., Dardiri, A. H., and Abdalla, S. K.,** Experimental transmission of African horse sickness virus by means of dog tick, *Rhipicephalus sanguineus sanguineus, J. Egypt. Vet. Med.*, in press.
22. **Awad, F. I., Amin, M. M., Salama, S. A., and Samia Khidr,** The role played by *Hyalomma dromedarii* in the transmission of African horse sickness virus in Egypt, *Bull. Anim. Health Prod. Afr.*, 29, 337, 1981.
23. **Theiler, A.,** African horse sickness *(Pestis equorum), S. Afr. Dep. Agric. Bull.*, 99, 1921.
24. **Carpano, M.,** La peste equina con particolare ruquardo a quella osservata in Egitto ad in Eritrea (Equine plague, particularly the type observed in Egypt and Eritrea), *Clin. Vet. Milano*, 53, 471, 1930.
25. **Reid, N. R.,** African horse sickness, *Br. Vet. J.*, 118, 137, 1961.
26. **Montilla, D. R. and Marti, P.,** Epizootologia de la peste equina en Espana, *Bull. Off. Int. Epiz.*, 68, 705, 1967.
27. **Matthews, R. E. F.,** Classification and nomenclature of viruses, *Intervirology*, 17, 1, 1982.
28. **Knudson, D. L. and Shope, R. E.,** Overview of the orbiviruses, in *Bluetongue and Related Orbiviruses*, Alan R. Liss, New York, 1985, 255.
29. **Della-Porta, A. J.,** Classification of obiviruses: a need for supergroups or genera, in *Bluetongue and Related Orbiviruses*, Alan R. Liss, New York, 1985, 267.
30. **Verwoerd, D. W. and Huismans, H.,** On the relationship between bluetongue African horsesickness, and reoviruses: hybridization studies, *Onderstepoort J. Vet. Res.*, 36, 175, 1969.
31. **McIntosh, B. M.,** Complement fixation with horsesickness viruses, *Onderstepoort J. Vet. Res.*, 27, 165, 1956.
32. **McIntosh, B. M.,** Immunological types of horsesickness virus and their significance in immunization, *Onderstepoort J. Vet. Res.*, 27, 465, 1958.
33. **Howell, P. G.,** The isolation and identification of further antigenic types of African horsesickness virus, *Onderstepoort J. Vet. Res.*, 29, 139, 1962.
34. **Theiler, A.,** *African Horsesickness. A System of Bacteriology in Relation to Medicine*, Vol. 7, Her Majesty's Stationery Office, London, 1930, 362.
35. **Theiler, A.,** The susceptibility of the dog to African Horsesickness, *J. Comp. Pathol. Ther.*, 23, 315, 1910.
36. **M'Fadyean, J.,** The susceptibility of the dog to African horse-sickness, *J. Comp. Pathol. Ther.*, 23, 27, 1910.
37. **Bevan, L. E. W.,** The transmission of African horse-sickness to the dog by feeding, *Vet. J.*, 67, 402, 1911.
38. **Van Rensburg, I. B. J., De Clerk, J., Groenewald, H. B., and Botha, W. S.,** An outbreak of African horsesickness in dogs, *J. S. Afr. Vet. Assoc.*, 52, 323, 1981.
39. **Mirchamsy, H. and Hazrati, M.,** A review of aetiology and pathogeny of African horsesickness, *Arch. Inst. Razi*, 25, 23, 1973.
40. **Davies, F. G. and Lund, L. J.,** The application of fluorescent antibody techniques to the virus of African horsesickness, *Res. Vet. Sci.* 17, 128, 1974.
41. **Erasmus, B. J., Young, E., Pieterse, L. M., and Boshoff, S. T.,** The susceptibility of zebra and elephants to African horsesickness virus, in *Proc. 4th Int. Conf. Equine Infect. Dis.*, Lyon, 1976, 409.
42. **Davies, F. G. and Otieno, S.,** Elephants and zebras as possible reservoir hosts for African horsesickness virus, *Vet. Rev.*, 100, 291, 1977.
43. **Awad, F. I., Amin, M. M., Salama, S. A., and Aly, M. M.,** The incidence of African horsesickness antibodies in animals of various species in Egypt, *Bull. Anim. Health Prod. Afr.*, 29, 285, 1981.
44. **Alexander, R. A.,** Studies on neurotropic virus of horsesickness. I. to IV., *Ondersterpoort J. Vet. Sci. Anim. Ind.*, 4, 291, 1935.
45. **Theiler, A.,** The immunication of mules with polyvalent serum and virus, *Ann. Rep. Gov. Vet. Bact.*, 1906/1907, 192, 1908.
46. **Erasmus, B. J.,** A new approach to polyvalent immunzation against African horsesickness, in *Proc. 4th Int. Conf. Equine Infect. Dis.*, Lyon, 1976, 401.
47. **Ozawa, Y., Salama, S. A., and Dardiri, A. H.,** Methods for recovering African horsesickness virus from horse blood, in *Proc. 3rd Int. Conf. Equine Infect. Dis.*, S. Karger, Basel, 1973, 58.
48. **Mirchamsy, H. and Taslimi, H.,** Adaptation of horsesickness virus to tissue culture, *Nature (London)*, 198, 704, 1963.
49. **Ozawa, Y. and Hazrati, A.,** Growth of African horse-sickness in monkey cell cultures, *Am. J. Vet. Res.*, 25, 505, 1964.
50. **Erasmus, B. J.,** The pathogenesis of African horsesickness, in *Proc. 3rd Int. Conf. Equine Infect. Dis.*, S. Karger, Basel, 1973, 1.

51. **Newsholme, S. J.,** A morphological study of the lesions of African horsesickness, *Onderstepoort J. Vet. Res.*, 50, 7. 1983.
52. **Hazrati, A. and Mirchamsy, H.,** Preparation and characterization of a soluble precipitating antigen from African horsesickness virus propagated in cell cultures, in *Proc. 3rd Int. Conf. Equine Infect. Dis.*, S. Karger, Basel, 1973, 38.
53. **Salama, S. A.,** Epidemiological and immunological studies on South African horsesickness and trials of production of inactivated vaccine. Final technical report, *Minst. Agric. Vet. Serum Vacc. Res. Inst. Cairo*, 1984.
54. **Ozawa, Y.,** Bluetongue and related orbiviruses: overview of the world situation, in *Bluetongue and Related Orbiviruses*, Alan R. Liss, New York, 1985, 13.
55. **Bourdin, P.,** Ecology of African horsesickness, in *Proc. 3rd Int. Conf. Equine Inf. Dis.*, S. Karger, Basel, 1973, 12.
56. **Maurice, Y. and Provost, A.,** La peste equine a type 9 en Afrique central. Enquete serologique, *Rev. Elev. Med. Vet. Pays Trop.*, 20, 21, 1967.
57. **Gilbert, Y.,** *Rapports Annuels du Laboratoire National*, Curasson, Dakar, Senegal, 1954, 1957.
58. **Mellor, P. S., Boorman, J. P. T., Wilkinson, P. J., and Martinez-Gomez, F.,** Potential vectors of bluetongue and African horse sickness viruses in Spain, *Vet. Rec.*, 112, 229, 1983.
59. **Henning, M. W.,** *Animal Diseases in South Africa*, 3rd ed., Central News Agency, Johannesburg, South Africa, 1956, 785.
60. **Nevill, E. M.,** A significant new breeding site of *Culicoides pallidipennis* Carter, Ingram and Macfie (Diptera:Ceratopogonidae), *J. S. Afr. Vet. Med. Assoc.*, 39, 61, 1968.
61. **Braverman, Y. Linley, J. R., Marcus, R., and Frish, K.,** Seasonal survival and expectation of infective life of *Culicoides* spp. (Diptera:Ceratopogonidae) in Israel, with implications for bluetongue virus transmission and a comparison of the parous rate in *C. imicola* from Israel and Zimbabwe, *J. Med. Entomol.*, 22, 476, 1985.
62. **Becker, P.,** Observations on the life cycle and immature stages of *Culicoides circumscriptus* Kieff. (Diptera:Ceratopogonidae), *Proc. R. Soc. Edinburgh*, B67, 363, 1961.
63. **Campbell, M. M. and Kettle, D. S.,** Oogenesis in *Culicoides brevitarsis* Kieffer (Diptera:Ceratopogonidae) and the development of a plastron-like layer on the egg, *Aust. J. Zool.*, 23, 203, 1975.
64. **Hoogstraal, H.,** African Ixodoidea. I. Ticks of the Sudan (with special reference to Equatoria Province and with preliminary reviews of the genera *Boophilus, Margaropus, and Hyalomma*, Res. Rep. NM 005 050.29.07, Bureau of Medicine and Surgery, Department of the Navy, Washington, D.C., 1956, 420.
65. **Delpy, L. P. and Gouchey, S. H.,** Biologie de *Hyalomma dromedarii* (Koch 1844), *Ann. Parasitol. Hum. Comp.*, 15, 487, 1937.
66. **Nevill, E. M.,** Cattle and *Culicoides* biting midges as possible overwintering hosts of bluetongue virus, *Onderstepoort J. Vet. Res.*, 38, 65, 1971.
67. **Nevill, E. M. and Anderson, D.,** Host preferences of *Culicoides* midges (Diptera:Ceratopogonidae) in South Africa as determined by precipitin tests and light trap catches, *Onderstepoort J. Vet. Res.*, 39, 147, 1972.
68. **Kettle, D. S.,** The spatial distribution of *Culicoides impunctatus* Goet. under woodland and moorland conditions and its flight range through woodland, *Bull. Entomol. Res.*, 42, 239, 1951.
69. **Sellers, R. F.,** Weather, host and vector — their interplay in the spread of insect-borne animal virus diseases, *J. Hyg.*, 85, 65, 1980.

Chapter 15

AFRICAN SWINE FEVER

William R. Hess

TABLE OF CONTENTS

I. HISTORICAL BACKGROUND

A. Discovery of Agent and Vectors

African swine fever (ASF) was first described by Montgomery[1] in a report published in 1921 based on his observations in Kenya from 1909 to 1915 when 15 outbreaks occurred involving 1366 pigs, of which 1352 (98.9%) died. He established the viral nature of the disease, and studied the host range, mode of transmission, and the stability of the virus under a variety of environmental conditions and explored methods of immunization. Because outbreaks in domestic pigs usually occurred in areas frequented by warthogs, he suggested that the wild pig was perhaps the reservoir of the virus in nature. It was soon found that both warthogs and bush pigs often carried the virus without signs of illness.[2-4] However, the mode of transmission from the wild pig to the domestic pig was not readily apparent. Because pigs placed in contact with infected warthogs failed to become infected, the possible involvement of bloodsucking arthropods as vectors was explored. Fleas and lice failed to transmit the virus experimentally.[1,5-7] Several hundreds of lice *(Haematopinus phacochoeri* and *Haematomyzus hopkinsi)* collected from infected warthogs in Kenya were free of virus.[8] The fact that domestic pigs could be protected from acquiring the disease merely by confining them to paddocks that excluded wild pigs ruled out the possibility that flying insects were involved in the transmission.

The first attempts to incriminate ticks as vectors of ASF virus in Africa failed.[9] Ixodid ticks commonly found on warthogs or in their burrows were free of the virus and were also unable to experimentally transmit the virus when fed sequentially on infected and healthy susceptible pigs. Others attempted to demonstrate transmission of ASF virus with species of hard ticks and also obtained negative results.[7,10] In 1963, Sanchez Botija[11] announced that the virus was recovered from argasid ticks (*Ornithodoros erraticus)* found on farms in Spain where ASF outbreaks had occurred. Efforts to find a tick vector for the disease in Africa were renewed. *O. moubata porcinus,* a tick that was often found in warthog burrows, became the main object of the ASF vector studies in Africa. Experimental transmission of ASF virus from infected to healthy susceptible pigs by *O. moubata porcinus* was soon demonstrated,[7] and in a series of reports by Plowright and coworkers,[6,12-14] the full potential of *O. moubata porcinus* as a vector of ASF virus was revealed. The arthropod vector may be essential for the maintenance of the virus in nature and may have been responsible in most instances for its spread to domestic pigs in Africa. However, once ASF virus has entered a domestic pig population, it readily spreads among the pigs by contact.

B. History of Epidemics and Their Economic Impact

The virus of ASF was undoubtedly present in soft ticks and the wild indigenous pigs of Africa for a very long time, but it did not emerge as a disease agent until breeds of domestic swine were imported from Europe. That which was present as an inapparent infection in wild pigs then appeared as an acute, highly contagious and lethal febrile disease in the domestic pigs. Soon after Montgomery described the disease, outbreaks were reported in a number of other areas south of the Sahara. Because swine production was sparse, outbreaks were usually isolated and self limiting, and each one represented a new incursion of the virus from its natural reservoir. However, in South Africa in 1933, ASF appeared in domestic pigs in the Northern Transvaal and spread to Witwatersrand and on to the Western Cape Province. For the first time, an extensive spread of ASF involving a large number of pigs was reported.[3] Of the 11,000 pigs affected, only 8% survived. Most of the survivors were in poor condition, and many continued to carry the virus in their blood for several months. As is true today, a satisfactory vaccine was not available, and further spread of the disease could only be halted by a strict quarantine and a slaughter program that eliminated all surviving pigs in the affected area. The disease was first clearly described in the Portuguese colony of Angola in 1947, but there is some evidence that it was already present in 1932.[15] Domestic pigs were allowed to range freely, and although there are no estimates of the number of pigs lost to ASF, it was stated that the disease had become enzootic, and the development of a viable swine industry in Angola was virtually impossible.

In 1957, ASF appeared in Portugal. It was contracted by pigs fed on food wastes from the airport in Lisbon and was believed to have come from Angola. This first incursion of the disease outside the continent of Africa was said to be eradicated in 1958 after 6103 pigs died of the infection and 10,354 more were slaughtered as probable contacts.[16] However, in 1960, the disease appeared once again near Lisbon and during the same year spread to Spain. Since then, ASF has been enzootic on the Iberian Peninsula, and losses of 2 to 3% of the national swine populations have been reported annually in both Spain and Portugal. Millions of dollars have been spent annually in efforts to control the disease, and losses of millions more are attributed to the loss of export markets for their agricultural products.

Presumably as extension from the Iberian Peninsula, outbreaks of ASF have occurred in island possessions and neighboring countries in the Mediterranean area. Outbreaks occurred in France in 1964, 1967, and 1974.[17,18] Fortunately, in each instance the disease was quickly recognized and eradicated by the application of a drastic slaughter program, and losses were not great. The disease also spread to Italy in 1967, and during that year 100,000 pigs died or were slaughtered.[19] It was finally eradicated in 1969 after estimated losses of more than $5 million were sustained. The disease was reported in Madeira in 1965. The full extent of the losses was not reported.

The first incursion of ASF in the Western Hemisphere occurred in Cuba in 1971.[20] With the help of advisers from Russia, France, and Canada, the disease was eradicated, but only after more than 400,000 pigs died or were slaughtered.

In Spain and Portugal, the number of outbreaks has varied in a cyclic fashion over the years, with 1977 being a year of particularly high incidence. During the following year, ASF appeared in a number of widely separated areas of the world. In the Mediterranean, it appeared in Malta and Sardinia, and in the Western Hemisphere, it appeared in Brazil and the Dominican Republic.[21-23] In 1979, it was reported in Haiti. Although the origins of these outbreaks may never be known with certainty, the viruses isolated have usually been of lower virulence producing infections similar to those that were occurring on the Iberian Peninsula. The disease was eliminated in Malta by the death or slaughter of the island's entire population of 80,000 pigs. Costly eradication programs were similarly conducted in the Dominican Republic and Haiti. In Haiti, e.g., more than 384,000 pigs were systematically elminated in about 1 year at the cost of $9,548,860 for indemnities alone.

In 1980, ASF reappeared in Cuba.[24] Food brought in by refugees from Haiti was thought to be the source of the outbreak. Again, substantial losses were sustained before the disease was eradicated. After more than 6 years of testing, slaughtering, and surveillance, Brazil claims to be free of ASF and hopes to regain her lost export markets. In parts of West Africa where ASF had not previously occurred, large outbreaks have been reported in recent years. In Cameroon, an estimated 80% of the sizable swine population of the nation was destroyed by the disease in 1982. Most of the commercial producers lost their entire herds. Belgium had an outbreak of ASF in 1985. It was their first experience with ASF, and nearly 20,000 pigs were lost before the country was declared free of the disease. The outbreak lasted only a few months. Lesser developed countries lacking adequate veterinary services do not fare as well. A new disease usually becomes well established in the animal population before it is diagnosed and steps are taken to combat it. The losses, whether they are due to the disease or to the measures taken to control it, usually are most damaging to the small farmer. In Haiti, the pig is not only one of the few sources of dietary protein, it is often the only source of income or purchasing power for the small farmer. Without substantial funds and assistance given to Haiti for eradication of the disease, ASF would have been an unbearable disaster for many of the small farmers.

II. THE VIRUS

A. Antigenic Relationships and Strain Variations

As a large icosahedral cytoplasmic DNA virus, ASFV has been placed in the family Iridoviridae.[25] Although the viruses of this family are morphologically similar and have certain other characteristics in common, most of them have separate and exclusive host ranges and are not serologically related. African swine fever virus is the only member of the group that infects a warm-blooded animal, and it is the only animal DNA virus presently known to satisfy the criteria for classification as an arbovirus. In the pig, ASF virus attacks cells of the reticuloendothelial system or the mononuclear phagocyte system of van Furth et al.[26] Cultures of these cells extracted from bone marrow or blood of uninfected domestic pigs are readily infected with ASF virus and are used to isolate and titrate the virus.[27] Depending on how the cultures are prepared, erythrocytes present in or added to the cultures adsorb to the surfaces of the infected mononuclear cells to form rosettes that serve as distinctive indicators of an infected culture. Infected lymphocytes and neutrophils have also been reported; however, this has been disputed by others. In any case, some of the cells that are involved in immune responses are targets of the virus. Despite this knowledge, the immunology of ASF is poorly understood. Chronically infected pigs and pigs that have survived infection frequently have high levels of precipitating, complement-fixing, and hemadsorption-inhibiting antibodies in their blood, and at the same time, virus may also be present. Virus neutralizing antibodies have not been convincingly demonstrated.

Numerous isolates of ASF virus from widely separated geographical locations have been available for study and comparison. When it has been possible to carry out cross-immunity tests, it has often been found that pigs refractory to one virus isolate may not withstand challenge inoculation with another isolate. Serums of surviving pigs are usually able to inhibit the hemadsorption reaction produced in cell cultures by the homologous virus, but may fail to inhibit the reaction produced by another virus isolate.[27] At the molecular level, restriction endonuclease cleavage patterns of the ASF viral DNA from different isolates have revealed distinct differences in the genomes of the isolates.[30] From these findings, it would appear that several immunological types of the virus exist. However, a number of antigens are common to all of the isolates, and none of several serological tests that may be used to identify the disease is able to distinguish one isolate from another. While this may be advantageous for detection of the disease, the strain differences appear to be of sufficient magnitude to preclude the possibility of producing a single vaccine that will protect against all strains of the virus.

B. Host Range

Natural infections with ASF virus have been found only in porcine species and ticks of the genus *Ornithodoros*. In Africa, the virus has been isolated several times from warthogs *(Phacochoerus aethiopicus)* and bush pigs *(Potamochoerus porcus)*, and both species have been infected experimentally.[1,9] A single isolation was made from a giant forest hog *(Hylochoerus meineizhageni)*.[31] Isolations from a hippopotamus,[9,34] a porcupine, and a hyena have also been claimed,[32] but these findings have not been confirmed by additional isolations, nor have these species been experimentally infected. The European wild boar *(Sus scrofa ferus)* in Spain and Sardinia has acquired the infection from domestic pigs and reacts similarly.[34,35] A survey in Spain has indicated that 5.8% of the outbreaks in domestic pigs resulted from contacts with wild boars. Feral pigs from the southeastern U.S. experimentally infected with ASF virus were fully susceptible, and the disease produced was similar to that seen in the domestic pig.[36] The American white-collared peccary belongs to the Tayassuidae family and is susceptible to a number of virus diseases that affect domestic pigs, but is resistant to ASF.[37]

Among other animals tested were cattle, horses, sheep, goats, dogs, cats, guinea pigs, rabbits, hedgehogs, hamsters, rats, mice, and various fowl.[1,38] With the exception of kids less than 4 months old, all were refractory to ASF virus. Through a series of alternating passages in pigs and rabbits, strains of ASF virus were derived that could be propagated in rabbits.[39] However, it appears that members of the family Suidae are the only vertebrates that may be infected naturally.

C. Methods for Assay

Strains of ASF virus which grow well in a variety of cell culture systems have been derived from several field isolates. These strains have been used extensively in research on the virus and for the production of diagnostic reagents. They are readily assayed by cytopathic effect or plaque counting in the particular cell lines to which they have been adapted. Field isolates from pigs or ticks must be assayed in pig monocyte-macrophage cultures. Preferably, the cells are cultured in plastic plates having flat-bottom wells (24, 48, or 96 wells per plate). In the acutely infected pig, virus is detectable in all tissues that have a good blood supply, but here we are concerned mainly with the level of viremia at the time of vector feeding. For this purpose, blood drawn from the anterior vena cava is heparinized, and tenfold serial dilutions are made in a cell culture medium such as Eagle's minimum essential medium (MEM). Eight wells are inoculated per dilution. To assay the virus in vectors, individual ticks are weighed, surface sterilized as described elsewhere,[40] triturated in Ten Broeck grinders, and serially diluted in MEM. In 96-well plates, eight wells per dilution are inoculated using 0.01 mℓ per well. The cultures are incubated at 37°C in a 3 to 4% CO_2 atmosphere for at least 7 days, during which time they are periodically shaken sufficiently to resuspend the unadsorbed erythrocytes and are examined for hemadsorption using an inverted microscope. The titers are calculated by the method of Karber.[41]

III. DISEASE ASSOCIATIONS

A. Humans

Concurrent events in the epidemiology of ASF and the acquired immune deficiency syndrome (AIDS) in humans led to the suggestion that ASF virus might be the causative agent of the syndrome.[42] Recently, through the application of cell culture and serological methods ordinarily used in diagnosing ASF in pigs, a number of apparently positive results were obtained with the blood of AIDS patients.[43] If the results are truly positive, based on the tests used, isolation and identification of the virus should be relatively easy. Until this is done, an association between ASF and AIDS has not been definitely proven.

B. Domestic Animals

Thus far, the pig is the only domestic animal known to contract ASF. It occurs in clinical forms ranging from peracute and acute to subacute, chronic, and inapparent. In the peracute form of the disease, death may occur before any clinical signs appear. When this kind of infection was produced experimentally, a febrile response occurred within 3 days, and death occurred 3 days thereafter. Fever was the only indication of illness. In acute ASF, the first symptom is a fever that may persist for 3 or 4 days and may exceed 106°F. During this time, the total leukocyte count may fall to 40% of normal.[9] Affected animals may appear quite normal until their temperatures begin to fall. They then stop eating and tend to lie huddled together. If forced to rise, they appear weak and unstable. Red areas often appear in the skin on the ears and flanks, and the pulse and respiration accelerate as death approaches. Other signs occasionally encountered are vomiting, bloody diarrhea, and mucopurulent conjunctival and nasal discharges. Death usually occurs by the 7th day after the onset of fever.

A subacute infection is one that runs a course ending in death or recovery in 3 or 4 weeks. A high fever marks the onset of the disease and usually persists for several days or fluctuates irregularly throughout the course of the disease. Abortions are common and may be the only sign of the disease. Survivors may harbor the virus for the rest of their lives.

Chronic ASF is extremely variable in its appearance, and may mimic a variety of illnesses. It may persist for several months with stunting and emaciation being the only signs of illness. Swollen joints, lameness, and skin ulcerations are other signs that may appear. The chronically infected pig may be the survivor of an acute infection and may continue to harbor the virus for the rest of its life. Pneumonia is often the ultimate cause of death.

None of the forms of ASF is unique enought to be diagnosed clinically. Other swine diseases, especially hog cholera, may display similar signs. In most instances, the same is true for the pathology of the disease. In some instances when death occurs before there are any signs of disease, gross lesions may be minimal. The most typical lesions of peracute and acute ASF are an enlarged dark-colored spleen and hemorrhages may occur in any internal organ. Petechiae are often found in the renal cortex. They may also occur on the mucosa of the intestines and urinary bladder, in the myocardium, and on the subendocardial and epicardial surfaces of the heart. Nearly all lymph nodes are enlarged and show peripheral reddening. The renal and especially the gastrohepatic nodes are most severely involved and often resemble blood clots. The gallbladder may be distended with bile, and the wall may be edematous. Some lesions occur more frequently than others, and a few, when present, are strongly suggestive of ASF. An engorged spleen that may be several times its normal size and the severely involved gastrohepatic and renal lymph nodes are the most distinct features of acute ASF.

Subacute ASF has become common in recent years, and in some outbreaks has been the predominant form of the disease. Again, the reticuloendothelial system is involved. Hemorrhages are most pronounced in the lymph nodes and kidneys. If the spleen is enlarged, it is due to hyperplasia of cellular elements rather than engorgement with blood. The lungs are often involved with areas of lobular consolidation and, in some case, there may be a diffuse interstitial pneumonia.

The lesions of chronic ASF are not unique. In pigs that die after one of the recurring cycles of pyrexia, the lesions seen are similar to those of the subacute disease. Hemorrhages may be prominent, but any enlargement of the lymph nodes or spleen is due to hyperplasia rather than engorgement. In fact, hyperplasia of the lymphoreticular tissues is one of the most prominent features of chronic ASF. Fibrinous pericarditis and pleuritis are often seen. Pneumonia accounts for more than half of the deaths in the chronic disease.

C. Wildlife

As stated previously, warthogs, bush pigs, and the giant forest hog have been identified as carriers of the virus in Africa. None of these appear to suffer from the infection, and no lesions have been reported. However, the European wild boar has become infected in Spain and Sardinia and expresses illness similar to that seen in the domestic pig.[34,35] Feral pigs from the southeastern U.S. have been shown experimentally to be fully susceptible to ASF.[36] They, too, react the same as domestic pigs.

D. Diagnostic Procedures

Although there are some clinical signs and gross pathology that might lead one to suspect ASF, a positive diagnosis of the disease in the field is virtually impossible. Samples must be submitted to a laboratory for detection of the virus or demonstration of ASF-specific antibodies. To detect virus, the following tests may be used:

1. Inoculation of hog cholera-immune and -susceptible pigs[31]
2. Hemadsorption (HAd) reaction[27,44]
3. Direct immunofluorescence (DIF)[45,46]
4. Agar gel diffusion precipitin test (AGDP)[47]
5. Complement fixation (CF)[48,49]
6. Radioimmunoassay (RIA)[50]

The inoculation of hog cholera-immune pigs has been regarded as the definitive test for distinguishing ASF from hog cholera, but the test is relatively slow and costly and is now used only to confirm the first positive diagnosis in a newly infected country. The HAd test is now generally regarded as a definitive differential test for ASF and is the most widely used means of isolating virus from field specimens. If the required pig leukocyte or bone marrow cultures are available for immediate inoculation, the virus can often be detected by the HAd reaction within 12 hr. With the DIF test performed on tissue smears or preferably on frozen tissue sections, the virus may be detected in less than 1 hr. This test is usually not applicable to tissues from protracted cases, for apparently enough antibody may be present to block the DIF reaction. The AGDP, CF, and RIA are generally considered to be too slow or inconvenient to be used as routine diagnostic tests.

ASF antibodies may be detected by the following tests:

1. AGDP test[51]
2. Indirect immunofluorescence (IIF)[52]
3. Immunoelectroosmophoresis (IEOP)[53]
4. Reverse radial immunodiffusion (RID)[54]
5. CF[48]
6. Enzyme-linked immunosorbent assay (ELISA)[55-58]
7. Indirect immunoperoxidase plaque staining (IIPS)[58]
8. RIA[50]

The IIF, IEOP, ELISA, and IIPS are the tests of choice to detect antibody. Because no single test can be expected to detect the disease under all conditions, an array of tests may be conducted if necessary. Spleen, liver, lymph nodes, whole blood, and serum should be submitted when an initial diagnosis is being sought. All tissue samples and serums should be kept on ice or refrigerated or 4°C until tested.

E. Effects of the Virus on the Vector

Ornithodoros moubata porcinus and *O. erraticus* are the only arthropod species that have thus far been found to harbor ASF virus in nature, and neither species has been reported to be adversely affected by the virus. However, there is no indication that anyone has looked at that aspect of the vector-virus relationship in ASF. In an ongoing survey to identify arthropods indigenous to North America and the Caribbean Basin which might serve as vectors and perhaps reservoirs of ASF virus if the disease were to enter the U.S., *O. coriaceus*, a tick found along the coastal range of California, was found capable of transmitting the virus.[10] In the first trials, an elevated mortality rate in the infected ticks was noted, and it was thought that it might be due to the virus. In a later study using a much larger number of ticks, the vector capability of *O. coriaceus* was more thoroughly explored.[40] Second- and third-stage nymphs were infected by feeding on ASF-infected pigs. Three different isolates of ASF virus were used: one was from a domestic pig in Malawi (Tengani); another was from a domestic pig in the Dominican Republic (DR2); and the third isolate was from *O. moubata porcinus* taken from a warthog burrow in Zimbabwe (Z1). The ticks were fed periodically on normal pigs to determine the duration of their ability to transmit. At 502 days after feeding on the infected pig, the surviving DR2-infected ticks were still able to transmit. During an observation period of 463 days, they had molted at least 3 times before becoming adults, and 454 or 40% of the 1141 that originally fed had died. The levels of the Tengani and Z1 infections steadily decreased until virus was no longer detectable after 220 and 463 days, respectively. With the Z1 virus, 124 out of 344 (36%) died, and 69 of the 425 (19%) infected with Tengani virus died. Most deaths occurred at about the time of molting. In nearly every instance, the dead ticks had virus titers comparable to or higher than surviving ticks sampled at about the same time. Mortality was only 3% in a control lot of 892 ticks fed on normal pigs during the same period of observation. It appears that the elevated mortality rates in ASF-infected *O. coriaceus* may be due to the virus and may vary depending on the isolate involved. It would be of considerable interest to learn if *O. moubata porcinus* and *O. erraticus* are also adversely affected by ASF virus infection.

IV. EPIDEMIOLOGY

A. Georgaphic Distribution

Most countries south of the Sahara in Africa have had outbreaks of ASF. In Kenya, Tanzania, Uganda, Zimbabwe, Malawi, Zaire, and the Republic of South Africa, the wild indigenous pigs and argasid ticks of the genus *Ornithodoros* are often carriers and reservoirs of the virus. The disease is enzootic in domestic pigs in Angola and Mozambique, and the virus is probably also present in ticks and wild pigs. It now appears that ASF has become enzootic in domestic pigs in Cameroon.

In Europe, the disease continues to be enzootic in Portugal and Spain and must now also be regarded as enzootic in Sardinia. The disease has been successfully eradicated in France, the Italian mainland, Malta, and Belgium.

It appears likely that ASF has been eradicated in the Dominican Republic, Haiti, and Cuba, and no further outbreaks have been reported in Brazil since 1981. It is hoped that the Western Hemisphere is once again free of the disease.

B. Incidence

ASF outbreaks, wherever they occur, are regularly reported in the *Bulletin of the Office of International Epizootics*. From these reports, it is apparent that the incidence of ASF varies considerably from one year to the next. On the Iberian Peninsula where the disease is enzootic, there is a cyclic variation in the number of outbreaks with peaks occurring about every 7 years. Cyclic variations in the incidence of other infectious diseases have often been found to coincide with the variations in the antibody level of the population. Because ASF virus-neutralizing antibodies have not been convincingly demonstrated, an explanation for the cyclic variations in the incidence of the disease on the Iberian Peninsula is not readily apparent.

C. Seasonal Distribution

There is no apparent seasonal variation in the distribution of the disease. However, generalization on this point is not possible, for there has never been a protracted incursion of ASF in an area subject to severe seasonal changes in climate.

D. Risk Factors

Although domestic pigs of both sexes and all ages and breeds are susceptible to ASF, in some recent outbreaks attributed to viruses of low virulence, young pigs were said to be most vulnerable and had the highest mortality rate. Pregnant sows were particularly at risk, and abortions were common.[59]

E. Serologic Epidemiology

The presence of ASF antibodies in a domestic pig indicates that the animal was infected and, in all probability, is still carrying the virus. The basis for this assumption is strong enough to warrant using antibody detection to determine the bounds of an outbreak and follow the progress of an eradication program. Because of its devastating potential and the lack of a satisfactory vaccine, quarantine and slaughter are resorted to as the only means of halting the spread and achieving the eradication of the disease. Once an outbreak has been definitely identified as ASF, a positive antibody test in a single animal is considered sufficient to condemn the rest of the herd. These practices appear to have been successful in controlling and eradicating ASF in the Western Hemisphere and several locations in Europe. It might be argued that such drastic measures are unnecessary if the ASF virus strain causing the outbreak is of low virulence. Perhaps in that case, only positive reactors need to be eliminated, and subseqeunt serological testing would yield the epidemiological information needed to enable lesser developed nations to deal with such outbreaks in more economical fashion. However, without a vaccine, few veterinary officials would care to see this premise tested in their own country or elsewhere.

In Africa, the indigenous wild pigs, especially warthogs and bushpigs, are distributed widely and, in some areas, have often been found to be infected with ASF virus. Because these animals are free living and do not show signs of illness when infected with the virus, serological testing of them is of considerable interest. With warthogs, such studies are difficult at best, but with bushpigs, they are virtually impossible because of their nocturnal habits and the dense cover of their habitat. The animals must be hunted and killed or immobilized with anesthetizing darts. Nevertheless, reasonably good estimates of the prevalence of ASF antibodies in warthog populations in a number of different locations in eastern and southern Africa have been obtained.[8,60] In the same studies, viremia levels in the warthogs, the presence of *O. moubata porcinus*, and the occurrence of ASF outbreaks in domestic pigs in the areas were recorded. The significance of the antibody data only becomes apparent when examined in relation to these other findings. In areas where antibody prevalence in warthogs was high, the number of warthog burrows harboring *O. moubata porcinus* was also high. In such areas, infection rates in ticks, as well as warthogs, were usually high, and outbreaks in domestic pigs occurred. Warthogs with relatively high viremias also had high antibody levels. Again, as in the domestic pig, the presence of antibodies in the warthog also signals the presence of infectious virus in the animal.

Although the epidemiology of ASF in Africa is not understood completely, it is quite clear that the wild indigenous pigs in association with ticks of the *O. moubata porcinus* complex play a major role in maintaining ASF virus in its sylvatic state. From the work of Plowright[8] in East Africa and Thomson[60] in South Africa, it is apparent that a serological survey of the free-living hosts of ASF virus in any particular area could serve as an accurate means of determining the risks involved and the preventive measures that must be applied to assure the viability of a swine industry in that area. However, collection of a statistically significant sample to determine the prevalence of antibodies in a wild animal population is not a task that many countries in Africa would be willing or able to undertake.

The inability to demonstrate neutralizing antibodies and the prevalence of common antigens have made it virtually impossible to accurately identify virus strains and trace their origins. Although the hemadsorption inhibition test has been used to sort a few virus strains into groups, it has never been applied in an extensive survey to determine its value as a tool for studying the epidemiology of ASF.[61] Instead, efforts have been made recently to identify and differentiate virus isolates on the basis of the patterns obtained by restriction endonuclease cleavages of their DNA genomes.[30]

V. TRANSMISSION CYCLES

A. Evidence from Field Studies

1. Vectors

Before proceeding with the discussions on the two vectors that are known to be carrying ASF virus in the field, their taxonomic positions should be clarified. *Ornithodoros moubata* Murray 1877 was studied for many years by Walton[62] who was concerned with the epidemiology of human relapsing fever in Africa. He decided that the species was in fact a complex consisting of four species and two subspecies. One species was associated with porcupines in Kenya and another with tortoises in South Africa. A third species for which he reserved the name *O. moubata* is the vector of *Borrelia duttoni*, the cause of human relapsing fever in Africa. It is found in huts and the lairs of wild animals in many of the arid regions of Africa. A fourth species that was found in large numbers in burrows occupied by warthogs and other wild animals was designated *O. porcinus* with the subspecies *O. porcinus porcinus* and *O. porcinus domesticus*. In reviewing Walton's complex, van der Merwe[63] concluded that *O. porcinus porcinus* should be designated *O. moubata porcinus*. The nomenclature of van der Merwe is used in this presentation.

The ASF virus-infected tick found in the Spanish provinces of Andalusia and Extremadura and identified as *O. erraticus* by Spanish investigators is believed by some authorities to be *O. marocanus* Velu, a species closely related to *O. erraticus*. The Spanish identification is accepted here because their studies are the only ones, thus far, that relate specifically to the ASF vector in Spain.[64]

When seeking information on the biology of these vectors, it is often difficult to determine from which member of their respective complexes the information was derived.

It was early recognized in Africa that ASF outbreaks usually occurred in areas where infected warthogs existed, but the mode of transmission from the warthog to the domestic pig was not clearly established. Then it was discovered that *O. moubata porcinus* found in warthog burrows often carried the virus and could transmit it by bite to domestic pigs, and transstadial, sexual, and transovarial transmission of the virus in the tick were established.[6,8] Prior to the African studies, *O. erraticus* found in pig pens where outbreaks had occurred in Spain were often infected and could serve as the source of a new outbreak when the pens were restocked 6 to 12 months later.[65]

2. Vertebrate Hosts

As stated previously, natural infections with ASF virus have been found only in porcine species. Although the ASF virus-*O. moubata porcinus*-warthog complex is probably the most important natural reservoir of ASF in Africa, it does not account for many of the outbreaks on the African continent. In addition to incursions of ASF which result from human activities such as the movement of infected animals and the feeding of ASF virus-contaminated swill, there are outbreaks that appear to originate from unknown or undefined reservoirs in nature. In an area of central Kenya at altitudes of more than 6200 ft above sea level, no ticks were found in any of the more than 100 warthog burrows examined, yet ASF virus infection in the warthogs was universal, and outbreaks in domestic pigs occurred.[8] On the other hand, there are parts of South Africa where warthog populations appear to be entirely free of ASF virus infection.[8] In other areas where *O. moubata porcinus* and the warthog coexist and both populations are carrying the virus, there may be substantial differences from one geographical location to another in the overall infection rates in the two populations, the infection rates in various age groups of the warthogs, the maximum virus titers found in each group, and the percentage of warthog burrows yielding infected ticks.[8] There has been much study and speculation concerning the reasons for these differences. However, if we consider the number of ways that ASF virus isolates may differ from one another and the differences that may exist in a single tick species collected from different geographical locations, variations in the ASF virus-*O. moubata porcinus*-warthog complex and its epidemiological manifestations are to be expected. It was thought that the warthog did not play an important role in the maintenance of the virus in nature because it was never found to have a high enough level of viremia to establish a persistent infection in the feeding tick.[8] However, more recent studies show that 3-month-old warthogs experimentally infected with ASF virus had viremias that persisted up to 11 days after infection and were high enought to persistently infect some of the ticks fed on them, mainly large nymphs and adults.[66] Later, Thomson and co-workers[67] found 6 out of 20 2-week-old warthogs captured in Namibia to also have sufficiently high titers to infect feeding ticks. Although *O. moubata porcinus* is a fast feeding tick, it is sometimes found on the warthog outside its burrow during the day. At the very least, the warthog probably serves to disseminate infected ticks.

B. Evidence from Experimental Infection Studies

1. Vectors

O. savignyi, a widely distributed soft tick species in Africa, has been found to be experimentally capable of acting as a biological vector of ASF virus.[68] Because it is found in many areas where ASF occurs, it may actually be serving as a natural vector of the disease. Recent studies indicate that most, if not all, *Ornithodoros* species that will feed on mammals may be capable of acting as vectors of ASF virus. Thus far, three species in North America and the Caribbean Basin, *O. coriaceus*, *O. turicata*, and *O. puertoricensis*, have been shown experimentally to be susceptible to ASF virus infection and to be capable of transmitting the virus by bite to healthy domestic pigs.[40] Transstadial transmission of ASF virus has been demonstrated in all three species, and transovarial transmission has been experimentally produced in *O. puertoricensis*.

2. Vertebrate Hosts

Experimental infections with ASF virus have been produced in goats and rabbits. In goats, infection could be initiated only in kids less than 4 months old.[38] A series of alternating passages in pigs and rabbits was required before a virus strain was obtained that could regularly infect rabbits.[39] The feral pigs of the southeastern U.S. have proved experimentally to be fully susceptible.[36] Again, it appears that only true porcine species are naturally susceptible to ASF.

There have been several experimental efforts to determine the mode of transmission involved in the rapid spread of ASF among domestic pigs. Montgomery[1] found that a susceptible pig confined behind wire mesh fences that kept it separated by only a few inches from an ASF-infected pig did not contract the disease. He also showed that susceptible pigs in direct contact with infected pigs did not become infected if they were muzzled and unable to ingest excretions of the infected pigs. Although later experiments indicated that airborne transmission can occur,[69] it is apparently possible only over very short distances. Therefore, it is assumed that virus present in the secretions and excretions of acutely infected pigs is transmitted to other pigs by nuzzling or ingestion and that primary invasion occurs in the upper respiratory or alimentary tracts.[70] Based on that assumption, studies on the pathogenesis of ASF have been conducted on pigs infected by intranasal and oral instillation of virus or by contact. Pigs of different sizes and ages and virus isolates of different degrees of virulence were used, and the results obtained in each study differed somewhat but in most respects were quite similar.[71-74]

Whether the virus was inoculated orally or intranasally, the primary infection usually occurred in the upper respiratory tract and rapidly spread to lymph nodes in the cephalic region as early as 1 day postinoculation (DPI). In mature animals, there was rarely evidence of virus invasion by way of the alimentary tract. In one study using a highly virulent isolate, the virus was first detected in the tonsils and mandibular lymph nodes.[71] In another study using the same route of infection and the same virus isolate, the initial infection was detected in the retropharyngeal mucosa and the retropharyngeal nodes.[72] From these primary sites, the virus apparently spread via the lymph ducts and the bloodstream to practically all tissues of the body. In some instances, viremia occurred 1 or 2 days before the onset of fever. The infection was usually generalized after 3 DPI, and pyrexia usually occurred about the same time or within the next 24 hr. In acute ASF, the virus titer in the blood may quickly rise to 10^7 to 10^8 50% HAd units/mℓ and remain at that level until death occurs 3 or 4 days later. Because the virus has a predilection for cells of the reticuloendothelial system, it was found in highest concentrations in tissues having large components of reticuloendothelial cells. The lymphoid tissues involved in the primary infection harbored large quantities of virus throughout the course of the disease. The spleen, bone marrow, liver, and lungs were the main secondary sites of virus multiplication.

The infection usually followed a similar but more rapid course in newborn pigs after oral administration of ASF virus.[73] The tonsils were the most frequent site of initial entry of the virus. Viremia occurred as early as 8 hr after inoculation, and generalization was well advanced by 30 hr. Occasionally, primary infection was found in the mesenteric lymph nodes indicating that the virus may have entered through the small intestine.

With low or moderately virulent viruses, the invasion, primary infection, and subsequent generalization followed much the same time sequence as seen in the acute and fatal infection. However, after the initial febrile responses that varied in duration, many of the animals appeared to recover completely.[74] Their temperatures returned to normal and, after 5 or 6 weeks, viremia was no longer demonstrable. Virus was found in the lymph nodes and the tonsils for more than 90 DPI. Other internal organs were usually free of virus.

Having established the upper respiratory tract as the most probable portal of entry of the virus in transmission by direct contact, the routes of excretion that made the virus available for such transmission were determined. In a study using a highly virulent strain of virus,[75] excretion was first detected from the nasopharynx, and it occurred as early as 1 or 2 days before the onset of fever in many cases. The titers rose rapidly and by 48 to 72 hr after the onset of fever, nasal and pharyngeal swabs picked up about 10^4 to 10^5 50% HAd units of virus. Virus appeared in the secretions of the conjunctiva and the lower urogenital tract later and was of lower titer. The feces and urine contained insignificant amounts of virus. With a slower-acting virus, excretion in the feces and urine was considerably higher. The feces proved to be the major route of virus excretion in another study in which a moderately virulent virus isolate was used.[74]

Montgomery[1] and Plowright et al.[72] found that the infection was not contagious during the first 12 to 24 hr of fever. At first it was thought that no virus was being excreted that early in the infection. However, when fairly high concentrations of virus were detected in nasopharyneal secretions often prior to the onset of pyrexia, it could only be concluded that viral contamination of the environment did not reach a level high enough to assure transmission until sometime later.[75] Anyone who has observed numerous ASFV-infected pigs is likely to agree that the excretion of blood-tinged urine and feces is not an uncommon occurrence, especially during advanced stages of the infection. The amount of virus in such excretions is likely to be very high, and a pig exposed to them could become infected through inhalation, ocular instillation, or an abrasion in the skin or mucus membrane.

All attempts to demonstrate transmission from infected warthogs to domestic pigs by contact have failed.[1,9,76] Although it was possible to isolate virus from the blood of an experimentally infected warthog for at least 54 DPI, there was no evidence of virus excretion to an extent sufficient to infect a domestic pig in the same stall. In fact, chronic ASFV infection and a prolonged carrier state in the warthog have not been proved. Despite the fact that virus has been isolated from the blood and lymph nodes of many warthogs in a number of locations in Africa, it is quite possible that they are the victims of frequent reinfection from the arthropod reservoir of the virus. In any case, the epidemiological importance of the mature warthog as a carrier of ASF is obscure because it has never been found to have a level of viremia high enough to support arthropod transmission, and it has never been found capable of transmission by contact.

3. Maintenance

Both *O. moubata porcinus* and *O. erraticus* are found only in areas where the climate is warm or mild throughout the year. They are active year-round.

VI. ECOLOGICAL DYNAMICS

A. Environment

Like many of the argasid ticks, *O. moubata porcinus* and *O. erraticus* live in close proximity to their favorite hosts. *O. moubata porcinus* lives in the warthog's burrow which may be several feet deep thus affording a broad variety of microenvironments graded in temperature and humidity. *O. erraticus* is found in the pens and sheds occupied by domestic pigs in Spain. During the day, it retreats into the loose soil and crevices in the stone walls of the pens and sheds.

B. Climate and Weather

The climate and weather to which ASF virus-infected *O. moubata porcinus* and *O. erraticus* and their hosts are exposed are quite similar. The seasonal temperatures do not vary greatly and are generally quite warm. However, there are extreme seasonal variations in rainfall giving rise to alternating wet and very dry periods.

C. Biting Activity and Host Preference

Like many *Ornithodoros* species, *O. moubata porcinus* and *O. erraticus* are fast feeders that require less than 1 hr to fully engorge; they feed at night and are only occasionally found on their hosts during the day or away from their shelters. While both species appear to prefer feeding on porcine species, they have been known to feed on any animal species available to them. It is perhaps indicative of its host preference that the fecundity and rate of development of *O. moubata porcinus* are substantially affected by the source of its blood meal.[77] Females fed on porcine blood produced 1.23 eggs per milligram of blood taken as opposed to 0.76 eggs produced per milligram of a bovine blood meal. It was also found that nymphs fed on porcine blood developed more rapidly and required fewer molts in reaching the adult stage than nymphs fed on bovine blood. Comparable studies have not been made with *O. erraticus*. However, the association of *O. erraticus* with pigs in Extremadura has undoubtedly existed for many generations.

O. moubata porcinus emerges from the egg as the first-stage nymph, and both tick species have several nymphal stages that usually molt after each feeding. All instars of both species are rapid feeders, and all, including the larvae of *O. erraticus*, will feed on pigs.

D. Vector Oviposition

The adults of both species may feed to full engorgement at least several times, and the females may produce egg batches after each feeding. The eggs are deposited in the soil or in cracks and crevices in walls. They are coated with a waxy material that is impervious to water and enables them to survive under arid conditions.

E. Vector Density, Fecundity, and Longevity

In his studies on the distribution and ecology of *O. moubata procinus* in animal burrows in East Africa, Peirce[78] found tick populations ranging from a few to as many 250,000 per burrow. The larger populations were generally found in regions with neutral soils, high relative humidity, and an altitude of about 3000 to 5000 ft above sea level which insured an optimal temperature of about 24°C throughout the year. Vegetation surrounding the burrow and the presence of burrow-inhabiting animals were other factors that favored large tick populations.[78] There is a dearth of quantitative information on the biology of *O. erraticus* which might apply to the species infesting the pig sties in the Extremadura region of Spain. However, some cursory observations made by Gladney, a tick expert from the U.S. Department of Agriculture, and the writer during a tour of the Extremadura region in 1975 may be somewhat indicative of the density of the ticks in that area. In talking to a farmer and a local veterinarian in the region of the oak forests near Jerez de los Caballeros, we were told that the pigs were often restless at night, and when examined with a light were found to be heavily infested with ticks. To lend credence to this, several ticks were promptly collected from a crevice in the stone wall of the sty.

As many as eight egg batches may be produced by a single female of the *O. moubata* complex. Although the total number of eggs produced by a female during her lifetime may vary among the different subspecies of *O. moubata*, all fall within the range of 500 to 1200.[79] There is a steady decrease in the number of eggs per batch. Starting at 140 for the first oviposition, it may subsequently decline to about 30 for the 7th batch. This is accompanied by a decline in egg fertility which may be from about 90% for the first egg batch to about 70% for the last batch. These conditions prevail even if mating takes place after each blood meal. The declines may be greater if mating does not occur after each blood meal. However, frequent mating shortens the life of the female.

O. erraticus females fed on a pig and held with males at room temperature (25°C) without controlled humidity oviposited 6 times. The number of eggs per batch varied from 77 to 237. Egg fertility varied from 85 to 100%.[64]

In general, ticks of the genus *Ornithodoros* are long lived. Under favorable conditions of temperature and humidity, adults of *O. moubata* species have been known to survive up to 5 years in absence of hosts.[8] Three years and 49 days after their last blood meal, 20 out of 100 *O. erraticus* nymphs were still alive.[64]

F. Vector Competence

O. moubata porcinus from warthog burrows in various areas of East Africa had overall ASF infection rates estimated to be 0.23 to 1.35%.[8] The infection rate in unfed first-stage nymphs was 0.15% in one area, thus indicating transovarial transmission of the virus in nature. There was a gradual increase in infection rates in successive nymphal stages. However, the rate increased in the adults by about four- to sixfold over that of the final nymphal stage. It was thought that this substantial increase might have been due partially to larger blood meals which increased the chances of acquiring the virus, but perhaps of greater importance was the ability of the male to transmit virus to the female during mating.[8] While only a small percentage of ticks are infected, a single infected tick is capable of infecting a pig and may do so each time it feeds.

G. Movements and Migrations of Vectors and Hosts

The warthog is distributed widely in Africa. It may forage over a considerable area, but it is not a migratory animal. *O. moubata procinus* live in burrows occupied by warthogs and other wild animals. Although it is a fast-feeding tick, it is occasionally found on the warthog during the day and is presumably disseminated by that means. In Spain, the situation is quite different; *O. erraticus* became infected by feeding on the domestic pigs. It became a reservoir of the virus and has been the source of infection in pigs brought in to restock piggeries that were previously depopulated by ASF. The tick has not been responsible for spreading the disease to other parts of the country.

H. Human Element in the Ecology of ASF

Pigs and pork products are items of commerce distributed worldwide. Whenever ASF has spread over long distances or across oceans, it has been through the movement of infected pigs or infected pork products. The first outbreaks in Portugal in 1957 and Brazil in 1978 are clear-cut examples. In both instances, the outbreaks occurred on farms where food wastes from international airports were being fed to pigs. A number of other examples could be cited. Like other infectious animal diseases, ASF is sometimes unwittingly spread from one farm to another on the clothes and implements of the very people who are attempting to diagnose and control animal diseases.

VII. PREVENTION AND CONTROL

A. Vector Control

Because of the vastness and expense of the task, no serious consideration has ever been given to the control *O. moubata porcinus* anywhere in Africa. Furthermore, control of the tick may be of limited value because there are areas where warthogs are infected, but ticks are not present. A variety of methods were tried in efforts to rid piggeries in Spain of *O. erraticus*. Among them were steaming, flaming, and spraying with a number of different insecticides. All of the treatments were ineffective because they failed to penetrate to the retreats of the ticks.

B. Control of Vertebrate Hosts

Thus far, the control of vertebrate hosts has been the most effective means of preventing the spread of ASF virus. In several countries in East and South Africa where the wild indigenous pig and *O. moubata porcinus* are known to be carrying the virus, outbreaks in domestic pigs are now quite rare. This has been achieved by insisting that domestic pigs be raised within double-fenced enclosures that protect them from contact with wild pigs.[80] Although most countries still do not allow importation of pigs or pork products from countries where the virus is known to exist, these African countries are at least able to produce enough for their own consumption. Despite the fact that ASF is enzootic in Spain, more pigs are now being raised there than before the incursion of the disease. Again, this has been achieved to a large extent by raising pigs in isolation. Other contributing factors have been early detection and prompt application of eradication procedures. In addition to prohibiting the importation of pigs and pork products from countries that have ASF, most countries that are free of the disease require that all food wastes discarded at ports of entry be incinerated.

C. Environmental Modification

Since it is difficult and expensive to eliminate tick populations, and it is no longer acceptable to destroy a wild animal species, adoption of modern swine husbandry methods presently offers the best means of establishing a viable swine industry in an area where infected ticks and wild pigs exist. These methods include strict isolation from wild or free-ranging pigs and confinement to shelters with concrete floors and accesses with disinfectant foot baths. Facilities of this kind are being used successfully by commercial swine producers in Spain and parts of Africa. Such facilities are beyond the reach of the small subsistence farmer in the lesser developed countries. Until a satisfactory and inexpensive vaccine becomes available, they will have to depend on animals other than pigs as a source of animal protein.

VIII. FUTURE RESEARCH

A. Field Studies

Studies should be undertaken to determine what arthropods or other factors account for the transmission of ASF among wild pigs and to domestic pigs in areas of Africa where *O. moubata porcinus* does not exist.

B. Experimental Studies

Efforts to clarify the obscure immunology of ASF continue. Monoclonal antibodies that react with certain ASF viral proteins are being produced.[81] They are being used to pinpoint the locations of some of the structural proteins of the virus and identify antibody binding sites or epitopes. Restriction endonucleases have been used to cleave ASF viral DNA producing fragments that account for about 98% of the DNA molecule, and the positions of these fragments on the viral genome have been mapped.[82,83] Using this and other information being developed at the molecular level, efforts will be made to engineer a safe and effective vaccine.

To increase the level of preparedness to deal with a possible incursion of ASF in the U.S., continued efforts will be made to identify potential vectors of ASF among arthropods indigenous to the Americas.

REFERENCES

1. **Montgomery, R. E.,** On a form of swine fever occurring in British East Africa (Kenya Colony), *J. Comp. Pathol.,* 34, 159 and 243, 1921.
2. **Steyn, D. G.,** East African virus disease in pigs. Union of South Africa, *18th Rep. Dir. Vet. Ser. Anim. Ind.,* 99, 1932.
3. **De Kock, G., Robinson, E. M., and Keppel, J. J. G.,** Swine fever in South Africa (East African swine fever), *Onderstepoort J. Vet. Sci. Anim. Husb.,* 14, 31, 1940.
4. **Hammond, R. A. and DeTray, D. E.,** A recent case of Africa swine fever in Kenya, East Africa, *J. Am. Vet. Med. Assoc.,* 126, 389, 1955.
5. **Kovalenko, J. R., Sidorov, M. A., and Burba, L. G., II.,** Pasture ticks and Haematopinus as possible reservoirs and vectors of African swine fever, *Trudy Vses. Inst. Eksp. Vet.,* 33, 91, 1967 (in Russian).
6. **Plowright, W., Parker, J., and Peirce, M. A.,** The epizootiology of African swine fever in Africa, *Vet. Rec.,* 85, 668, 1969.
7. **Heuschele, W. P. and Coggins, L.,** Studies on the transmission of African swine fever by arthropods, *Proc. U.S. Livestock Sanit. Assoc.,* 69, 94, 1965.
8. **Plowright, W.,** Vector transmission of African swine fever virus, in *Agricultural Research Seminar on Hog Cholera/Classical Swine Fever and African Swine Fever: CEC,* Publ. No. 5904EN, Luxembourg, 1977, 575.
9. **DeTray, D. E.,** African swine fever, *Adv. Vet. Sci.,* 8, 299, 1963.
10. **Groocock, C. M., Hess, W. R., and Gladney, W. J.,** Experimental transmission of ASFV by *Ornithodoros coriaceus,* an Argasid tick indigenous to the U.S., *Am. J. Vet. Res.,* 44, 591, 1980.
11. **Sanchez Botija, C.,** Reservorios del virus de la peste porcina africana. Investigacion de virus de la P.P.A. en los arthropodos mediante la prueba de la hemoadsorcion, *Bull. Off. Int. Epiz.,* 60, 895, 1963.

12. **Plowright, W., Parker, J., and Peirce, M. A.,** African swine fever virus in ticks *(Ornithodoros moubata,* Murray) collected from animal burrows in Tanzania, *Nature (London),* 221, 1071, 1969.
13. **Plowright, W., Perry, C. T., Peirce, M. A., and Parker, J.,** Experimental infection of the argasid tick, *Ornithodoros moubata porcinus,* with African swine fever virus, *Arch. Ges. Virusforsch.,* 31, 33, 1970.
14. **Plowright, W., Perry, C. T., and Peirce, M. A.,** Transovarial infection with African swine fever virus in the Argasid tick, *Ornithodoros moubata porcinus,* Walton, *Res. Vet. Sci.,* 11, 582, 1970.
15. **Conceicao, J. M.,** Estude das zoonoses porcinas de Angola, premiere relatorio. A zoonoses porcine africana de virus filtravel, *Pecuaria,* 1, 217, 1947.
16. **Manso Ribeiro, J., Rosa Azevedo, J. A., Teixeira, M. J. O., Braco Forte, M. C., Rodrigues Ribeiro, A. M., Oliveira, E., Noronha, F., Grave Pereira, C., and Dias Vigario, J.,** Peste porcine provoquee par une souche differente (Souche L) de la souche classique, *Bull. Off. Int. Epiz.,* 50, 516, 1958.
17. **Larenaudie, B., Haag, J., and Lacaze, B.,** Identification en France metropolitaine de la peste porcine africaine ou maladie de Montgomery, *Bull. Acad. Vet. Fr.,* 37, 257, 1964.
18. **Gayot, G., Carnero, R., Costes, C., and Plateau, F.,** Peste porcine africaine, *Bull. Acad. Vet. Fr.,* 47, 91, 1974.
19. **Mazzaracchio, V.,** L'episodio di peste suina africana in Italia, *Ann. Ist. Super. Sanit.,* 4, 650, 1968.
20. **Oropesa, P. R.,** Reporte Preliminar del Brote de Fiebre Porcina Africana en Cuba, Instituto de Medicina Veterinario, Havana, Cuba, 1971.
21. **Wilkinson, P. J., Lawman, M. J. P., and Johnston, R. S.,** African swine fever in Malta, *Vet. Rec.,* 106, 94, 1980.
22. **Contini, A., Cossu, P., Rutili, D., and Firinu, A.,** African swine fever in Sardinia, in *African Swine Fever,* CEC/FAO Res. Semin. Sardinia, Wilkinson, P. J., Ed., 1983, 1.
23. **Mebus, C. A., Dardiri, A. H., Hamdy, F. M., Ferris, D. H., Hess, W. R., and Callis, J. J.,** Some characteristics of African swine fever viruses isolated from Brazil and the Dominican Republic, *Proc. U.S. Anim. Health Assoc.,* p. 232, 1978.
24. **Mussman, H. C.,** African Swine Fever Newsl. No. 10, Food and Agriculture Organization Regional Office for Latin America, Santiago, Chile, 1980.
25. **Fenner, F.,** The classification and nomenclature of viruses, *Intervirology,* 7, 1, 1976.
26. **Van Furth, R., Cohn, Z. A., Hirsch, J. G., Humphrey, J. H., Spector, W. G., and Langevoort, H. L.,** The mononuclear phagocyte system: a new classification of macrophages, monocytes and their precursor cells, *Bull. WHO,* 46, 845, 1972.
27. **Malmquist, W. A. and Hay, D.,** Hemadsorption and cytopathic effect produced by African swine fever virus in swine bone marrow and buffy coat cultures, *Am. J. Vet. Res.,* 24, 450, 1963.
28. **Maurer, F. D., Griesmer, R. A., and Jones, T. C.,** The pathology of African swine fever. A comparison with hog cholera, *Am. J. Vet. Res.,* 19, 517, 1958.
29. **Moulton, J. and Coggins, L.,** Comparison of lesions in acute and chronic African swine fever, *Cornell Vet.,* 58, 364, 1968.
30. **Wesley, R. D. and Pan, I. C.,** African swine fever virus DNA: restriction endonuclease cleavage patterns of wild-type, Vero cell-adapted and plaque-purified virus, *J. Gen. Virol.,* 63, 383, 1982.
31. **Heuschele, W. P. and Coggins, L.,** Isolation of African swine fever virus from a giant forest hog, *Bull. Epiz. Dis. Afr.,* 13, 255, 1965.
32. **Cox, B. F.,** African swine fever, *Bull. Epiz. Dis. Afr.,* 11, 147, 1963.
33. **Stone, S. S. and Heuschele, W. P.,** The role of the hippopotamus in the epizootiology of African swine fever, *Bull. Epiz. Dis. Afr.,* 13, 23, 1965.
34. **Polo Jover, F. and Sanchez Botija, C.,** African swine fever in Spain, *Bull. Off. Int. Epiz.,* 107, 1961.
35. **Ravaioli, L., Palliola, E., and Ioppoto, A.,** La peste swina Africana dei cinghiali. I. Possibilita d'infezione eperimentale da inoculazione, *Vet. Ital.,* 18, 499, 1967.
36. **McVicar, J. W., Mebus, C. A., Becker, H. M., Belden, R. C., and Gibbs, E. P. J.,** Induced African swine fever in feral pigs, *J. Am. Vet. Med. Assoc.,* 179, 441, 1981.
37. **Dardiri, A. H., Yedloutschnig, R. J., and Taylor, W. D.,** Clinical and serological response of American white collared peccaries to African swine fever, foot-and-mouth disease, vesicular stomatitis, vesicular exanthema of swine, hog cholera, and rinderpest viruses, *Proc. U.S. Anim. Health Assoc.,* 73, 437, 1969.
38. **Kovalenko, J. R., Sidorov, M. A., and Burba, L. G.,** Experimental investigations on African swine fever, *Bull. Off. Int. Epiz.,* 63, 169, 1965.
39. **Mendes, A. M.,** The lapinization of the virus of African swine fever, *Bull. Off. Int. Epiz.,* 58, 699, 1962.
40. **Hess, W. R., Endris, R. G., Haslett, T. M., Monahan, M. J., and McCoy, J. P.,** Potential arthropod vectors of African swine fever in North America and the Caribbean Basin, *Vet. Parasitol.,* 26, 145, 1987.
41. **Karber, G.,** Beitrag zur Kollektiven Behandlung Pharmakologischer Reihenversuche, *Arch. Exp. Pathol. Pharmakol.,* 162, 480, 1931.
42. **Teas, J.,** Could AIDS be a new variant of African swine fever, *Lancet,* 1, 923, 1983.
43. **Beldekas, J., Teas, J., and Hebert, J. R.,** African swine fever and AIDS, *Lancet,* 1(8480), 564, 1986.

44. **Hess, W. R. and DeTray, D. E.,** The use of leukocyte cultures for diagnosing African swine fever (ASF), *Bull. Epiz. Dis. Afr.*, 8, 317, 1960.
45. **Heuschele, W. P., Coggins, L., and Stone, S. S.,** Fluorescent antibody studies on African swine fever, *Am. J. Vet. Res.*, 27, 477, 1966.
46. **Boulanger, P., Bannister, G. L., Greig, A. S., Gray, D. P., Ruckerbauer, G. M., and Willis, N. C.,** African swine fever. IV. Demonstration of the viral antigen by means of immunofluorescence, *Can. J. Comp. Med. Vet. Sci.*, 31, 16, 1967.
47. **Coggins, L. and Heuschele, W. P.,** Use of agar diffusion precipitation test in the diagnosis of African swine fever, *Am. J. Vet. Res.*, 27, 485, 1966.
48. **Cowan, K. M.,** Immunological studies on African swine fever virus. I. Elimination of the procomplementary activity of swine serum with formalin, *J. Immunol.*, 86, 465, 1961.
49. **Boulanger, P., Bannister, G. L., Gray, D. P., Ruckerbauer, G. M., and Willis, N. C.,** African swine fever. II. Detection of the virus in swine tissues by means of the modified direct complement fixation test, *Can. J. Comp. Med. Vet. Sci.*, 31, 7, 1967.
50. **Crowther, J. R., Wardley, R. C., and Wilkinson, P. J.,** Solid-phase radioimmunoassay techniques for the detection of ASF antigen and antibody, *J. Hyg.*, 83, 353, 1979.
51. **Malmquist, W. A.,** Serologic and immunologic studies with African swine fever virus, *Am. J. Vet. Res.*, 21, 450, 1963.
52. **Bool, P. H., Ordas, A., and Sanchez Botija, C. S.,** Diagnosis of ASFV by immunofluorescence, *Bull. Off. Int. Epiz.*, 72, 819, 1969.
53. **Pan, I. C., De Boer, C. J., and Hess, W. R.,** African swine fever; application of immunoelectroosmophoresis for the detection of antibody, *Can. J. Comp. Med.*, 36, 309, 1972.
54. **Pan, I. C., Trautman, R., Hess, W. R., De Boer, C. J., and Tessler, J.,** African swine fever: detection of antibody by reverse single radial immunodiffusion, *Am. J. Vet. Res.*, 35, 351, 1974.
55. **Hamdy, F. M., Colgrove, G. S., Rodriguez, E. M., Snyder, M. L., and Stewart, W. C.,** Field evaluation of enzyme-linked immunosorbent assay for detection of antibody to African swine fever virus, *Am. J. Vet. Res.*, 42, 1441, 1981.
56. **Sanchez-Viscaino, J. M., Martin, O. L., and Ordas, A.,** Adaptation and evaluation of immunoenzyme assay for detection of African swine fever antibodies, *Laboratorio (Granada, Spain)*, 67, 311, 1979.
57. **Wardley, R. C., Abu Eizein, E. M. E., Crowther, J. R., and Wilkinson, P. J.,** A solid-phase enzyme-linked immunosorbent assay for the detection of African swine fever virus antigen and antibody, *J. Hyg.*, 83, 363, 1979.
58. **Pan, I. C., Huang, T. S., and Hess, W. R.,** New method of antibody detection by indirect immunoperoxidase plaque staining for serodiagnosis of African swine fever, *J. Clin. Microbiol.*, 16, 650, 1982.
59. **McDaniel, H. A.,** African swine fever, in *Proc. 21st Annu. Meet. Am. Assoc. Vet. Lab. Diagnost.*, 1978, 391.
60. **Thomson, G. R.,** The epidemiology of African swine fever: the role of free-living hosts in Africa, *Onderstepoort J. Vet. Res.*, 52, 201, 1985.
61. **Vigario, J. D., Terrinha, A. M., and Moura Nunes, J. F.,** Antigenic relationships among strains of ASFV, *Arch Ges. Virusforsch.*, 45, 272, 1974.
62. **Walton, G. A.,** The *Ornithodoros moubata* superspecies problem in relation to human relapsing fever epidemiology, *Symp. Zool. Soc. London*, 6, 83, 1962.
63. **Van der Merwe, S.,** Some remarks on the "tampans" of the *Ornithodoros moubata* complex in southern Africa, *Zool. Anz.*, 181, 280, 1968.
64. **Fernandez Garcia, J. M.,** Aportaciones al conocimiento de la biologia de *Ornithodoros erraticus, An. Fac. Vet. Leon*, 16, 195, 1970.
65. **Sanchez Botija, C.,** Reservorios del virus de la peste porcina africana. Investigacion del virus de la P.O.A. en los artropodos mediante la prueba de la hemadosorcion, *Bull. Off. Int. Epiz.*, 60, 895, 1963.
66. **Thomson, G. R., Gainaru, M. D., and van Dellen, A. F.,** Experimental infection of warthog *(Phacochoerus aethiopicus)* with African swine fever virus, *Onderstepoort J. Vet. Res.*, 47, 19, 1980.
67. **Thomson, G., Gainaru, M., Lewis, A., Biggs, H., Nevill, E., van der Pyjpekamp, H., Gerber, L., Esterhuysen, J., Bengis, R., Bezuidenhout, D., and Condy, J.,** The relationship between African swine fever virus, the warthog and *Ornithodoros* species in southern Africa, Proc. CEC/FAO Res. Semin. Sardinia, Wilkinson, P. J., Ed., 1983, 85.
68. **Mellor, P. S. and Wilkinson, P. J.,** Experimental transmission of African swine fever virus by *Ornithodoros savignyi* (Audouin), *Res. Vet. Sci.*, 39, 353, 1985.
69. **Wilkinson, P. J., Donaldson, A. I., Greig, A., and Bruce, W.,** Transmission studies with African swine fever virus. Infection of pigs by airborne virus, *J. Comp. Pathol.*, 87, 487, 1977.
70. **Scott, G. R.,** African swine fever virus, *Vet. Rec.*, 77, 1421, 1967.
71. **Heuschele, W. P.,** Studies on the pathogenesis of African swine fever. I. Quantitative studies on the sequential development of virus in pig tissue, *Arch. Ges. Virusforsch.*, 21, 349, 1967.

72. **Plowright, W., Parker, J., and Staple, R. F.,** Growth of a virulent strain of African swine fever virus in domestic pigs, *J. Hyg.*, 66, 117, 1968.
73. **Colgrove, G. S., Haelterman, E. O., and Coggins, L.,** Pathogenesis of African swine fever in young pigs, *Am. J. Vet. Res.*, 30, 1343, 1969.
74. **McVicar, J. W.,** Quantitative aspects of the transmission of African swine fever, *Am. J. Vet. Res.*, 45, 1535, 1984.
75. **Grieg, A. and Plowright, W.,** The excretion of two virulent strains of African swine fever virus by domestic pigs, *J. Hyg.*, 68, 673, 1970.
76. **Walker, J.,** East African Swine Fever, D.V.M. thesis, Zurich. Balliere, Tindall and Cox, 1933.
77. **Mango, C. K. A. and Galun, R.,** *Ornithodoros moubata:* breeding *in vitro*, *Exp. Parasitol.*, 42, 282, 1977.
78. **Peirce, M. A.,** Distribution and ecology of *Ornithodoros moubata porcinus* Walton (Acarina) in animal burrows in East Africa, *Bull. Entomol. Res.*, 64, 605, 1974.
79. **Aeschlimann, A. and Grandjean, O.,** Observations on fecundity in *Ornitodoros moubata*, Murray (Ixodoidea:Argasidae), *Acarologia*, 15, 206, 1973.
80. **Dorman, A. E.,** The control of African swine fever in Kenya, *Bull. Off. Int. Epiz.*, 63, 807, 1965.
81. **Sanz, A., Garcia-Barreno, B., Nogal, M. L., Venuela, E., and Enjuanes, L.,** Monoclonal antibodies specific for African swine fever virus proteins, *J. Virol.*, 54, 199, 1985.
82. **Ley, V., Almendral, J. M., Carbonero, P., Beloso, A., and Venuela, E., and Talavera, A.,** Molecular cloning of African swine fever virus, *Virology*, 133, 249, 1984.
83. **Almendral, J. M., Blasco, R., Ley, V., Beloso, A., Talavera, A., and Venuela, E.,** Restriction site map of African swine fever virus DNA, *Virology*, 133, 258, 1984.

Chapter 16

BLUETONGUE AND EPIZOOTIC HEMORRHAGIC DISEASE

E. Paul J. Gibbs and Ellis C. Greiner

TABLE OF CONTENTS

I. HISTORICAL BACKGROUND

Bluetongue is an arthropod–borne viral disease of ruminants characterized by congestion of the buccal and nasal mucosa and the coronary tissue of the hooves, stiffness due to muscle degeneration, and, on occasion, edema of the head and neck; congenital abnormalities may occur in the fetuses of animals infected during pregnancy. Although all ruminant species are probably susceptible to infection with bluetongue virus, clinical disease is usually more severe in sheep. Extensive epidemics of bluetongue have been recorded in which many thousands of sheep have died, thus the disease is of particular concern to countries in Western Europe and Australasia which have large sheep industries to protect.

Bluetongue is caused by a virus that is classified within the *Orbivirus* genus in the family *Reoviridae*. Also within this genus, and antigenically closely related to bluetongue virus, is epizootic hemorrhagic disease virus. This virus can cause disease that is clinically indistinguishable from bluetongue. In view of the close relationship of these viruses and the similarity of the diseases they cause, they are discussed together.

The importance of bluetongue as a clinical disease of domestic livestock is reflected in the availability of several reviews on the orbiviruses and the diseases associated with them, to which the reader is referred for details outside the scope of the text.[1-12]

A. Discovery of Bluetongue Virus and History of Epidemics

Although bluetongue had been observed as a clinical disease in sheep in South Africa in the 19th century, the first detailed clinical descriptions were not published until 1902.[13] Early studies demonstrated that the agent was filterable[14] and an attenuated vaccine, against what now emerges to be bluetongue virus serotype 4, had been developed by Theiler[15] as early as 1908.

Bluetongue has traditionally been regarded as an African disease and for the early part of this century was believed to be confined to Africa, but in 1943, a major outbreak of the disease occurred in sheep on the island of Cyprus.[16] This outbreak was caused by a particularly virulent strain of virus which focused attention on the importance of the disease and subsequently led to the recognition of bluetongue in other countries in the region, e.g., Israel in 1951.[17]

In retrospect, bluetongue possibly existed in North America prior to World War II, but it was not recognized until the 1940s and — perhaps because of its sporadic occurrence — it was 1952 before the disease was confirmed by virus isolation in sheep in California.[18]

In 1956, a major epidemic of bluetongue began in Portugal[19] and soon spread into Spain,[20] causing disease in approximately 300,000 sheep. This epidemic, the first recorded in Europe, is of particular interest because, in contrast with bluetongue in North America, the infection did not become established and eventually disappeared after about 4 years. The appearance of bluetongue in Europe, however, generated a steady escalation in recognizing bluetongue as an "emerging disease"[1] likely to cause severe economic loss if introduced to countries with large sheep populations. The confirmation of bluetongue in Pakistan in 1959[21] and India in 1963[22] caused further concern. Gee[23] has remarked that Australia — which was considered at that time to be free of bluetongue virus — with its large and valuable sheep population and its constant attention to bluetongue prevention, helped focus and maintain world interest; protocols for international movement of breeding livestock in the 1960s and 1970s reflected the Australian "obsession" with this disease.[24,25]

The increasing use in the early 1970s of simple serological tests to detect antibody in the sera of previously infected animals revealed that bluetongue virus was more widely distributed in the countries of the tropics and subtropics than had previously been thought;[11] during the 1970s, the virus was isolated in several countries (among them, ironically, Australia[26]) without clinical disease being recognized in domestic livestock. Today, it is probable that most, if not all, countries in the tropics and subtropics have livestock infected either with bluetongue virus or closely related viruses.

In the period since the epidemic in the Iberian Peninsula, the impact of the virus worldwide as a cause of disease has not been as serious as feared. Clearly, the epidemic potential of the virus remains, but the concept of bluetongue as an "emerging disease" — with the implication that it originates in Africa — has been revised.

While bluetongue is traditionally regarded as a disease of sheep, particularly those of European stock, Spreull[14] had demonstrated as early as 1905 the susceptibility of calves to bluetongue virus. Subsequently, natural outbreaks of disease have been confirmed in cattle in South Africa[27] and other countries.[2,17,20] As a disease in adult cattle, bluetongue is similar to that in sheep, but it is uncommon, clinically less severe, and generally affects less than 5% of a herd.[28] Because cattle are regarded by some as reservoirs of infection and the virus is known to be associated with reproductive problems, countries with large cattle industries have become concerned about bluetongue. It may emerge that bluetongue virus is an important cause of reproductive disease in cattle in agricultural systems, such as those in the U.S., which are heavily committed to cattle raising.

B. Discovery of Epizootic Hemorrhagic Disease Virus and History of Epidemics

There are reports in the U.S. of diseases in deer similar to epizootic hemorrhagic disease (EHD) dating back to the 19th century, but it was not until 1955 that a virus was isolated during an epizootic in white-tailed deer *(Odocoileus virginianus)* in New Jersey.[29] EHD is considered by many to be the most important disease of deer in North America. Similar viruses have been isolated from *Culicoides* in West Africa[30] and from subclinical infections in cattle in Australia.[31]

In 1959, an orbivirus was isolated from an epidemic of a "bluetongue-like" disease in cattle in Japan.[32] This virus, which caused approximately 4000 deaths among 39,000 cases, was named Ibaraki after the prefecture from where the first isolate originated. It is now recognized to be closely related to EHD virus.

It has been proposed that cattle are natural hosts of both bluetongue and EHD viruses,[33] a concept implicit in this review of the epidemiology of the two viruses.

C. Economic Impact

It is surprising that there are very few studies on the economic impact of bluetongue and EHD, despite the obvious importance of both diseases in agriculture and wildlife conservation and management.

When considering the financial loss attributable to bluetongue virus being present in a country, it is important to differentiate between the cost of clinical disease and loss of exports of breeding livestock and germplasm. The latter is clearly important only to countries with an export trade.

The major clinical impact of bluetongue is within the sheep industry; death losses may be high (>30%) and indirect losses due to the protracted convalescence are particularly important. For countries that produce high-quality wool, fleece loss due to "wool break", which can occur as a sequel of bluetongue, is also important. In cattle, a recent study[34] estimated that direct losses in an epidemic in Mississippi and surrounding states were over

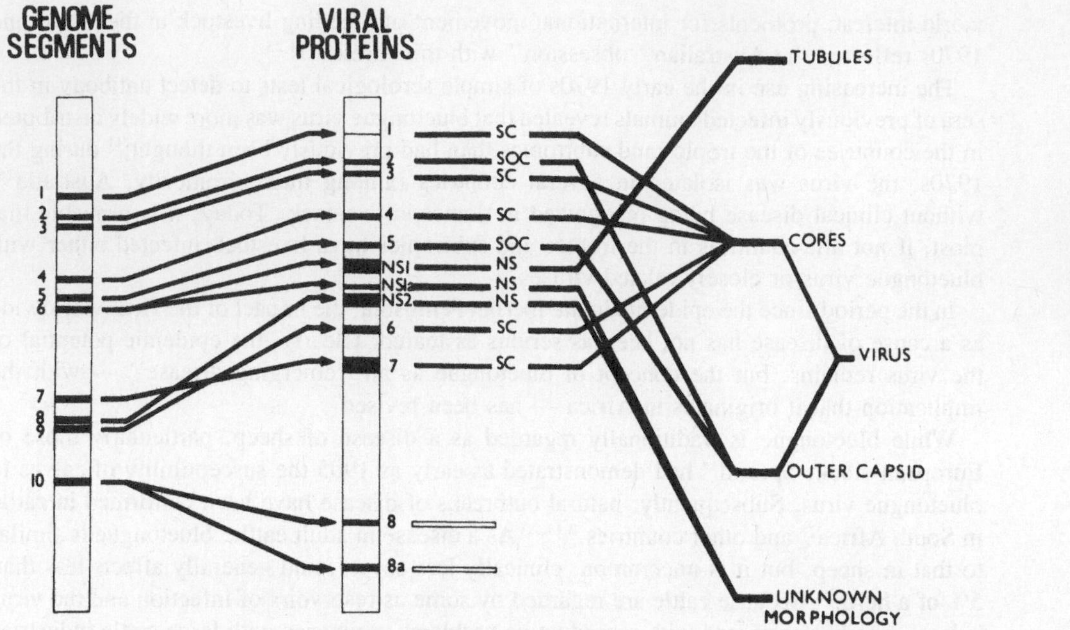

FIGURE 1. Assignment of the genome segments of bluetongue virus to the proteins which they encode (based on bluetongue virus type 1). The structural core proteins (SC), structural outer capsid proteins (SOC), and nonstructural proteins (NS) are indicated. Genome segment 2 encodes viral protein 2, the major serotype-specific antigen, and genome segment 7 encodes viral protein 7, the serogroup-specific antigen. (From Mertens, P. P. C., Brown, F., and Sanger, D. V., *Virology*, 135, 207, 1984. With permission.)

$6 million. However, the major impact of bluetongue on the cattle industry is as a constraint on exports. In the U.S., it has been estimated that the ban on cattle semen exports to the U.K., Australia, and New Zealand, when in effect in the late 1970s, resulted in an annual loss of approximately $24 million.[35]

II. BLUETONGUE AND EPIZOOTIC HEMORRHAGIC DISEASE VIRUSES

A. Classification

Bluetongue and EHD viruses are classified as orbiviruses, which is a genus within the family Reoviridae. The name orbivirus was proposed by Borden and colleagues[36] to describe a number of arthropod–borne viruses that differed from most of the recognized orbiviruses in (1) their relative resistance to inactivation by lipid solvents and detergents and (2) their characteristic appearance when examined by electron microscopy. Indeed, it is the characteristic appearance of the capsomers of these viruses which led to the selection of "orbi" to denote the genus (orbis: Latin, meaning ring or circle).

The orbiviruses have a genome of double-stranded RNA, which has 10 segments ranging in molecular weight from $2.5 \times 10^{6.0}$ for the slowest-migrating segment to $0.3 \times 10^{6.0}$ for the fastest-migrating segment when fractionated by polyacrylamide gel electrophoresis.[6] Each genome segment is considered to be a gene coding for either structural or nonstructural proteins (Figure 1). Further details of the morphology, physicochemical characteristics, and strategy of replication may be found in several reviews published recently.[8,10-12,37]

B. Antigenic Relationships

To epidemiologists and clinicians, the following discussion of antigenic variation and gene function and their relevance to classification may appear at first to be of esoteric interest, but an appreciation of genetic diversity within the serotypes of bluetongue and EHD viruses and their relationships to other viruses is of fundamental importance in understanding their epidemiology.

On present evidence, there are 12 serological groups of orbiviruses most of which are not associated with disease (Table 1). No common generic antigen has been identified, but viruses within a serogroup share antigens detectable by immunodiffusion, complement fixation, and fluorescent-antibody tests. Within each group, serotypes are distinguishable by cross-protection tests in animals and in vitro neutralization tests.

Low-level cross reactions among orbiviruses — normally considered members of distinct serogroups — do occur; e.g., viruses within the Eubenangee and EHD serogroups may cross-react with bluetongue viruses. Netivot virus — a recently discovered virus isolated from mosquitoes collected in the Negev Desert in Israel — also exhibits low-level cross reactions.[38]

Neitz[39] in 1948 was the first to identify different antigenic strains of bluetongue virus. Using viruses isolated from clinical outbreaks of bluetongue in sheep in South Africa, he found in a series of tests in sheep that each of ten strains produced solid immunity against itself, but variable degrees of protection against heterologous viruses. In later work in South Africa, another 6 antigenic strains were identified such that, by 1970, 16 antigenic strains were known.[3] The six additional strains had been identified by using the serum neutralization test and had not been correlated directly to cross-protection tests in sheep; however, each additional serotype — as they are now known — was an isolate associated with clinical disease.

Since 1970, many isolates of bluetongue virus have been obtained from arthropod vectors or from animals in countries where bluetongue has not been clinically diagnosed. Several of these isolates have been classified as new serotypes of bluetongue virus without reference to cross-protection studies in sheep.

The history of the emergence of several serotypes of EHD virus is similar. The significance of the current system of classification of bluetongue and EHD viruses into distinct serotypes has been questioned, given the virtually continuous nature of biological variation and the discovery of genetic reassortment within the orbiviruses.[8] Gorman and colleagues[8] and Della-Porta[40] have argued that the recognized serogroups of orbiviruses should be clustered into higher taxonomic units to accommodate comparison of antigenic relationships. Further, Gorman and colleagues,[8,41-43] in an elegant series of papers and reviews addressing the evolution of orbiviruses and the concept of viral species, have concluded that the exclusive use of immunological reactions in defining orbivirus groups underestimates the degree of genetic similarity among viruses.

As mentioned above, the ten RNA genome segments of orbiviruses are considered functionally distinct inside the virion and each codes for a single protein (Figure 1). This arrangement facilitates genetic reassortment between different viruses; in cells simultaneously infected with two or more related viruses, recombinants are derived by independent reassortment of parental genes. The process is analogous to sexual reproduction in higher organisms and can generate genetic diversity.

As yet, genetic reassortment has been demonstrated in orbiviruses only between serotypes of the defined serogroups.[42,43] However, the implications are clear; serological tests alone cannot determine the relationships between viruses, nor are they necessarily informative in determining the potential importance of virus isolates from infected vectors or infected animals that are clinically normal.

Bluetongue is the paradigm of this dilemma in the classification of orbiviruses. As mentioned earlier, a serotype was originally defined as an isolate that did not confer immunity in sheep against challenge with other serotypes. As the number of serotypes has increased, serological and biological characterization has, for pragmatic reasons, supplanted systematic

Table 1
SEROGROUPS WITHIN THE GENUS *ORBIVIRUS*: THEIR GEOGRAPHICAL DISTRIBUTION AND ARTHROPOD VECTOR[37]

Serogroup	No. of serotypes	Geographic distribution: Africa	Europe	Asia	Australia	North America	Central & South America	Vector
Bluetongue	24	+	+	+	+	+	+	*Culicoides*
EHD of deer	7	+		+	+	+	+	*Culicoides*
Eubenangee	3	+			+			Mosquitoes
African horse sickness	9	+	+	+				*Culicoides*
Palyam	10	+		+	+			Mosquitoes, ticks, *Culicoides*
Changuinola	12						+	*Lutzomyia*, mosquitoes
Corriparta	4	+			+		+	Mosquitoes
Warrego	2				+			Mosquitoes, *Culicoides*
Wallal	2				+			Mosquitoes, *Culicoides*
Equine encephalosis	5	+						*Culicoides*
Kemerovo	22	+	+	+	+	+	+	Ticks
Umatilla	2					+		Mosquitoes
Ungrouped								
Ife		+						NR[a] (bat)
Japanaut					+			Mosquitoes
Lebombo		+						Mosquitoes
Orungo		+						Mosquitoes
Paroo River					+			Mosquitoes
T5-0616						+		NR (skunk)
Netivot				+				Mosquitoes

NR, none recognized (isolated from).

in vivo testing. It is conceivable that some of the recent orbivirus isolates, currently classified as bluetongue viruses and originating from countries where bluetongue has not been reported, may not possess the gene(s) necessary to cause clinical disease. The implications of this for the licensed international exchange of breeding livestock and germplasm are discussed in Section VIII.

C. In Vitro Growth of Bluetongue Virus and Epizootic Hemorrhagic Disease Virus

Although bluetongue virus is difficult to isolate in cell cultures, once adapted to in vitro growth, the virus produces cytopathic effect in a wide variety of continuous cell lines such as BHK 21, VERO, and L cells.[6] Development of serological assays and detailed biochemical studies frequently use these cell lines.[10] In general, and for reasons that are not understood, bluetongue virus is difficult to isolate by inoculation of either unestablished or continuous cell cultures. Most diagnostic laboratories have found that intravenous (i.v.) inoculation of 10-day-old embryonated chicken eggs incubated at 33°C remains the most sensitive technique.[44] Infected embryos die from day 3 after inoculation; when the egg is opened, the embryo is seen to be hemorrhagic. Routine inoculation of organ suspensions from infected embryos to cell cultures is usually all that is necessary to adapt the virus to in vitro growth.

In contrast with bluetongue virus, EHD virus grows directly upon inoculation to BHK cultures.[45] Insect cell cultures, particularly the C6/36 mosquito cell line, have been used to isolate several different types of orbivirus including EHD virus.[46] Both bluetongue and EHD virus can be isolated by intracerebral (i.c.) inoculation of neonatal mice or hamsters. Other laboratory animals are considered refractory to infection.

III. DISEASE ASSOCIATIONS

A. Domestic Animals and Wildlife Species

Although there are several orbiviruses that have been associated with disease in man and there are anecdotal reports of bluetongue virus causing infections in laboratory workers, neither bluetongue nor EHD virus has been conclusively shown to cause clinical disease in man. To our knowledge, there are no published reports of any serological surveys to detect antibody in man.

The incubation period of bluetongue in sheep is between 5 and 12 days with the majority of sheep developing clinical signs between the 6th and 8th days. The disease is characterized by an initial fever which may last several days before hyperemia, excessive salivation, and frothing are noticed in the mouth. The muzzle and lips are often edematous and a nasal discharge, initially serous but often mucopurulent, is common. The tongue may become cyanosed, although this is uncommon despite the name "bluetongue". At this stage of the disease, sheep may become recumbent and die in shock. In most animals, there is a marked loss of condition. Some animals appear to have difficulty in swallowing; others may vomit which can lead to aspiration pneumonia, and death. The coronary bands of all four feet can be hyperemic and so painful that affected animals are often reluctant to walk and tend to be recumbent. Close examination of the skin may reveal that this is also hyperemic, which can lead to "wool break" some weeks later. Muscle degeneration occurs leading to a protracted convalescence of several weeks. Some animals develop torticollis.

The extent of the clinical signs varies from subclinical infection to severe clinical disease depending upon the strain of virus, the breed of sheep, and the geographical area. In endemic areas, disease is rarely present except in animals introduced from areas free of the virus; however, in some outbreaks of bluetongue in geographical areas where the virus is not normally active, morbidity can be as high as 80% of the flock and mortality as high as 50%.

Bluetongue in cattle is milder and, even when a major epidemic is occurring in sheep in an area, relatively few cattle will be affected. The clinical signs are similar to bluetongue in sheep.[47] The first signs are fever, hyperemia of the mucous membranes in the mouth, and edema of the muzzle; affected cattle often salivate excessively. Some animals develop coronitis and lameness.

Although disease is rare in goats, it has been reported.[48]

The susceptibility of wild animals to bluetongue virus was first established by Neitz[49] who demonstrated the susceptibility of South African blesbok *(Damaliscus albifrons)*. It is now recognized that most, possibly all, ruminant species are susceptible to infection with bluetongue virus. In general, African antelope do not develop clinical disease; in contrast, the wildlife species found in North America, such as the white-tailed deer *(Odocoileus virginianus)*, the pronghorn *(Antilocapra americana)*, and the desert bighorn sheep *(Ovis canadensis)*, are susceptible to infection and may develop severe clinical disease similar to bluetongue in sheep.[5,50] While there are no reports of bluetongue in wildlife species on other continents, outbreaks of bluetongue in zoological collections in North America have indicated that some species of Asiatic deer [e.g., Muntjac deer *(Muntiacus reevsei)*] are susceptible to infection and may develop severe clinical disease.[51]

Bluetongue virus can infect the fetus leading to abortion, congenital malformations, and growth retardation. Affected calves and lambs display a range of physical and functional abnormalities depending upon the gestational age at which infection occurred. Offspring may be spastic, blind, and deaf, but otherwise physically normal; however, some calves and lambs have obvious abnormalities such as hydranencephaly, arthrogryposis, and prognathia. Such severely affected calves and lambs rarely survive.[52,53]

EHD virus has not been associated with naturally occurring clinical disease in sheep, nor has it been possible to produce obvious disease under experimental conditions.[33] In the North American white-tailed deer, the disease caused by EHD virus is indistinguishable from bluetongue.[5] In cattle in North America and Australia, EHD viruses have not been found to cause clinical disease.[31,54] The outbreaks of Ibaraki disease in Japan were clinically similar to bluetongue in cattle.[55]

B. Pathogenesis

A detailed description of the pathogenesis of bluetongue and EHD virus infections in ruminants is beyond the scope of this review, but there are several features of epidemiological significance.

Surprisingly, there is still uncertainty about (1) how these viruses cause clinical disease in their mammalian hosts, (2) whether the viruses can induce persistent infection in immunologically competent animals, and (3) whether *in utero* infection can induce immunotolerance and persistent infection in the fetus (reviewed by Osburn[56]).

Upon introduction into the animal by the bite of an infected *Culicoides* insect, virus first replicates in the local lymph nodes, subsequently inducing a primary viremia which seeds other lymph nodes, spleen, and bone marrow, where further replication occurs.[57] The virus also replicates in endothelial cells lining the blood vessels and capillaries. It is logical to assume that replication of the virus in monocytes, macrophages, and endothelial cells can induce sufficient physical and functional damage to lead to hemorrhage and edema of tissues, but no adequate explanation is available as to why subclinical infections are so common. In most animals a viremia can be detected between 5 and 12 days after infection with both EHD and bluetongue viruses (maximum titer approximately $10^{5.0}$ $\log_{10}$ tissue culture infective doses per milliliter); the viruses are closely cell associated — predominantly with the buffy coat cells. Beyond 12 days after infection, it becomes progressively more difficult to isolate virus; however, if the cell fraction of the blood is washed to remove antibody present in the plasma and disrupted by ultrasonics, occasional isolates can be made up to 100 days after infection. Although it has not been demonstrated by electron microscopy, it is probable that during the replication of virus in stem cells in the bone marrow, some virus particles become "incarcerated" in the maturing erythrocytes and remain there until the erythrocyte is removed from circulation. Since the half-life of bovine erythrocytes, e.g., is estimated at 120 days, isolation of virus at 100 days after infection is consistent with this theory. While the presence of virus beyond 28 days after infection can be considered to technically constitute a persistent infection,[24] if such a theory is correct, the recovery of virus does not indicate continuing replication of virus in the mammalian host. Nevertheless, this feature of bluetongue virus, and probably EHD virus, may be of epidemiological significance for it potentially extends the period over which a mammalian host is infective for an arthropod vector.

The early belief that bluetongue virus infections in cattle could frequently lead to a persistent or latent infection in bulls,[58] with periodic excretion of virus in their semen over several years, is now questioned. Recent studies with bulls infected with bluetongue virus have shown that bulls only shed virus in their semen during the period when they are viremic.[59]

In this respect, bluetongue virus is not different from many other viruses that cause viremia; however, the systems of low temperature storage used in artificial insemination and embryo-transfer centers for semen and embryos are also ideal for virus preservation. Thus, in view of the risk — albeit small — that the international movement of cattle semen and germplasm could lead to an epidemic of bluetongue in a country previously free of infection with the virus, there have been several experimental studies on the effects of (1) inseminating susceptible heifers with semen infected with bluetongue virus and (2) implanting recipient heifers with embryos collected from donor heifers that either had been inseminated with infected semen or were viremic at the time of ovulation (reviewed by Thomas et al.[60]). The general consensus of opinion at this time is that while infected semen can infect susceptible cows and heifers and produce a viremia, bluetongue virus is not present in embryos resulting from fertilization of ova with infected semen or from heifers viremic at ovulation. Pending the results of field trials with larger numbers of embryo transfers in areas where bluetongue viruses are active, it appears that embryo transfer, if conducted correctly,[61] substantially reduces the risk of bluetongue virus transmission when compared with international movement of breeding livestock.

If infection of pregnant ruminants occurs later in gestation, both bluetongue and EHD viruses can cross the placenta and establish infection in the fetus; the results of such infections depend upon the stage of gestation and range from abortion, through congenital abnormalities, to growth retardation. The ontogeny of the immune response is intricately linked to the outcome of fetal infection, and several authors have reported studies involving experimental infections of pregnant ruminants with bluetongue and EHD viruses. Although early work[52,62] suggested that transplacental infection of cattle could lead to latent infections that could be activated later in life by the bite of *Culicoides* species, it has not been possible to confirm these observations[63] and currently they are considered to be of historical interest only. There is sufficient evidence to conclude, however, that fetuses infected mid-gestation or later may be viremic at birth, but the titers of virus are low and within a few months these young animals have ceased to be viremic.[64] The relevance of these observations to "overwintering mechanisms" is discussed later.

C. Diagnostic Procedures

Bluetongue in sheep and EHD in deer are usually sufficiently characteristic for an experienced clinician to diagnose.[28] In cattle, diagnosis of bluetongue is more difficult and can be confused with bovine virus diarrhea/mucosal disease, infectious bovine rhinotracheitis, and malignant catarrh. In both cattle, and sheep, photosensitization should be excluded from the diagnosis. Apart from hemorrhage at the base of the pulmonary artery, which is not a consistent finding and for which there is no adequate explanation, there is no pathognomonic gross pathology.

As mentioned previously, bluetongue virus is often difficult to isolate in the laboratory and the techniques currently in general use are considered to be insensitive.[65] The probability of isolating virus is enhanced if blood is collected from animals in the early stages of disease. Some strains of bluetongue can be isolated directly into cell cultures, but others have to be isolated first by i.v. inoculation of washed and ultrasonicated buffy coat cells into 10- or 11-day-old embryonated chicken eggs.[44] Both techniques are time consuming and expensive. Apart from the immediate problems associated with the failure to confirm the clinical diagnosis or suspicion of disease, this feature creates difficulty in mapping the geographical distribution of the different serotypes of virus and in certifying livestock and germplasm free of infection for import/export.[25]

Epizootic hemorrhagic disease virus can be isolated directly in cell cultures of baby hamster kidney (BHK 21) and other established cell lines.[45]

Both bluetongue virus and EHD virus have been isolated directly from clinical tissues and *Culicoides* spp. by i.c. inoculation of day-old mice,[30,45] however, not all strains of bluetongue virus can be isolated in this manner.[26]

To overcome the problems in isolating bluetongue virus from clinical specimens and to assist in the rapid identification of isolates, the fluorescent antibody technique has been investigated by several authors to detect bluetongue virus in infected tissues and cell cultures,[64,65] but nonspecificity has limited its use for diagnosis.[66] An indirect immunoperoxidase procedure to overcome this nonspecificity has been reported.[67] More recently, genetic probes have been developed.[68,69] The probe that has been used most widely is a DNA copy of RNA segment 3 of an isolate of serotype 17 bluetongue virus.[68] Using Northern blot hybridization techniques, this probe has been found to hybridize to segment 3 of serotypes 1 to 17, 20, and 21 grown in cell cultures. When translated and biotinylated, the probe has been used successfully to detect viral RNA in blood from sheep experimentally infected with either bluetongue virus serotype 2 or 11 7 days previously. The sensitivity of the probe for detecting bluetongue virus in naturally infected animals and their germplasm remains to be evaluated, but the technique holds considerable promise for overcoming the diagnostic uncertainty that has restricted investigation of the epidemiology of bluetongue virus for so long. The probe does not hybridize with serotype 1 of EHD virus (the only serotype tested to date); currently, there are no DNA probes reported for EHD viruses.

Serological tests for the diagnosis of bluetongue also have limitations. The tests may be divided into two groups: those detecting antibody to the group antigen (protein 7, Figure 1), e.g., the agar-gel immunodiffusion and complement-fixation (CF) test, and those used to detect antibody to the different serotypes (protein 2, Figure 1), e.g., the serum neutralization and hemagglutination inhibition tests.

The problems in interpreting serological data due to cross reactions have been reviewed by several authors;[65,70,71] unless great caution is exercised in the interpretation of serological data, misleading conclusions can result. All the tests suffer from problems of standardization and specificity; some of the group-reactive tests detect cross reactions with EHD and Palyam virus serogroups, and multiple infections of ruminants with related viruses can produce serum antibody that will cross react in type-specific virus neutralization tests with bluetongue virus serotypes to which the animal has never been exposed.[72] These problems are brought into sharp focus when interpreting serological information obtained in countries where clinical disease has not been recorded and where there are no virus isolates available for critical study.

Of the various tests available, the agar-gel immunodiffusion test is preferred for detecting antibody to the group antigen and the plaque-inhibition virus neutralization test for detecting type-specific antibody.[71] Preliminary studies indicate that a newly developed "blocking" ELISA may prove to be a more specific and sensitive group-specific test than the agar-gel immunodiffusion test.[73] The test is based upon the interruption of the reaction between bluetongue virus antigen and a group-specific monoclonal antibody raised against bluetongue virus by the addition of serial dilutions of test sera.

IV. EPIDEMIOLOGY

It is important to remember, when reading the following sections on the epidemiology, transmission, and ecology of bluetongue and EHD viruses, that both viruses have evolved in the tropics and subtropics where they are well-adapted parasites of many mammalian species belonging to the suborders Ruminantia and Tylopoda. While infection of animals in the tropics and subtropics is common, clinical disease in indigenous species is unusual. As a general rule, clinical disease is noticed, or likely to be noticed, only when:

1. Nonindigenous species or breeds (particularly sheep) are introduced to endemically infected areas
2. Atypical climatic conditions permit infected arthropod vectors to spread beyond their normal geographical distribution and infect susceptible mammalian hosts
3. The virus is introduced — through international trade in animals and their products — into a region of the world where a suitable vector exists to transmit the disease to susceptible mammalian hosts

The limitation to the spread of bluetongue and EHD into most of the temperate areas of the world is the absence of suitable arthropod vectors, and not the availability of susceptible mammalian species.

A. Geographical Distribution

Historically, it was naturally disease that first attracted attention. It was not until the introduction of simple serological techniques in the 1970s to detect the presence of bluetongue and EHD virus infections in livestock that the true geographical distribution of the viruses began to be recognized (see also Section I.A).

The geographical distribution of bluetongue virus as presently known is summarized in Table 2 and Figure 2 from which it can be seen that infection is present, or has occurred, on all continents where livestock are reared. Only Europe is normally considered free of infection.

EHD viruses have been isolated in North America, Africa, Japan, and Australia. There are at least seven serotypes recognized and probably more exist because no detailed serological comparisons of isolates from different geographical areas have been made.

B. Incidence and Risk Factors

The incidence of clinical disease is highly variable and influenced by several factors, most importantly, vaccination history, geographical location, and climate.

Two areas of the world seem subject to repeated annual outbreaks of bluetongue in sheep: California and the Republic of South Africa. Elsewhere, the disease is periodic and several years may elapse between recorded outbreaks.

Both California and South Africa are in temperate zones, and there is a marked seasonal incidence of disease in late summer and early autumn.[13,14,74,75] Cases of bluetongue generally cease after the first frosts of winter.

The epidemics of bluetongue in the Iberian Peninsula, Cyprus, and Turkey have been similar. Where disease occurs infrequently, epidemics have usually been associated with only one serotype of virus (e.g., serotype 10 in Portugal in 1956 and serotype 4 in Cyprus in 1977), but where disease is common, a random distribution of serotypes may be isolated during the course of an epidemic.[74,75] Epidemics of EHD in North America and Ibaraki disease in Japan have followed much the same pattern.

Table 2
GEOGRAPHICAL DISTRIBUTION OF BLUETONGUE VIRUS

Continent/region	Geographical extent	Clinical disease reported	Serotypes of bluetongue virus isolated
Africa	Probably endemic in all countries except northwest Africa	Yes	1—16, 18, 19, and 22
Asia	Probably endemic in all countries from Turkey through Indian subcontinent to Indonesia; northern extent beyond Nepal not determined	Yes	2—4, 9, 10, 12, 16, 17, 23
Australia	Endemic in northern part of continent	No	1, 20, 21, (also 2 untyped viruses)
Europe	Epidemic in Iberian Peninsula (1956); serological evidence on Greek Islands (1979); not considered endemic on continent	Yes	4, 10
North America	Endemic in southern and western states of U.S. and Mexico	Yes	2, 10, 11, 13, 17
South America, Central America, and Caribbean	Endemic but southern extent not determined	No	No isolates; only serological diagnosis

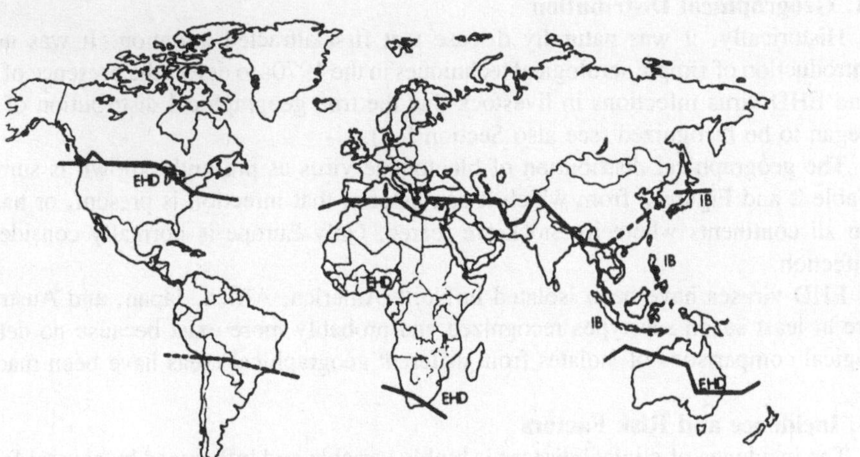

FIGURE 2. Geographical distribution of bluetongue, epizootic hemorrhagic, and Ibaraki disease viruses. The northern and southern limits of bluetongue are indicated by a solid line (dotted line indicates that limit is uncertain). EHD, epizootic hemorrhagic disease virus; IB, Ibaraki virus. (Adapted from Sellers, R. F., in *Virus Diseases of Food Animals*, Gibbs, E. P. J., Ed., Academic Press, New York, 1981, chap. 23. With permission.)

In South Africa, clinical disease is usually confined to the fine wool and European breeds of sheep; only occasionally do farmers report disease in Karakul and Black Head Persian sheep — breeds that have developed in areas that we now consider endemic areas of bluetongue virus activity.[75] However, while European breeds of sheep may be most susceptible, local breeds of sheep can exhibit clinical disease such as was reported in Cyprus in 1977.[76]

Management and other factors also determine the incidence and severity of clinical disease during an epidemic. Two examples may be given: (1) sheep that recently have been shorn and exposed to strong sunlight are reported to develop more severe disease and (2) most European breeds of sheep have restricted breeding seasons and lamb in the spring, thus a large susceptible population develops in autumn as passive immunity wanes and is coincident with the peak infectivity rates in insect vectors.[75]

The incidence of bluetongue in sheep in a geographical area may also be modified by the frequency of vaccination as is discussed in Section VIII.

C. Serologic Epidemiology

Despite the limitation of the currently available serological tests, serological surveys using the agar-gel immunodiffusion test (detects group-specific antibody) and the neutralization test (detects type-specific antibody) have provided radical new insights into the epidemiology of bluetongue and EHD viruses.

Serological surveys using the agar-gel immunodiffusion test have established the ubiquity of bluetongue virus infections — or at least infections caused by closely related viruses — in many parts of the world where clinical disease has not been recognized.[77,78] In conjunction with sentinel cattle herds, the test has been employed to indicate seasonal activity of virus transmission and to focus attempts to isolate virus from subclinical infections.[79,80]

There are several ways to establish the serotypes of bluetongue and EHD viruses active in a given area. The preferred way is to isolate viruses, but in many countries and even regions the existing veterinary diagnostic laboratories are not equipped for isolation of viruses from either ruminants or arthropod vectors. Further, in the absence of any perceived clinical problems in livestock in most tropical and subtropical countries, it is unreasonable to expect diagnostic laboratories to undertake a serendipitous search for viruses that are notoriously difficult to isolate. Thus, serological studies involving the detection of neutralizing antibody in the sera of ruminants have been used as indicators to the serotypes active in a country or region.

As mentioned earlier, multiple infections of ruminants with related viruses can produce serum antibody that will cross-react in type-specific virus neutralization tests with bluetongue virus serotypes to which the animal has never been exposed.[72] To overcome this problem, the concept of "cluster analysis" has been introduced.[81] Data are examined for the frequency of neutralizing antibody against all serotypes and, if a high frequency or "cluster" appears of sera with antibody to one particular serotype, this suggests previous exposure to that serotype in the group of animals studied. While the conclusions drawn from such surveys should be considered indicative of the serotypes active in an area rather than definitive proof, their predictive value should not be underestimated. This was illustrated recently in Florida. Examination of sera from a group of cattle that were being bled at monthly intervals indicated that serotype 2 virus had caused previous infection. Initial attempts at the time of collection of the samples to isolate virus had been unsuccessful, but a reappraisal of the techniques in use led to the identification of bluetongue virus serotype 2 — the first time that this virus had been recorded in the Western Hemisphere.[80] Similar studies had indicated that bluetongue virus serotype 1 was present in Australia prior to its isolation.[82]

The value of using both group-reactive and type-specific tests for investigating the epidemiology of bluetongue and EHD viruses has been reported by several groups.[50,83-85] The studies initiated in Australia following the isolation of bluetongue virus from a pool of *Culicoides* in 1975 are a paradigm of this approach (reviewed by St.George[86] and Parsonson and Snowdon[87]).

V. TRANSMISSION CYCLES OF BLUETONGUE AND EPIZOOTIC HEMORRHAGIC DISEASE VIRUSES

Although venereal and congenital transmission of bluetongue and EHD viruses occurs in ruminant populations as discussed in Section III.B, the geographical restriction of bluetongue and EHD (Figure 2) indicates that these mechanisms are unimportant in the long-term perpetuation of the viruses. The viruses rely upon arthropod transmission for their survival in nature.

A. Field Studies on Arthropod Transmission

Du Toit,[88] in 1944, was the first to demonstrate that biting midges of the genus *Culicoides* were vectors of bluetongue virus. He isolated bluetongue virus from *C. imicola* collected in the Republic of South Africa and succeeded in transmitting the virus with triturated pools of this species. Since then, bluetongue viruses have been isolated from *C. imicola* (= *C. pallidipennis*) elsewhere in Africa, e.g., Sudan,[89] Kenya,[90] and Zimbabwe.[91]

Bluetongue virus has also been isolated from *C. imicola* in Israel,[92] but other species are generally involved in transmission outside Africa. Mellor and Pitzolis[93] recovered bluetongue virus from *C. obsoletus* from Cyprus; in Australia, bluetongue viruses have been detected in wild populations of *C. brevitarsis, C. fulvus, C. oxystoma,* and *C. peregrinus.*[94]

The studies on bluetongue in North America in the 1950s quickly identified that *C. variipennis* was transmitting the virus.[95] Subsequently, collections from Colorado,[96,97] Montana,[98] and California[99] also yielded bluetongue virus. Other speices of *Culicoides* are possibly vectors in North American since bluetongue virus was isolated from *C. insignis* near cattle in Florida.[100]

Additional vectors of bluetongue have been sought in many places where bluetongue viruses exist. Surveys of *Culicoides* associated with ruminant livestock have been made in locations where virus isolation could not be attempted logistically. Others have attempted to correlate which of these potential vectors were present during the time of seroconversion in sentinel flocks and herds. Some of these studies include Puerto Rico where *C. insignis* and *C. pusillus* were the primary species near three cattle herds[101] and Nigeria where *C. imicola* and *C. schultzei* groups were present at their peak population sizes during the time of seroconversion.[102]

The present state of knowledge of *Culicoides* spp. as proven vectors (biological transmission has been made by fly bite), suspected vectors [virus has been isolated from either naturally (N) or experimentally (E) infected flies], and potential vectors (species which are ruminant associated but for which no virus work has been done) of bluetongue virus is summarized in Table 3.

B. Experimental Infections of Arthropods

In the U.S., early work on *C. variipennis* using colonized females resulted in the transmission of bluetongue virus from sheep to sheep and between sheep and cattle.[114,115] This species recently has been shown to be capable of transmitting bluetongue serotype 2 virus from Florida, although it was less efficient with type 2 than other serotypes in North America.[116] EHD virus was shown to replicate in colonized *C. variipennis;*[117] 3 of 22 pools of free-flying *C. variipennis* females from Kentucky contained EHD; and transmission to deer was accomplished with the same species.[118]

The most complete nationwide examination of vectors of bluetongue virus is the recent work done in Australia. This was precipitated by the realization that bluetongue viruses occurred in Australia as reported in 1978.[26] Standfast et al.[94] summarized the results from those studies. Experimental transmission of the virus from sheep to sheep has been completed with *C. fulvus* and *C. actoni.* Four other species have been infected in the laboratory by

Table 3
PROVEN, SUSPECTED, OR POTENTIAL VECTORS OF BLUETONGUE VIRUS

	Vector categorization		
Location	**Proven**	**Suspected**[a]	**Potential**
Africa	*C. imicola* (= *C. pallidipennis*)[88]	*C. milnei* (N)	*C. cornutus*
		C. gulbenkiani (N)[103]	*C. grahami*
			C. magnus
			C. schultzei
			C. zulnensis
			C. krameri
			C. austeni
			C. kingi[30,89,91,102]
Mediterranean		*C. obsoletus (N)*[93]	*C. newsteadi*
			C. pulicaris
			C. punctatus
			C. circumscriptus
			C. puncticollis
			C. kingi[92,104]
Australia	*C. fulvus*	*C. brevitarsis* (N)	
	C. actoni[94]	*C. wadai* (E)	
		C. peregrinus (E)	
		C. oxystoma (E,N)	
		C. brevipalpis (E)[94]	
Eastern Asia			*C. nipponensis*
			C. obsoletus
			C. oxystona
			C. pulicaris
			C. sigaensis
			C. sinanoensis
			C. toyamuruae
			C. oxystoma
			C. similis
			C. shortti[105-107]
			C. actoni
			C. orientalis
			C. jacobsoni
			C. raripalpis
			C. peregrinus
			C. arakawai[108]
Caribbean			*C. insignis*
			C. pusillus
			C. filariferus[109]
South America			*C. insignis*
			C. pusillus
			C. maruim[107]
North America	*C. variipennis*	*C. insignis*	*C. stellifer*
		C. debilipalpis[109,110]	*C. paraensis*
			C. obsoletus
			C. biguttatus
			C. venustus[b]
			C. cockerelli
			C. freeborni
			C. neomontanus
			C. hieroglyphicus
			C. owhyeensis[97,98,111,112]

[a] N, natural infection; E, experimental infection.

[b] Jones et al.[113] demonstrated a very low infection rate for experimental infections of bluetongue virus and EHD virus.

feeding on viremic sheep: *C. wadai, C. peregrinus, C. oxystoma,* and *C. brevitarsis.* Vector competence as measured by infection rates (IR) suggests that *C. fulvus* was the most efficient species with a 62% IR followed by *C. wadai* with 42%, and the remainder with less than 1%.

Because of the difficulty in inducing many *Culicoides* spp. to take a blood meal in the laboratory, a system of intrathoracic inoculation has been used to assess vector competence. Several species from the eastern Mediterranean area (*C. puncticollis, C. circumscriptus, C. schultzei,* and *C. obsoletus*) have been inoculated with bluetongue virus via the intrathoracic route,[119] and in North America, *C. stellifer* has been inoculated.[110] In each species virus apparently replicated, but no transmission of infection to ruminants was attempted. EHD virus replicated following intrathoracic inoculation into *C. riethi, C. nubeculosis,* and *C. impunctatus.*[117]

The relevance of intrathoracic inoculation as a meaningful indicator of vector competence is difficult to assess, but a recent observation casts the technique in a new light. Mellor and colleagues[119,120] established that two European species, *C. puncticollis* and *C. nubeculosus,* were refractory to infection when fed on infected blood meals, but were susceptible when inoculated intrathoracically; however, when microfilariae from *Onchocerca cervicalis* were added to the infective blood meal and fed to *C. nubeculosus,* the virus infected the insects. This suggests that an initial "barrier" exists in the gut of these *Culicoides* species. Because microfilarial infections of *Culicoides* are common, this synergism warrants further attention.

Although mosquitoes have not been identified as vectors of bluetongue virus, intrathoracic inoculation has been used to propagate bluetongue viruses in several species.[121,122]

Arthropods other than biting midges have been examined as potential biological vectors of bluetongue. *Stomoxys calcitrans,* the stablefly, and *Aedes aegypti,* the yellow fever mosquito, were tried unsuccessfully.[114] The potential for sheep keds *(Melophagus ovinus)* to transmit bluetongue virus has also been examined.[123,124] Although the virus can be transmitted by placing keds from infected sheep onto susceptible sheep, it is not known whether the ked is a mechanical or biological vector. More recently, a soft tick, *Ornithodorus coriacius,* was shown to effect transmission of bluetongue.[125] Virus could be detected in this argasid tick for up to 105 days after feeding on an infective meal.

In contrast to argasid ticks such as *O. coriacius,* which can live up to 5 years, *Culicoides* live a matter of days. To further investigate the vector potential of colonized *C. variipennis,* females have been infected and dissected at various time intervals to determine by direct immunofluorescence the timing and sites of virus replication.[126,127] By day 5 postfeeding, bluetongue virus was detected in the cells of the midgut. Other organs/tissues such as the ovaries, salivary glands, fat bodies, hindgut, and thoracic muscles contained small amounts of virus by day 5 and the percentage of flies containing virus in these secondary sites increased markedly until day 14 postfeeding. This is the time of maximal virus titer in *C. variipennis.*[128]

C. Maintenance and Overwintering Mechanisms

Arthropod vectors have been suggested as mechanisms by which bluetongue virus may overwinter. Nevill[129] hypothesized that because small numbers of *Culicoides* adults are active during the warmer winter nights in South Africa and feed upon susceptible cattle, the *Culicoides* could be a major means of maintaining the virus through short winters.

Another proposed component in the overwintering could be the birth of calves, lambs, and possibly other young ruminants that are viremic at birth as a result of a previous infection *in utero.*[52,64] Although the levels of viremia generally have been low and do not persist, this feature may be of epidemiological significance in regions with short winters. In studies with infected lambs, there was a maximum time span of 145 days from infection of the dam to the last detectable viremia in her lambs.[64] Equally intriguing is a phenomenon observed in a bull reported to be persistently infected with bluetongue virus.[62] When colonized, uninfected *C. variipennis* females were allowed to feed on this bull, and a viremia was detectable several hours later, thus providing an infective blood meal for other *Culicoides* feeding at this time.

Transovarian transmission of the virus in *Culicoides* is another potential mechanism for overwintering. Transovarian transmission of bluetongue virus has been examined, but no virus was found in the progeny of infected females.[130] In studying the growth of bluetongue virus in *Culicoides,* virus particles were found in the ovarian sheath of *C. variipennis,* but could not be detected in ovarian follicles or eggs.[127] This suggests that transovarian transmission does not occur with this vector-agent system.

The epidemiological significance of argasid tick transmission of bluetongue virus has not been established.[125] However, it is easy to visualize that a hematophagous arthropod with a life expectancy of over 5 years and a proven capability of transmitting these viruses could be a significant feature in overwintering of the virus. Whether other soft ticks have the potential to transmit bluetongue viruses is unknown.

VI. ECOLOGICAL DYNAMICS OF VECTOR-HOST RELATIONSHIPS

A. Seasonality and Flight Periodicity of Vector Species

Seasonality and diel periodicity of adult *Culicoides* species vary regionally. In Colorado, *C. variipennis* adults emerge from mid-March to October, depending upon the year,[131] with peak populations from late June to early September.[132] Flight activity is greatest around sunset, but shifts from late afternoon activity during the cooler months to throughout the nocturnal hours in the summer.[132] Carbon dioxide-baited traps in California suggest the peak biting activity is at sunset.[133] In Virginia, this species is present from early May through late September.[134] In Florida, *C. variipennis* can be caught in light traps throughout the year with peak populations in spring and summer. The second important species in Florida, *C. insignis,* increases in numbers inversely with *C. variipennis* populations (where the two are sympatric, Figure 3) and peaks occur in late summer to late autumn.[135] This species is present in parts of Florida year-round, unless the winter months are extremely cold. Parous females have been collected in all months during mild winters.[136] *C. insignis* is a nocturnal species in Florida.[109] Similarly, *C. imicola* and *C. schultzei* are nocturnal species and are present throughout the year in Kenya.[137]

B. Vector Oviposition

The genus *Culicoides* utilizes a wide variety of moist to aquatic habitats for oviposition and larval and pupal development. Species with very large geographic ranges have been documented from a range of larval/pupal habitats. Some of these variations have led to the suggestion that we are actually dealing with species complexes rather than a single taxon. A good example of this is *C. variipennis* which has been discussed recently.[138] Presently, five subspecies are recognized: (1) *C. v. sonorensis,* which breeds in sewage effluents with high organic loads and occurs in the more arid western and southwestern U.S., (2) *C. v. variipennis,* immature stages of which occur in fresh water aquatic sites in the more wooded areas of northern and eastern U.S. and southern Canada, (3) *C. v. albertenis,* which inhabits portions of the Great Plains and the Rocky Mountains and uses alkaline lakes for its immature stages, (4) *C. var. occidentalis,* which also uses alkaline and saline habitats and occurs in the western coastal states, and (5) *C. v. australis,* a subspecies found in the southeastern U.S. which develops in saline habitats. In the southeastern-most collections of the species in Florida, larvae have only been collected from dairy sewage lagoons.[139]

C. insignis larvae have been obtained from muddy areas in pastures and edges of fresh water ponds in Florida.[139] The important point is that the Florida *Culicoides* larval collections of both *C. v. australis* and *C. insignis* were taken near cattle.

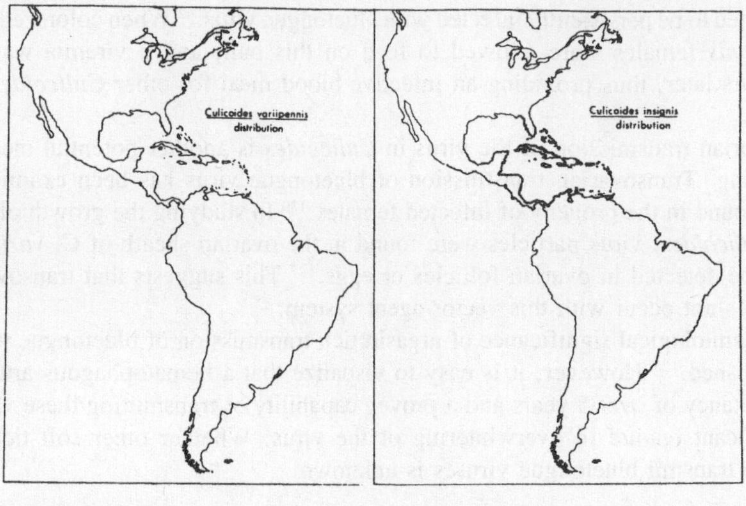

FIGURE 3. The general distribution of *C. variipennis* and *C. insignis*. Both species feed on ruminants and breed in areas where cattle are raised. *C. variipennis* is a proven biological vector of bluetongue virus. Although bluetongue virus has been isolated from *C. insignis*, experimental studies to demonstrate that it is a biological vector have not been completed. (From Kramer, W. L., Greiner, E. C., and Gibbs, E. P. J., *J. Med. Entomol.*, 22, 163, 1985. With permission.)

C. Vector Density, Fecundity, and Longevity

No study has examined adequately the critical population size of bluetongue vectors and the ability of that population to transmit the virus. Seemingly similar populations of *Culicoides* from one year to another effect transmission one year but not another.[109]

Population sizes do vary from year to year depending upon meteorological conditions. Too little rain inhibits larval development and too much rain apparently floods out prime habitat and results in population suppression. Ambient temperature also modifies the time of development and the size the flies attain.[136,140] Fecundity in free-flying *C. insignis* was highest during the cooler months of the year when larval and pupal development took their longest. At that time, they contained up to 100 eggs per gravid female; during the warmer months, the mean egg counts were as low as 80 eggs per gravid female.[136]

D. Biting Activity and Host Preference

Resource partitioning occurs within the biting habits of species of *Culicoides*. Some species feed during the daylight, others mainly at the crepuscular times, and still others are nocturnal. They also feed on different portions of the body of the host.[141] Different species are present in different seasons.[135] Possibly more important than the aforementioned resource partitioning is the fact that there are host feeding preferences as well. Widely distributed species such as *C. variipennis* may behave differently in terms of host selection as *C. v. sonorensis* usually feeds on livestock and rarely humans, whereas *C. v. occidentalis* is a serious man biter.

The Australian setting is an excellent example of the latter point. As stated elsewhere, clinical bluetongue has not been seen in Australia. The distribution of vectors and their relative vector competences and the geographical distribution of sheep and cattle production have been major factors in keeping bluetongue virus-sheep interactions to a minimum. The most efficient vectors in Australia are *C. fulvus* and *C. wadai* which are limited in distribution to the most northerly regions and do not overlap with the sheep production regions (Figure 4). Only the inefficient vector *C. brevitarsis* overlaps with sheep in distribution, but it prefers

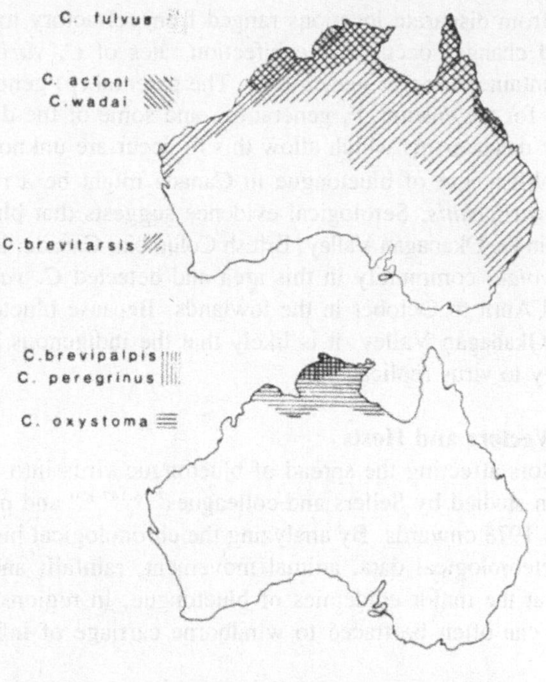

FIGURE 4. The geographical distribution of *Culicoides* species in Australia which have been demonstrated by experimental studies to be biological vectors of bluetongue virus. *C. fulvus* is an efficient vector, but is restricted to a narrow subcoastal strip in the tropics; it is the dominant vector species with transmission assisted in times of epidemics by less efficient vectors, namely *C. wadai*, *C. actoni*, and *C. brevitarsis*, in order of importance. In dry areas inland, *C. brevitarsis* is the only vector — albeit an inefficient one — present. In general, sheep are raised in areas where only *C. brevitarsis* occurs. The other vector species, *C. brevipalpis*, *C. oxystoma*, and *C. peregrinus*, are restricted in distribution to the far north and are closely associated with cattle. (Adapted from Standfast, H. A., Dyce, A. L., and Muller, M. J., in *Bluetongue and Related Orbiviruses*, Barber, T. L., Jochim, M. M., and Osburn, B. I., Eds., Alan R. Liss, New York, 1985, 177. With permission.)

cattle to sheep. Because *C. brevitarsis* breeds in cattle feces, it only occurs where cattle are present in moderate to high numbers. Thus, sheep raised in the same region are somewhat protected by the feeding preference. Unless there are major southerly expansions in the breeding range of the efficient vectors or sheep are moved into the more tropical range of *C. fulvus* or *C. wadai*, this protection of the sheep industry should continue. Serological data suggest that sheep in Australia are not usually exposed to infection with bluetongue viruses.[86,142]

E. Vector Competence

Intraspecific vector competence can vary within the geographical distribution of widely distributed species as can other aspects of midge biology. Vector competence of some of the Australian bluetongue vectors has already been discussed, but more work has been done to examine the vector competence of *C. variipennis* in North America than any other species. Through selective matings of a heterogenous colony of *C. variipennis*, strains both highly susceptible and resistant to bluetongue replication have been developed and field collections of *C. variipennis* from disparate locations ranged from refractory to highly susceptible.[143] Furthermore, rapid changes occur in the infection rates of *C. variipennis* taken into the laboratory and maintained for two generations. The parental (p) generation infection rate is usually lower than for the ensuing F_1 generation, and some of the differences are substantial.[144] The genetic mechanisms which allow this to occur are unknown.

The lack of establishment of bluetongue in Canada might be a reflection of refractory populations of *C. variipennis*. Serological evidence suggests that bluetongue viruses were circulating in cattle in the Okanagan Valley, British Columbia, Canada, in 1976.[145] McMullen[146] surveyed the *Culicoides* community in this area and detected *C. variipennis* (= *C. occidentalis)* from mid-April to October in the lowlands. Because bluetongue did not become established in the Okanagan Valley, it is likely that the indigenous population of *C. variipennis* is refractory to virus replication.

F. Movement of Vectors and Hosts

The various factors affecting the spread of bluetongue virus into areas normally free of infection have been studied by Sellers and colleagues[7,76,147-150] and published as a series of papers dating from 1978 onwards. By analyzing the chronological history of epidemics and relating this to meteorological data, animal movement, rainfall, and other factors, it has been postulated that the major epidemics of bluetongue, in regions where disease occurs only sporadically, can often be traced to windborne carriage of infected *Culicoides* from distant areas.

On the continents of America, Africa, Asia, and Australia where bluetongue virus is enzootic, there is a gradation from continuous bluetongue virus activity to complete absence of bluetongue virus transmission. To facilitate analysis, a number of zones have been designated. The zones are climate dependent (which in turn affects the presence of host and vector) and vary with latitude and altitude. The system is dynamic, however, and the boundaries of any zone are diffuse, hence it is difficult to define them on maps. In a simplified presentation, those areas where infection can be found most of the year have been defined as the endemic zone; where there are outbreaks every few years as the epidemic zone; and where disease occurs only sporadically as the incursive zone (Figure 5).

Studies over several years in West Africa have established that in the enzootic zone, there may be widespread virus activity for most of the year, but the main peaks relate to the carriage of infected *Culicoides* on winds associated with the north- or southward movements of the Inter-Tropical Convergence Zone (ITCZ). However, the movement of the ITCZ also brings rains to parched areas which leads to better grazing; thus, the vector is carried forward by the ITCZ as its source of food — the host — concurrently moves in search of food. Where cattle herds are settled, in contrast to the majority of herds, which are nomadic, movement of the vector still continues.

Outside the limits of the ITCZ, i.e., in the epizootic zone, virus activity appears to be highly focal. Outbreaks of disease are seen in these areas when climatic conditions appear to favor more widespread breeding of vectors. On occasion, windborne carriage of infected *Culicoides* occurs from this region into areas where suitable vectors do not normally exist. This phenomenon has been proposed to explain incursions of disease into Portugal in 1956 and Cyprus in 1977. (The analysis of these epidemics of disease was simplified by the separation of both countries from known sources of infection by the Straits of Gibraltar and the Eastern Mediterranean Sea, respectively.) A similar situation probably occurs in North America, which would explain the introduction of the virus into Canada in 1976.[145] In areas such as Portugal, Cyprus, and Canada, the virus has not become endemic, indicating that suitable vectors (i.e., susceptible species, or subspecies, or even populations; see Section VI.E) are not present for the long-term perpetuation of virus.[76,145]

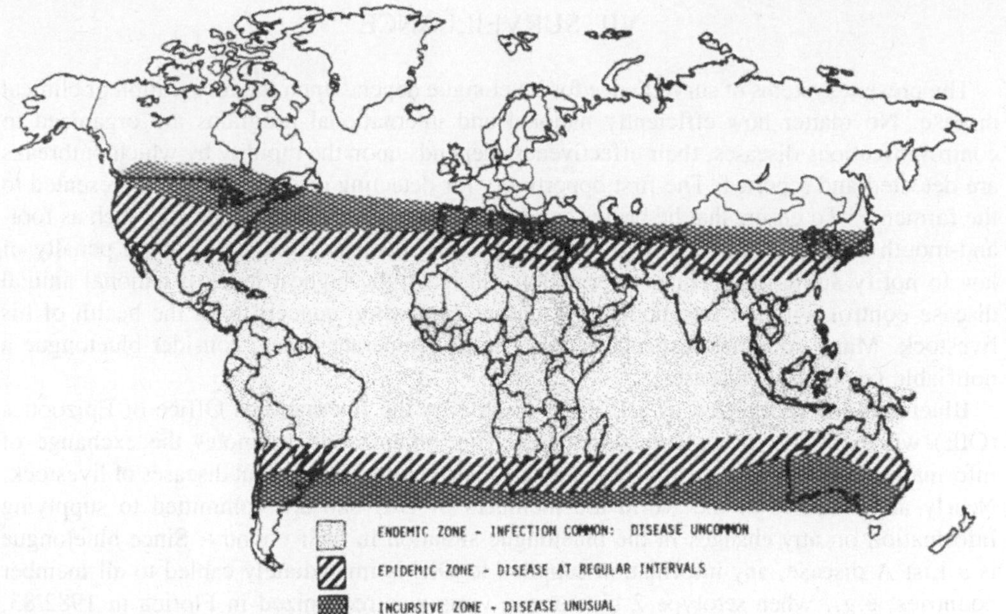

FIGURE 5. Geographical distribution of bluetongue virus and clinical disease. The three zones are depicted to simplify analysis of the epidemiology; they are dynamic and represent parts of a spectrum of virus activity dependent upon climate, altitude, and other factors. They should not be considered geographically accurate; e.g., no clinical disease has been recorded in Australia or South America. One would normally expect that in the endemic zone, most ruminants have antibody to the virus; in the epidemic zone, the percentage of animals with antibody to the virus varies and is focal, although herds/flocks with antibody can be found; and in the incursive zone, no ruminants with antibody can be found. (Adapted from Gibbs, E. P. J., in *Viral and Bacterial Diseases of Sheep*, Fehri, F., Ed., Association of French Speaking Universities, Paris, 1986. With permission.)

G. Human Element in Disease Ecology

The early concept that bluetongue virus was restricted to Africa, from where it "escaped" in the 1940s to cause disease first in Cyprus and subsequently in other parts of the world, has now been challenged by the discovery that infection of ruminants with the virus is common in the tropics and subtropics even though clinical disease is uncommon. Molecular comparisons of isolates from different continents where the virus is endemic, e.g., southern Africa and Australia, using such techniques as cross hybridization of the RNA segments and peptide mapping of virus proteins, are consistent with the evolution of bluetongue virus in continental isolation.[12,151]

This information necessitates a review of the importance of man's influence on the inadvertent spread of bluetongue virus through trade in live animals and their products. In contrast with diseases such as foot-and-mouth disease, we are not aware of any substantiated example of the intercontinental, or even international, introduction of bluetongue virus leading to clinical disease in a country that has imported livestock or animal products from an infected country. This statement is not a justification for relaxation of controls on the international movement of livestock, germplasm, and animal products; it simply highlights a previous misconception that trade has promoted the widespread distribution of bluetongue virus since its discovery in South Africa at the turn of the century.

Apart from land improvement, which in some areas has increased the supply of available ruminants and in others has decreased the supply, man has not had a profound influence on the ecology of the virus. Even vaccination, which is restricted to certain parts of the world and available only for sheep, probably has little effect on the perpetuation of the virus in an area.

VII. SURVEILLANCE

The present systems of surveillance for bluetongue depend upon the recognition of clinical disease. No matter how efficiently national and international programs are organized to control infectious diseases, their effectiveness depends upon the rapidity by which outbreaks are detected and reported. The first opportunity for detecting disease is usually presented to the farmer.[152] To ensure that he brings any suspicion of one of several diseases such as foot-and-mouth disease or rinderpest to veterinary authorities, he is expected under penalty of law to notify authorities. Thus, information can reach the highest level of national animal disease control within a few hours of a farmer expressing concern over the health of his livestock. Many countries, especially those in the temperate zones, consider bluetongue a notifiable (reportable) disease.

Bluetongue is recognized as a List A* disease by the International Office of Epizootics (OIE) which has its headquarters in Paris. This organization promotes the exchange of information on the incidence and geographical distribution of important diseases of livestock. Nearly all countries of the world are members of OIE and are committed to supplying information on any changes in the bluetongue situation in their country. Since bluetongue is a List A disease, any information supplied to OIE is immediately cabled to all member countries; e.g., when serotype 2 bluetongue virus was recognized in Florida in 1982/83, this discovery, which represented the first recovery of this serotype in the western hemisphere,[80] was reported to OIE, which in turn informed each of its member countries. This system provides governments with the opportunity to review the appropriateness of their international trade agreements and to examine critically the health status of their domestic livestock.

This system has worked well for several diseases such as rinderpest, which are rarely if ever associated with subclinical infection. Declaration by a country that it has evidence of a notifiable disease invariably invites embargoes on animal and animal product exports to some countries; however, in general, the benefits of a free exchange of information outweigh short-term disruptions in trade patterns.

The realization that bluetongue virus could be infecting livestock in a country without any evidence of clinical disease, however, has caused some countries to question the present system for the exchange of information through OIE.[153] The trade embargoes placed on Australia when it declared that it had incidentally detected bluetongue virus in a pool of *Culicoides* in the Northern territory had considerable effect on the Australian economy.[23] The dilemma of how to (or even whether to) report the isolation of viruses serologically related to those known to cause disease to international organizations, such as OIE, is unresolved. With the sensitivity of modern diagnostic tests revealing hitherto unrecognized relationships, the problem is likely to increase in importance. The definition of bluetongue virus and EHD virus becomes a critical issue in this debate as previously outlined in Section II.

VIII. PREVENTION AND CONTROL

There are three components to the prevention and control of bluetongue and EHD viruses: control of animal movement, vaccination, and vector control. Eradication is not a feasible proposition at present.

* The OIE definition of List A diseases describes them as diseases which have the potential for very serious and rapid spread, irrespective of national borders; which are of serious socioeconomic or public health consequence; and which are of major importance in the international trade of livestock and livestock products.

A. Control of Animal Movement

In contrast with humans, the movements of animals and germplasm between, and sometimes within, countries is closely regulated. Particularly for intercontinental shipment of animals destined for breeding and zoological collections, extensive laboratory testing to ensure freedom from specified viruses usually precedes movement; further, most animals are quarantined upon arrival in the importing country. To minimize the risk of introducing bluetongue and EHD viruses, a policy of restricting animal importation has been adopted by most countries with livestock presently free of infection with bluetongue and EHD viruses. Several countries such as the U.S. and Australia, which are known to have livestock infected with these viruses, also maintain a similar policy to minimize the risk of introducing additional types of bluetongue and EHD viruses.

Restrictions of animal movement within a country have been imposed on occasion during epidemics, e.g., in Portugal in 1956 and Cyprus in 1977.[76] When bluetongue virus was recognized in Australia in 1977, restrictions were initially placed on the shipment of cattle from the north into the predominantly sheep-raising areas in the south, but these were lifted once it was realized that infection had been present in Australia for several years, if not more, prior to the discovery of the virus. Further discussion of legislation and testing protocols is outside the scope of this review. Information is available in the OIE Sanitary Code[154] and other publications.

B. Vaccination

Once bluetongue virus has become established in an area, complete eradication seems impossible.[155] In areas such as South Africa, Israel, and California, this means living with the disease while attempting to minimize losses. Since the major effect of bluetongue is on the sheep industry, the use of vaccines to protect sheep has been studied from the early days of research on bluetongue. Vaccines against bluetongue are used widely in South Africa and to a lesser extent in other areas of the world. The early development of the attenuated live virus vaccines currently in use in South Africa has been described by Howell.[156] No commercially produced inactivated vaccines are available and problems exist with attenuated vaccines. Use in the 1950s of an attenuated vaccine prepared from serotype 10 in the U.S. led to the discovery that attenuated vaccines can cause congenital abnormalities in lambs born to ewes vaccinated when pregnant.[157] Later, it was discovered that the vaccine virus could be transmitted between sheep by *Culicoides*.[158]

As mentioned earlier, epidemics of bluetongue in the incursion zone (Figure 5) are usually associated with one serotype of the virus, but in the epizootic zone, a random distribution of serotypes may cause disease in any particular year. Howell[156] reported that during the course of an outbreak of bluetongue on two adjoining farms in South Africa, 9 serotypes were isolated over a 29-day sampling period. This plurality of serotypes has led to the use of multivalent vaccines to the point that in South Africa, sheep are inoculated with pentavalent vaccines 3 times, at weekly intervals; thus, each animal is exposed to viruses of 15 different serotypes within a 2-week period (B. J. Erasmus, personal communication, cited in Reference 11).

The effectiveness of polyvalent vaccination programs is debatable. It has been established that the administration of a single potent vaccine strain is followed by a durable lifelong immunity, but when combined in a polyvalent vaccine, young sheep vaccinated once or even twice may still develop clinical bluetongue and die.[155] However, despite these limitations, the large-scale use of attenuated vaccines in South Africa and Israel has undoubtedly minimized the losses due to bluetongue and has made sheep farming possible in regions where is was previously uneconomical.[155,159]

To avoid the necessity of vaccinating sheep with an attenuated strain of each serotype of bluetongue virus — an increasingly complex task as the multiplicity of serotypes increases

— Jeggo and colleagues have explored the humoral and cell-mediated immune responses of sheep to bluetongue vaccines, with a view to identifying a vaccine regimen that would promote heterotypic protection (reviewed by Jeggo and Wardley[160]). From the results of a series of experiments, some involving transfer of antibody and/or immune cells between monozygotic sheep, they questioned the current vaccination programs using polyvalent attenuated vaccines. The practice of repeat inoculations within 1 to 2 weeks is contraindicated because cytotoxic T lymphocytes induced by the initial vaccination, although short lived, are cross reactive and lyse cells infected with other virus serotypes. The research also revealed that the in vitro assay of neutralizing antibody induced in sheep vaccinated with polyvalent vaccines is a poor indicator of protection. Jeggo and Wardley concluded that for many years, the development and use of attenuated vaccines for the control of bluetongue has been on an empirical basis and would appear to have been incorrect; they propose that heterologous protection against bluetongue in areas where virus activity is common but seasonal can be generated by inoculation of attenuated vaccine (one type only) immediately prior to the anticipated period of virus activity followed by vaccination with a second serotype 1 month later.

Epidemiological observations are recorded in the published literature to support this iconoclastic proposal. Two are of particular interest:

1. Although reported,[161] rarely are multiple serotypes isolated from an individual animal during an epidemic; e.g., naturally infected sheep in South Africa yielded 40 isolates of 9 serotypes of bluetongue virus over a 29-day sampling period, but never was there any evidence to show that individual sheep became infected with more than one strain at a time.[156]
2. The epidemic of bluetongue in the Iberian Peninsula, referred to earlier, caused the death of 179,000 sheep within the first 4 months of the epidemic. A dramatic reduction in the incidence of disease followed a campaign of quarantine, slaughter, and compulsory vaccination. The value of vaccination in controlling the disease generally has been dismissed because it was subsequently shown that the vaccine used contained only serotypes 3, 11, and 12, whereas the outbreak was due to serotype 10. In light of the research findings of Jeggo and co-workers, the vaccine may have had a significant contribution.

Although a more logical and effective application of attenuated vaccines is now heralded for areas where disease is a recurring problem, the use of attenuated vaccines is condoned rather than encouraged because of the lack of alternatives. Apart from teratogenicity and insect transmission, attenuated vaccines — especially polyvalent vaccines — can theoretically give rise to virulent recombinants of new serotypes. The FAO/WHO Working Team[162] on Molecular Virology of Bluetongue has outlined the position succinctly.

In summary, the team concluded that the occurrence of reassortment does not in itself preclude the use of attenuated vaccines. If virulence is determined by a single gene and a strain is attenuated, then apart from reversion, its use in a polyvalent vaccine would not present problems in reassortment. However, if two genome segments are involved in virulence and only one is altered to produce attenuation in each of two strains, the reassortment could result in a virus particle with a set of the original genes; there would be a reassortment of virulence.

Since the basis of attenuation is unknown at this time, the Working Team has proposed that the application of polyvalent vaccines should be based on careful consideration of the possibility of genetic reassortment.

From the above description of the problems associated with attenuated vaccines, it is obvious that effective inactivated or subunit vaccines would be a great asset world-wide for the prevention and control of bluetongue, but especially for those countries in the epidemic and incursive zones. Unfortunately, despite research for several years on the development of such vaccines, none is commercially available at this time. β-Propiolactone and binary ethylenimine have shown promise as effective inactivants, but vaccines prepared using these inactivants have not progressed to field trials.[163,164] Laboratory studies with bluetongue virus inactivated with γ-irradiation indicate that this approach also warrants further study.[165] However, Stott and colleagues[164] have reported hypersensitization when sheep previously vaccinated with inactivated virus are challenged with virulent virus; thus, a subunit vaccine developed through genetic engineering may well be the most promising approach for novel vaccines against bluetongue. Studies are currently in progress on (1) the insertion of gene sequences of bluetongue into vaccinia virus as a vaccine vector and (2) anti-idiotype vaccines. At the present time, there are no published reports on the success of these approaches.

There are no vaccines available for the control of EHD. Because vaccination of wildlife is generally impractical and EHD virus is of little economic significance to the cattle and sheep industries, there is little incentive to develop vaccines.

C. Vector Control

The use of insecticides to control *Culicoides* — without specific reference to bluetongue — has been attempted in a variety of ways. Ultralow-volume aerial application of naled (Dibrom 14, 1,2-dibromo-2,2-dichloroethyl dimethyl phosphate) was used against coastal marsh breeding species in South Carolina.[166] This procedure can be effective, if applied at the correct time and with good weather, but requires a knowledge of the species of biting midge involved and details of population dynamics and biology. A by-product of mosquito control using naled in southern Florida was the apparent decimation of a large *C. insignis* population.[109]

Usually, control has been aimed at the aquatic stages. Larvicides can be applied to specific sites known to contain larvae of the species concerned. Granular formulations of chlorpyrifos, fenthion, and temephos have been used with good success.[167] Another approach is to place the pesticide into or onto the ruminant, killing the biting midges as they feed. In experiments with ivermectin in Australia, this led to a 99% reduction in the population of *C. brevitarsis* for 10 days after dosing.[168] Because this species breeds in cow pats, it is possible that control of the population could be assisted by ivermectin being shed in the feces and thus killing the immature stages developing in the pat.

Although the feasibility of using insecticides against the vectors of bluetongue and EHD viruses has been shown to be effective in restricted localities, widespread attempts to control *Culicoides* may cause environmental problems and would certainly be too expensive for routine use.

D. Environmental Modification

Populations of species like *C. variipennis*, which breed in sewage lagoons, can be controlled by keeping the sides of the lagoon steep to minimize breeding sites provided by sloping margins. Elimination of marshy areas and wet spots from seepage around watering tanks can reduce populations of species which breed in these areas. It is questionable whether many farms are prepared to adopt such policies unless bluetongue is a recurring problem.

IX. FUTURE RESEARCH

Research over the last 10 years has provided radical insights into the epidemiology of bluetongue, EHD, and related orbiviruses. The recognition that bluetongue and EHD viruses are probably ubiquitous in the livestock populations of the tropics and subtropics has led to a reassessment of the risk of these viruses causing extensive epidemics in temperate areas of the world. Although dramatic and costly epizootics, such as the one that occurred in the Iberian Peninsula in the 1950s, are still possible, the specter that bluetongue could become endemic in those countries/states that have traditionally been most concerned, i.e., those of northern Europe and the sheep-rearing areas of Australia, is now viewed as unlikely. If one considers epizootics of bluetongue and EHD as imbalances in the ecological equilibria of the two viruses, then the focus of future research is simple to define. Two major areas require attention: (1) analysis of the factors — principally climatic and their influence on vector distribution — which contribute to changes in the ecological equilibria of both bluetongue and EHD viruses and (2) development of effective and safe vaccines for protection of livestock in the event of an epizootic.

The first of these research areas provides the challenge of developing computer models reflecting the complex interactions of climate, virus evolution, and vector competence; the second provides the excitement of modern genetic approaches to vaccine development.

"We are aware that, in developing this overview, personal bias may have resulted in an emphasis on some items of information in preference to other items, and these instances may appear to the cognoscenti as capricious or arbitrary selections of the cited work. But our consistent intention was to provide, as far as possible, a balanced and succinct survey of the biology of these agents."[169]

REFERENCES

1. **Howell, P. G.,** Bluetongue, in *Emerging Diseases of Animals,* FAO Agric. Stud. No. 61, Foodland Agriculture Organization, Rome, 1963, 111.
2. **Bowne, J. G.,** Bluetongue disease, *Adv. Vet. Sci. Comp. Med.,* 15, 1, 1971.
3. **Howell, P. G. and Verwoerd, D. W.,** Bluetongue virus, *Vir. Monogr.,* 9, 35, 1971.
4. Multiple Authors, Symposium on bluetongue, *Aust. Vet. J.,* 51, 165, 1975.
5. **Hoff, G. L. and Trainer, D. O.,** Bluetongue and epizootic hemorrhagic disease viruses: their relationship to wildlife species, *Adv. Vet. Sci. Comp. Med.,* 22, 111, 1978.
6. **Verwoerd, D. W., Huismans, H., and Erasmus, B. J.,** Orbiviruses, in *Comprehensive Virology,* Vol. 14, Fraenkel-Conrat, H. and Wagner, R. R., Eds., Plenum Press, New York, 1979, chap. 5.
7. **Sellers, R. F.,** Bluetongue and related diseases, in *Virus Diseases of Food Animals,* Vol. 2, Gibbs, E. P. J., Ed., Academic Press, London, 1981, chap. 23.
8. **Gorman, B. M., Taylor, J., and Walker, P. J.,** Orbiviruses, in *The Reoviridae,* Joklik, W. K., Ed., Plenum Press, New York, 1983, chap. 7.
9. **Sellers, R. F.,** Orbiviruses, in *Topley and Wilson's Principles of Bacteriology, Virology, and Immunity,* Vol. 4, 7th ed., Brown, F. and Wilson, G., Eds., Williams, & Wilkins, Baltimore, 1984, chap. 95.
10. **Campbell, C. H., and Grubman, M. J.,** Current knowledge on the biochemistry and immunology of bluetongue, *Prog. Vet. Microbiol. Immunol.,* 1, 58, 1985.
11. **Barber, T. L., Jochim, M. M., and Osburn, B. I., Ed.,** *Bluetongue and Related Orbiviruses,* Alan R. Liss, New York, 1985, 1.
12. **Gorman, B. M. and Taylor, J.,** Orbiviruses, in *Virology,* Fields, B. N., Knipe, D. M., Chanock, R. M., Melnick, J. L., Roizman, B., and Shope, R. E., Eds., Raven Press, New York, 1985, chap. 38.
13. **Hutcheon, D.,** Malarial catarrhal fever of sheep, *Vet. Rec.,* 14, 629, 1902.
14. **Spreull, J.,** Malarial catarrhal fever (bluetongue) of sheep in South Africa, *J. Comp. Pathol. Ther.,* 18, 321, 1905.
15. **Theiler, A.,** Inoculation of sheep against bluetongue and results in practice, *Vet. J. (Transvaal),* 64, 600, 1908.
16. **Gambles, R. M.,** Bluetongue of sheep in Cyprus, *J. Comp. Pathol. Ther.,* 59, 176, 1949.
17. **Kamorov, A. and Goldsmith, L.,** A bluetongue-like disease of cattle and sheep in Israel, *Refu. Vet.,* 8, 69, 1951.
18. **McKercher, D. G., McGowan, B., Howarth, J. A., and Saito, J. K.,** A preliminary report on the isolation and identification of the bluetongue virus from sheep in California, *J. Am. Vet. Med. Assoc.,* 122, 300, 1953.

19. **Manso-Ribeiro, J., Rosa-Azevedo, J. A., Norontia, F. O., Braco-Forte, M. C., Grave-Pereiro, C., and Vasco-Fernandes, M.,** Fievre catarrhale du mouton (bluetongue), *Bull. Off. Int. Epiz.*, 48, 350, 1957.
20. **Campano, L.,** Rapport sur l'epizootie de fievre cattarrhale ovine "langue bleu" en Espagne, *Bull. Off. Int. Epiz.*, 48, 605, 1957.
21. **Sarwar, M. M.,** A note on bluetongue in sheep in West Pakistan, *Pak. J. Anim. Sci.*, 1, 1, 1962.
22. **Sapre, S. N.,** An outbreak of bluetongue in goats and sheep in Maharashtra State, India, *Vet. Rev.*, 15, 69, 1964.
23. **Gee, R. W.,** Bluetongue — effects on trade, in *Proc. 2nd Symp. Arbovirus Res. Aust.*, St.George, T. D. and French, E. L., Eds., Commonwealth Scientific and Industrial Research Organisation and Queensland Institute of Medical Research, Brisbane, 1979, 35.
24. **Gibbs, E. P. J.,** Persistent virus infections of food animals. Their relevance to international movement of livestock and germplasm, *Adv. Vet. Sci. Comp. Med.*, 25, 71, 1981.
25. **Gibbs, E. P. J.,** Bluetongue — an analysis of current problems with particular reference to importation of ruminants to the United States, *J. Am. Vet. Med. Assoc.*, 182, 1190, 1983.
26. **St.George, T. D., Standfast, H. A., Cybinski, D. H., Dyce, A. L., Muller, M. J., Doherty, R. L., Carley, J. G., Filippich, C., and Frazier, C. L.,** The isolation of a bluetongue virus from *Culicoides* collected in the Northern Territory of Australia, *Aust. Vet. J.*, 54, 153, 1978.
27. **Bekker, J. G., De Kock, G., and Quinlan, J. B.,** The occurrence and identification of bluetongue in cattle — the so-called psuedo foot and mouth disease in South Africa, *Onderstepoort J. Vet. Sci. Anim. Ind.*, 2, 393, 1934.
28. **Gibbs, E. P. J.,** Bluetongue disease, *Agripractice*, 4, 31, 1983.
29. **Shope, R. E., MacNamara, L. G., and Mangold, R.,** A virus-induced epizootic hemorrhagic disease of the Virginia white-tailed deer *(Odocoileus virginianus)*, *J. Exp. Med.*, 111, 155, 1960.
30. **Lee, V. H., Causey, O. R., and Moore, D. L.,** Bluetongue and related viruses in Ibadan Nigeria: isolation and preliminary identification of viruses, *Am. J. Vet. Res.*, 35, 1105, 1974.
31. **St.George, T. D., Cybinski, D. H., Standfast, H. A., Gard, G. P., and Della-Porta, A. J.,** The isolation of five different viruses of the epizootic hemorrhagic diseases of deer serogroup, *Aust. Vet. J.*, 60, 216, 1983.
32. **Inaba, Y.,** Ibaraki disease and its relationship to bluetongue, *Aust. Vet. J.*, 51, 178, 1975.
33. **Gibbs, E. P. J. and Lawman, M. J. P.,** Infection of British deer and farm animals with epizootic haemorrhagic disease of deer virus, *J. Comp. Pathol.*, 87, 335, 1977.
34. **Metcalf, H. E., Lomme, J., and Beal, V. C.,** Estimate of incidence and direct economic losses due to bluetongue in Mississippi cattle during 1979, in *Proc. 84th Annu. Meet. U.S. Anim. Health Assoc.*, 1980, 186.
35. U.S. Department of Agriculture, The riddle of bluetongue, *Agric. Res.*, 28, 4, 1980.
36. **Borden, E. C., Shope, R. E., and Murphy, F. A.,** Physicochemical and morphological relationships of some arthropod-borne viruses to bluetongue virus — a new taxonomic group, *J. Gen. Virol.*, 13, 261, 1971.
37. **Knudson, D. L. and Shope, R. E.,** Overview of the orbiviruses, in *Bluetongue and Related Orbiviruses*, Barber, T. L., Jochim, M. M, and Osburn, B. I., Eds., Alan R. Liss New York, 1985, 255.
38. **Tesh, R. B., Peleg, J., Samina, I., Margalit, J., Bodkin, D. K., Shope, R. E., and Knudson, D.,** Biological and antigenic characterization of Netivot virus, an unusual new *orbivirus* recovered from mosquitoes in Israel, *Am. J. Trop. Med. Hyg.*, 35, 418, 1986.
39. **Neitz, W. O.,** Immunological studies on bluetongue in sheep, *Onderstepoort J. Vet. Sci. Anim. Ind.*, 23, 93, 1948.
40. **Della-Porta, A. J.,** Classification of orbiviruses: a need for supergroups or genera, in *Bluetongue and Related Orbiviruses*, Barber, T. L., Jochim, M. M., and Osburn, B. I., Ed., Alan R. Liss, New York, 1985, 267.
41. **Gorman, B. M.,** Variation in orbiviruses, *J. Gen. Virol.*, 44, 1, 1979.
42. **Gorman, B. M.,** On the evolution of orbiviruses, *Intervirology*, 20, 169, 1983.
43. **Gorman, B. M.,** Speciation in orbiviruses, in *Bluetongue and Related Orbiviruses*, Barber, T. L, Jochim, M. M., and Osburn, B. I., Eds., Alan R. Liss New York, 1985, 275.
44. **Goldsmit, L. and Barzilai, E.,** An improved method for the isolation and identification of bluetongue virus by intravenous inoculation of embryonating chicken eggs, *J. Comp. Pathol. Ther.*, 78, 477, 1968.
45. **Lawman, M. J. P., Gibbs, E. P. J., and Davies, J. A.,** Epizootic haemorrhagic disease of deer virus in tissue culture and laboratory animals, *J. Comp. Pathol.*, 87, 345, 1977.
46. **Greiner, E. C., Barber, T. L., Pearson, J. E., Kramer, W. L., and Gibbs, E. P. J.,** Orbiviruses from *Culicoides* in Florida, in *Bluetongue and Related Orbiviruses*, Barber, T. L., Jochim, M. M., and Osburn, B. I., Eds., Alan R. Liss, New York, 1985, 195.
47. **Bowne, J. G.,** Is bluetongue an important disease in cattle?, *J. Am. Vet. Med. Assoc.*, 163, 911, 1973.
48. **Luedke, A. J. and Anakwenze, E. I.,** Bluetongue virus in goats, *Am. J. Vet. Res.*, 33, 1739, 1972.

49. **Neitz, W. D.,** The Blesbok *(Damaliscus albifrons)* as a carrier of heart-water and bluetongue, *J. S. Afr. Vet. Med. Assoc.*, 4, 24, 1933.
50. **Jessup, D. A.,** Epidemiology of two orbiviruses in California's native wild ruminants: preliminary report, in *Bluetongue and Related Orbiviruses*, Barber, T. L., Jochim, M. M., and Osburn, B. I., Eds., Alan R. Liss, New York, 1985, 53.
51. **Hoff, G. L., Griner, L. A., and Trainer, D. O.,** Bluetongue virus in exotic ruminants, *J. Am. Vet. Med. Assoc.*, 163, 565, 1973.
52. **Luedke, A. J., Jochim, M. M., and Jones, R. H.,** Bluetongue in cattle; effects of *Culicoides variipennis* transmitted bluetongue virus on pregnant heifers and their calves, *Am. J. Vet. Res.*, 38, 1687, 1977.
53. **Richardson, C., Taylor, W. P., Terlecki, S., and Gibbs, E. P. J.,** Observations on transplacental infection with bluetongue virus in sheep, *Am. J. Vet. Res.*, 46, 1912, 1985.
54. **Foster, N. M., Metcalf, H. E., Barber, T. L., Jones, R. H., and Luedke, A. J.,** Bluetongue and epizootic hemorrhagic disease virus isolations from vertebrate and invertebrate hosts at a common geographic site, *J. Am. Vet. Med. Assoc.*, 176, 126, 1980.
55. **Inaba, Y.,** Ibaraki disease and its relationship to bluetongue, *Aust. Vet. J.*, 51, 178, 1975.
56. **Osburn, B. I.,** Role of the immune system in bluetongue host-viral interactions, in *Bluetongue and Related Orbiviruses*, Barber, T. L., Jochim, M. M., and Osburn, B. I., Eds., Alan R. Liss, New York, 1985, 417.
57. **Pini, A.,** A study of the pathogenesis of bluetongue: replication of the virus in the organs of infected sheep, *Onderstepoort J. Vet. Res.*, 43, 159, 1976.
58. **Luedke, A. J.,** Effect of bluetongue virus on reproduction in sheep and cattle, *Bluetongue and Related Orbiviruses*, Barber, T. L., Jochim, M. M., and Osburn, B. I., Ed., Alan R. Liss, New York, 1985, 71.
59. **Bowen, R. A., Howard, T. H., Entwistle, K. W., and Pickett, B. W.,** Seminal shedding of bluetongue virus in experimentally infected mature bulls, *Am. J. Vet. Res.*, 44, 2268, 1983.
60. **Thomas, F. C., Singh, E. L., and Hare, W. C. D.,** Embryo transfer as a means of controlling viral infections. Bluetongue virus-free calves from infectious semen, *Theriogenology*, 24, 345, 1985.
61. **Hare, W. C. D.,** Diseases, transmissible by semen and embryo transfer techniques, Tech. Ser. No. 4, Office International des Epizootics, Paris, 1985, 1.
62. **Luedke, A. J., Jones, R. H., and Walton, T. E.,** Overwintering mechanism for bluetongue virus: biological recovery of latent virus from a bovine by bites of *Culicoides variipennis*, *Am. J. Trop. Med. Hyg.*, 26, 313, 1977.
63. **MacLachlan, N. J., Schore, C. E., and Osburn, B. I.,** Antiviral responses of bluetongue-virus inoculated bovine fetuses and their dams, *Am. J. Vet. Res.*, 45, 1469, 1984.
64. **Gibbs, E. P. J., Lawman, M. J. P., and Herniman, K. A. J.,** Preliminary observations on transplacental infection of bluetongue virus in sheep — a possible overwintering mechanism, *Res. Vet. Sci.*, 27, 118, 1979.
65. **Jochim, M. M.,** An overview of diagnostics for bluetongue, in *Bluetongue and Related Oribviruses*, Barber, T. L., Jochim, M. M., and Osburn, B. I., Eds., Alan R. Liss, New York, 1985, 423.
66. **Pini, A., Coackley, W., and Ohder, H.,** Studies on the fluorescent and neutralizing antibody to bluetongue virus in sheep, *Arch. Ges. Virusforsch.*, 18, 385, 1966.
67. **Cherrington, J. M., Ghalib, H. W., Sawyer, M. M., and Osburn, B. I.,** Detection of viral antigen in bluetongue virus-infected ovine tissues using the peroxidase-antiperoxidase technique, *Am. J. Vet. Res.*, 46, 2356, 1985.
68. **Roy, P., Ritter, G. D., Akashi, H., Collisson, E., and Inaba, Y.,** A genetic probe for identifying bluetongue virus infections *in vito* and *in vitro*, *J. Gen. Virol.*, 66, 163, 1985.
69. **Squire, K. R. E., Chuang, R. Y., Chuang, L. F., Doi, R. H., and Osburn, B. I.,** Detecting bluetongue virus RNA in cell culture by dot hybridization with a cloned genetic probe, *J. Virol. Methods*, 10, 59, 1985.
70. **Campbell, C. H.,** Bluetongue: diagnostic/antigenic interpretation, in *Bluetongue and Related Orbiviruses*, Barber, T. L., Jochim, M. M., and Osburn, B. I., Eds., Alan R. Liss, New York, 1985, 435.
71. **Della-Porta, A. J., Parsonson, I. M., and McPhee, D. A.,** Problems in the interpretation of diagnostic tests due to cross-reactions between orbiviruses and broad serological responses in animals, in *Bluetongue and Related Orbiviruses*, Barber, T. L., Jochim, M. M., and Osburn, B. I., Eds., Alan R. Liss, New York, 1985, 445.
72. **Jeggo, M. H., Gumm, I. D., and Taylor, W. P.,** Clinical and serological responses of sheep to serial challenge with different bluetongue virus types, *Res. Vet. Sci.*, 34, 205, 1983.
73. **Anderson, J.,** Use of monoclonal antibody in a blocking ELISA to detect group specific antibodies to bluetongue virus, *J. Immunol. Methods*, 74, 139, 1984.
74. **Osburn, B. I., McGowan, B., Heron, B., Loomis, E., Bushnell, R., Stott, J., and Utterback, W.,** Epizootiologic study of bluetongue: virologic and serologic results, *Am. J. Vet. Res.*, 42, 884, 1981.

75. **Howell, P. G.,** The epidemiology of bluetongue in South Africa, in *Proc. 2nd Symp. Arbovirus Res. Aust.*, St.George, T. D. and French, E. L., Eds., Commonwealth Scientific and Industrial Research Organisation and Queensland Institute for Medical Research, Brisbane, 1979, 3.
76. **Sellers, R. F., Gibbs, E. P. J., Herniman, K. A. J., Pedgley, D. E., and Tucker, M. R.,** Possible origin of the bluetongue epidemic in Cyprus, August 1977, *J. Hyg.*, 83, 547, 1979.
77. **Gibbs, E. P. J., Greiner, E. C., Alexander, F. C. M., King, T. H., and Roach, C. J.,** Serological survey of ruminant livestock in some countries of the Caribbean Region and South America for antibody to bluetongue virus, *Vet. Rec.*, 113, 446, 1983.
78. **Homan, E. J., Lorbacher de Ruiz, H., Donato, A. P., Taylor, W. P., and Yuill, T. M.,** A preliminary survey of the epidemiology of bluetongue in Costa Rica and Northern Colombia, *J. Hyg.*, 94, 357, 1985.
79. **Gibbs, E. P. J. and Greiner, E. C.,** Serological observations on the epidemiology of bluetongue virus infections in the Caribbean and Florida, in *Bluetongue and Related Orbiviruses,* Barber, T. L., Jochim, M. M., and Osburn, B. T., Eds., Alan R. Liss, New York, 1985, 563.
80. **Gibbs, E. P. J., Greiner, E. C., Taylor, W. P., Barber, T. L., House, J. A., and Pearson, J. E.,** Isolation of bluetongue virus serotype 2 from cattle in Florida: serotype of bluetongue virus hitherto unrecognized in the Western Hemisphere, *Am. J. Vet. Res.*, 44, 2226, 1983.
81. **Taylor, W. P., Gumm, I. D., Gibbs, E. P. J., and Homan, J.,** The use of serology in bluetongue epidemiology, in *Bluetongue and Related Orbiviruses,* Barber, T. L., Jochim, M. M., and Osburn, B. I., Eds., Alan R. Liss, New York, 1985, 461.
82. **Della-Porta, A. J., Sellers, R. F., Herniman, K. A. J., Littlejohns, I. R., Cybinski, D. H., St.George, T. D., McPhee, D. A., Snowdon, W. A., Campbell, J., Cargill, C., Corbould, A., Chung, Y. S., and Smith, V. W.,** Serological studies of Australian and Papua New Guinean cattle and Australian sheep for the presence of antibodies against bluetongue group viruses, *Vet. Microbiol.*, 8, 147, 1983.
83. **Taylor, W. P., Sellers, R. F., Gumm, I. D., Herniman, K. A. J., and Owen, L.,** Bluetongue epidemiology in the Middle East, in *Bluetongue and Related Orbiviruses,* Barber, T. L., Jochim, M. M., and Osburn, B. I., Eds., Alan R. Liss, New York, 1985, 527.
84. **Gumm, I. D., Taylor, W. P., Roach, J. C., Alexander, F. C. M., Greiner, E. C., and Gibbs, E. P. J.,** A serological survey of ruminant livestock in some countries of the Caribbean and South America for type specific antibody to bluetongue and epizootic hemorrhagic disease viruses, *Vet. Rec.*, 114, 635, 1984.
85. **Metcalf, H. E., Pearson, J. E., and Klingsporn, A. L.,** Bluetongue in cattle: a serologic survey of slaughter cattle in the United States, *Am. J. Vet. Res.*, 42, 1057, 1981.
86. **St.George, T. D.,** Epidemiology of bluetongue in Australia: the vertebrate hosts, in *Bluetongue and Related Orbiviruses,* Proc. Int. Symp., Barber, T. L., Jochim, M. M., and Osburn, B. I., Eds., Alan R. Liss, New York, 1985, 519.
87. **Parsonson, I. M. and Snowdon, W. A.,** Bluetongue, epizootic haemorrhagic disease of deer and related viruses, current situation in Australia, in *Bluetongue and Related Orbiviruses,* Barber, T. L., Jochim, M. M., and Osburn, B. I., Eds., Alan R. Liss, New York, 1985, 27.
88. **Du Toit, R. M.,** The transmission of bluetongue and horse sickness by *Culicoides, Onderstepoort J. Vet. Sci. Anim. Ind.*, 8, 129, 1944.
89. **Mellor, P. S., Osborne, R., and Jennings, D. M.,** Isolation of bluetongue and related viruses from *Culicoides* spp. in the Sudan, *J. Hyg.*, 93, 621, 1984.
90. **Davies, F. G., Walker, A. R., Ochieng, P., and Shaw, T.,** Arboviruses isolated from *Culicoides* midges in Kenya, *J. Comp. Pathol.*, 89, 587, 1979.
91. **Phelps, R. J., Blackburn, N. K., and Searle, L.,** *Culicoides* (Diptera:Ceratpogonidae) catches and virus isolations from them in the Mukwadzi Valley, Zimbabwe, *J. Entomol. Soc. S. Afr.*, 45, 195, 1982.
92. **Braverman, Y. and Galun, R.,** The occurrence of *Culicoides* in Israel with reference to the incidence of bluetongue, *Refu. Vet.*, 30, 121, 1973.
93. **Mellor, P. S. and Pitzolis, G.,** Observations on breeding sites and light trap collections of *Culicoides* during an outbreak of bluetongue in Cyprus, *Bull. Entomol. Res.*, 69, 229, 1979.
94. **Standfast, H. A., Dyce, A. L., and Muller, M. J.,** Vectors of bluetongue virus in Australia, in *Bluetongue and Related Orbiviruses,* Barber, T. L., Jochim, M. M., and Osburn, B. I., Eds., Alan R. Liss, New York, 1985, 177.
95. **Price, D. A. and Hardy, W. T.,** Isolation of the bluetongue virus from Texas sheep; *Culicoides* shown to be a vector, *J. Am. Vet. Med. Assoc.*, 121, 255, 1954.
96. **Jones, R. H.,** Some observations on biting flies attacking sheep, *Mosq. News,* 21, 113, 1961.
97. **Jones, R. H.,** Epidemiological notes: incidence of *Culicoides variipennis* in an outbreak of bluetongue disease, *Mosq. News,* 25, 217, 1965.
98. **Jones, R. H. and Luedke, A. J.,** Epidemiological notes: two bluetongue epizootics, *Mosq. News,* 29, 461, 1969.

99. **Stott, J. L., Bushnell, R. B., Lomis, E. C., Jessup, D., O'Rouke, M., Oliver, M. N., Osburn, B. I., and Walton, T. E.,** Overview of a longitudinal field study of bluetongue virus infection in resident populations of *Culicoides variipennis* and a multispecies sentinel herd, in *Double-Stranded RNA Viruses,* Compans, R. W. and Bishop, D. H. L., Eds., Elsevier, Amsterdam, 1983, 367.
100. **Greiner, E. C., Barber, T. L., Pearson, J. E., Kramer, W. L., and Gibbs, E. P. J.,** Orbiviruses from *Culicoides* in Florida, in *Bluetongue and Related Orbiviruses,* Barber, T. L., Jochim, M. M., and Osburn, B. I., Eds., Alan R. Liss, New York, 1985, 195.
101. **Greiner, E. C., Garris, G. I., Rollo, R. T., Knausenberger, W. I., Jones, J. E., and Gibbs, E. P. J.,** Preliminary studies on the *Culicoides* spp. as potential vectors of bluetongue in the Caribbean Region, *Prev. Vet. Med.,* 2, 389, 1984.
102. **Herniman, K. A. J., Boorman, J. P. T., and Taylor, W. P.,** Bluetongue virus in a Nigerian dairy cattle herd. Serological study and correlation of virus activity to vector population, *J. Hyg.,* 90, 177, 1983.
103. **Walker, A. R. and Davies, F. G.,** A preliminary survey of the epidemiology of bluetongue in Kenya, *J. Hyg.,* 69, 47, 1971.
104. **Boorman, J.,** *Culicoides* (Diptera, Ceratopogonidae) from Cyprus, *CAH Off. Rech. Sci. Tech. Outre-Mer Ser. Entomol. Med.,* 12, 7, 1974.
105. **Kitaoka, S.,** Notes on geographical distribution of *Culicoides* Latreille 1809 in Japan, *Bull. Natl. Inst. Anim. Health,* 46, 45, 1963.
106. **Smith, R. O. A. and Swamineth, C. S.,** Notes on some *Culicoides* from Assam, *Indian Med. Res. Mem.,* 25, 182, 1932.
107. **Wirth, W. W. and Dyce, A. L.,** The current taxonomic status of the *Culicoides* vectors of bluetongue viruses, in *Bluetongue and Related Orbiviruses,* Barber, T. L., Jochim, M. M., and Osburn, B. I., Eds., Alan R. Liss, New York, 1985, 151.
108. **Buckley, J. J. C.,** On *Culicoides* as a vector of *Onchocerca gibsoni* (Cleland and Johnson, 1910), *J. Helminthol.,* 16, 121, 1938.
109. **Greiner, E. C.,** unpublished data, 1979 to 1986.
110. **Mullen, G. R., Jones, R. H., Braverman, Y., and Nusbaum, K. E.,** Infections of *Culicoides debilipapis* and *C. stellifer* (Diptera:Ceratopogonidae) with bluetongue virus, in *Bluetongue and Related Orbiviruses,* Barber, T. L., Jochim, M. M., and Osburn, B. I., Eds., Alan R. Liss, New York, 1985, 239.
111. **Jorgensen, N. M.,** The systematics, occurrence and host preference of *Culicoides* (Diptera:Ceratopogonidae) in Southeastern Washington, *Melanderia,* 3, 1, 1969.
112. **Jones, R. H.,** Biting flies collected from recumbent bluetongue-infected sheep in Idaho, *Mosq. News,* 41, 183, 1981.
113. **Jones, R. H., Schmidtmann, E. T., and Foster, N. M.,** Vector-competence studies for bluetongue and epizootic hemorrhagic disease viruses with *Culicoides vensutus* (Ceratopogonidae), *Mosq. News,* 43, 184, 1983.
114. **Foster, N. M., Jones, R. H., and McCrory, B. R.,** Preliminary investigations on insect transmission of bluetongue virus in sheep, *Am. J. Vet. Res.,* 24, 1195, 1963.
115. **Luedke, A. J., Jones, R. H., and Jochim, M. M.,** Transmission of bluetongue between sheep and cattle by *Culicoides variipennis, Am. J. Vet. Res.,* 28, 457, 1967.
116. **Barber, T. L. and Jones, R. H.,** Bluetongue virus, serotype 2: vector transmission and pathogenicity for sheep, *Proc. 88th Annu. Meet. U.S. Anim. Health Assoc.,* 545, 1984.
117. **Boorman, J. and Gibbs, E. P. J.,** Multiplication of the virus of epizootic haemorrhagic disease of deer in *Culicoides* species (Diptera:Ceratopogonidae), *Arch. Ges. Virusforsch.,* 41, 259, 1973.
118. **Foster, N. M., Beckon, R. D., Luedke, A. J., Jones, R. H., and Metcalf, H. E.,** Transmission of two strains of epizootic hemorrhagic disease virus in deer by *Culicoides variipennis, J. Wildl. Dis.,* 13, 9, 1977.
119. **Mellor, P. S., Jennings, D. M., Braverman, Y., and Boorman, J.,** Infection of Israeli *Culicoides* with African horse sickness, bluetongue and Akabane viruses, *Acta Virol.,* 25, 401, 1981.
120. **Mellor, P. S. and Boorman, J.,** Multiplication of bluetongue virus in *Culicoides nubeculosus* (Meigan) simultaneously infected with the virus and the microfilariae of *Onchocerca cervicalis* (Railliet and Henry), *Ann. Trop. Med. Parasitol.,* 74, 463, 1980.
121. **Jennings, M., Boorman, J., and Mellor, P. S.,** Laboratory infection of the mosquito, *Toxorhynchites brevipalpis* (Diptera:Culicidae), with bluetongue virus, *Arch. Virol.,* 79, 79, 1984.
122. **St.George, T. D. and McCaughan, C. I.,** The transmission of the CSIRO 19 strain of bluetongue virus type 20 to sheep and cattle, *Aust. Vet. J.,* 55, 198, 1979.
123. **Gray, D. P. and Bannister, G. L.,** Studies on bluetongue. I. Infectivity of the virus in the sheep ked, *Melophagus ovinus* (L.), *Can. J. Comp. Med. Vet. Sci.,* 25, 230, 1961.
124. **Luedke, A. J., Jochim, M. M., and Bowne, J. G.,** Preliminary bluetongue transmission with the sheep ked, *Melophagus ovinus* (L.), *Can. J. Comp. Med. Vet. Sci.,* 29, 229, 1965.
125. **Stott, J. L., Osburn, B. I., and Alexander, L.,** *Ornithodoros coriacius* (Pajarello tick) as a vector of bluetongue virus, *Am. J. Vet. Res.,* 46, 1197, 1985.

126. **Chandler, L. J., Ballinger, M. E., Jones, R. H., and Beaty, E. J.,** The virogenesis of bluetongue virus in *Culicoides variipennis,* in *Bluetongue and Related Orbiviruses,* Barber, T. L., Jochim, M. M., and Osburn, B. I., Eds., Alan R. Liss, New York, 1985, 245.
127. **Ballinger, M. E., Jones, R. H., and Beaty, B. J.,** The comparative virogenesis of three serotypes of bluetongue virus in *Culicoides variipennis* (Diptera:Ceratopogonidae), *J. Med. Entomol.*, submitted.
128. **Foster, N. M. and Jones, R. H.,** Multiplication rate of bluetongue virus in the vector *Culicoides variipennis* (Diptera:Ceratopogonidae) infected orally, *J. Med. Entomol.*, 15, 302, 1979.
129. **Nevill, E. M.,** Cattle and *Culicoides* biting midges as possible overwintering host of bluetongue virus, *Onderstepoort J. Vet. Res.*, 38, 65, 1971.
130. **Jones, R. H. and Foster, N. M.,** Transovarian transmission of bluetongue virus unlikely for *Culicoides variipennis, Mosq. News,* 31, 434, 1971.
131. **Barnard, D. R. and Jones, R. H.,** *Culicoides variipennis:* seasonal abundance, overwintering, and voltinism in Northeastern Colorado, *Environ. Entomol.*, 9, 709, 1980.
132. **Barnard, D. R. and Jones, R. H.,** Diel and seasonal patterns of flight activity of Ceratopogonidae in Northeastern Colorado: *Culicoides, Environ. Entomol.*, 9, 446, 1980.
133. **Nelson, R. L. and Bellamy, R. E.,** Patterns of flight activity of *Culicoides variipennis* (Coquillett) (Diptera:Ceratopogonidae), *J. Med. Entomol.*, 8, 283, 1971.
134. **Zimmerman, R. H. and Turner, E. C., Jr.,** Seasonal abundance and parity of common *Culicoides* collected in blacklight traps in Virginia pastures, *Mosq. News,* 43, 63, 1983.
135. **Kramer, W. L., Greiner, E. C., and Gibbs, E. P. J.,** A survey of *Culicoides* midges (Diptera:Ceratopogonidae) associated with cattle operations in Florida, U.S.A., *J. Med. Entomol.*, 22, 153, 1985.
136. **Kramer, W. L., Greiner, E. C., and Gibbs, E. P. J.,** Seasonal variations in population size, fecundity, and parity rates of *Culicoides insignis* (Diptera:Ceratopogonidae) in Florida, U.S.A., *J. Med. Entomol.*, 22, 163, 1985.
137. **Walker, A. R.,** Seasonal fluctuations of *Culicoides* species (Diptera:Ceratopogonidae) in Kenya, *Bull. Entomol. Res.*, 67, 217, 1977.
138. **Wirth, W. W. and Morris, C.,** The taxonomic complex, *Culicoides variipennis,* in *Bluetongue and Related Orbiviruses,* Barber, T. L., Jochim, M. M., and Osburn, B. I., Eds., Alan R. Liss, New York, 1985, 165.
139. **Kline, D. L. and Greiner, E. C.,** Observations on larval habitat of suspected *Culicoides* vectors of bluetongue virus in Florida, in *Bluetongue and Related Orbiviruses,* Barber, T. L., Jochim, M. M., and Osburn, B. I., Eds., Alan R. Liss, New York, 1985, 221.
140. **Akey, D. H., Potter, H. W., and Jones, R. H.,** Effects of rearing temperature and larval density on longevity, size, and fecundity in the biting gnat *Culicoides variipennis, Ann. Entomol. Soc. Am.*, 71, 411, 1978.
141. **Schmidtmann, E. T., Jones, C. J., and Gollands, B.,** Comparative host-seeking activity of *Culicoides* (Diptera:Ceratopogonidae) attracted to pastured livestock in Central New York State, U.S.A., *J. Med. Entomol.*, 17, 221, 1980.
142. **Coackley, W., Smith, V. M., and Maker, D.,** A serological survey for bluetongue virus antibody in Western Australia, *Aust. Vet. J.*, 56, 487, 1980.
143. **Jones, R. H. and Foster, N. M.,** Heterogeneity of *Culicoides variipennis* field populations to oral infection with bluetongue virus, *Am. J. Trop. Med. Hyg.*, 27, 178, 1978.
144. **Jones, R. H.,** personal communication, 1985.
145. **Thomas, F. C., Skinner, D. J., and Samagh, B. S.,** Evidence of bluetongue virus in Canada: 1976 to 1979, *Can. J. Comp. Med.*, 46, 350, 1982.
146. **McMullen, R. D.,** *Culicoides* (Diptera:Ceratopogonidae) of the South Okanagan area of British Columbia, *Can. Entomol.*, 110, 1053, 1978.
147. **Sellers, R. G., Pedgley, D. E., and Tucker, M. R.,** Possible windborne spread of bluetongue to Portugal, June to July 1956, *J. Hyg.*, 81, 189, 1978.
148. **Sellers, R. F.,** Weather, host and vector — their interplay in the spread of insect-borne animal virus diseases, *J. Hyg.*, 85, 65, 1980.
149. **Sellers, R. F. and Pedgley, D. E.,** Possible windborne spread to Western Turkey of bluetongue virus in 1977 and of Akabane virus in 1979, *J. Hyg.*, 95, 149, 1985.
150. **Pedgley, D. E.,** *Windborne Pests and Diseases. Meteorology of Airborne Organisms,* Ellis Horwood, Chichester, England, 1982, 1.
151. **Whistler, T. and Newman, J. F. E.,** Peptide mapping of the group-specific antigens from the Australian bluetongue virus (BTV-20) and serotypes from Southern Africa and North America, *Vet. Microbiol.*, 11, 13, 1986.
152. **Henderson, W. M.,** Identification of existing and prospective problems of disease control, in *Virus Diseases of Food Animals,* Vol. 1, Gibbs, E. P. J., Ed., Academic Press, London, 1981, chap. 1.

153. **Morris, R. S. and Geering, W. A.,** Proposals for a modified system of international disease reporting, in *Proc. 2nd Int. Symp. Vet. Epidemiol. Econ.*, p. 290, 1980.
154. International Zoo-Sanitary Code, *Office International des Epizootics*, Paris, 1984.
155. **Erasmus, B. J.,** The control of bluetongue in an enzootic situation, *Aust. Vet. J.*, 51, 209, 1975.
156. **Howell, P. G.,** The epidemiology of bluetongue in South Africa, in *Proc. 2nd Symp. Arbovirus Res. Aust.*, St.George, T. D. and French, E. L., Eds., Commonwealth Scientific and Industrial Research Organisation and Queensland Istitute of Medical Research, Brisbane, 1979.
157. **Cordy, D. R. and Shultz, G.,** Congenital subcortical encephalopathies in lambs, *J. Neuropathol. Exp. Neurol.*, 20. 554, 1961.
158. **Foster, N. M., Jones, R. H., and Luedke, A. J.,** Transmission of attenuated and virulent bluetongue virus with *Culicoides variipennis* infected orally via sheep, *Am. J. Vet. Res.*, 29, 275, 1968.
159. **Barzilai, E. and Shimshony, A.,** Bluetongue: virological and epidemiological observations in Israel, in *Bluetongue and Related Orbiviruses,* Barber, T. L., Jochim, M. M., and Osburn, B. I., Eds., Alan R. Liss, New York, 1985, 545.
160. **Jeggo, M. H. and Wardley, R. C.,** Bluetongue vaccine: cells and/or antibodies, *Vaccine,* 3, 57, 1985.
161. **Stott, J. L., Osburn, B. I., and Barber, T. L.,** Recovery of dual serotypes of bluetongue viruses from infected sheep and cattle, *Vet. Microbiol.*, 1, 197, 1982.
162. WHO/FAO Working Team, Report: molecular virology, in *Bluetongue and Related Orbiviruses,* Barber, T. L., Jochim, M. M., and Osburn, B. I., Eds., Alan R. Liss, New York, 1985, 689.
163. **Parker, J., Herniman, K. A. J., Gibbs, E. P. J., and Sellers, R. F.,** An experimental inactivated vaccine against bluetongue, *Vet. Rec.*, 96, 284, 1975.
164. **Stott, J. L., Barber, T. L., and Osburn, B. I.,** Immunologic response of sheep to inactivated and virulent bluetongue virus, *Am. J. Vet. Res.*, 46, 1043, 1985.
165. **Campbell, C. H.,** Immunogenicity of bluetongue virus inactivated by gamma irradiation, *Vaccine,* 3, 401, 1985.
166. **Haile, D. G., Kline, D. I., Reinert, J. F., and Biery, T. L.,** Effects of aerial applications of naled on *Culicoides* biting midges, mosquitoes and tabanids on Parris Island, South Carolina, *Mosq. News,* 44, 178, 1984.
167. **Holbrook, F. R. and Agum, S. K.,** Field trials of pesticides to control larval *Culicoides variipennis* (Ceratopogonidae), *Mosq. News,* 44, 233, 1984.
168. **Standfast, H. A., Muller, M. J., and Wilson, D. D.,** Mortality of *Culicoides brevitarsis* (Diptera:Ceratopogonidae) fed on cattle treated with ivermectin, *J. Econ. Entomol.*, 77, 419, 1984.
169. **Dale, S. and Pogo, B. G. T.,** Biology of pox viruses, *Virol. Monogr.*, 18, 1, 1981.

Chapter 17

BOVINE EPHEMERAL FEVER

T. D. St.George and H. A. Standfast

TABLE OF CONTENTS

I. HISTORICAL BACKGROUND

A. Discovery of Agent and Vector(s)

The first detailed account of bovine ephemeral fever (BEF) appeared in the French literature.[1] All the essential clinical features of an epizootic of ephemeral fever in cattle were described as "dengue of cattle" in lower Egypt in 1895. Piot[1] drew attention to important similarities as well as differences from dengue in humans. The disease had not occurred previously in that part of Egypt within living memory. Somewhat earlier, Schweinfurth[2] had recorded the local name of ephemeral fever in the region of Africa which later became German East Africa, but this did not become general knowledge until 1910.[3] The probability exists that ephemeral fever is an ancient disease which the development of cattle industries or improved veterinary services has brought to prominence. In southern Africa, Bevan[4,5] described the disease quite clearly and mentioned its transmissibility by the intravenous (i.v.) injection of infective blood. It was much later that it was recorded in countries bordering the Pacific: Australia in 1936,[6] Japan in 1949,[7] China in 1954,[8] and in many countries of Africa and Asia. There is no suggestion that it spread from Africa to Asia in modern times.

The causative agent, BEF virus, was first transmitted successfully from cow to cow by Bevan[5] and the infection was characterized in comprehensive studies using cow inoculation as the assay system.[9] BEF virus was not isolated in another host until 1966 in South Africa.[10] That isolation was made from the leukocytes of febrile cattle passaged in other cattle since their original collection in 1958, and finally inoculated into the brains of suckling mice. This marked a critical turning point in the study of the disease by allowing for the characterization of the virus, development of diagnostic tests, production of vaccines, and treatment assessment.

Both the identification and the proving of vectors are in a poor state, even today.

B. History of Epizootics

The epizootiology of ephemeral fever has been described in detail in very few of the many countries[11] in which it occurs. The earlier history in Africa and Asia was reviewed by Curasson.[12] The disease appeared to be enzootic in many countries in the central third of the African continent, but produced isolated epizootics in the countries of southern Africa and Egypt. These epizootics moved in a north-south direction in Africa south of the Equator [4,5,13] and in a south-north direction north of the Equator in the Sudan and Egypt.[1,14] This movement away from the tropics occurred in China in 1983,[8] and many times in Australia (1936—

1937, 1955—1956, 1967—1968, 1970—1971, 1971—1972, 1972—1974)[6,15,16] and less distinctly thereafter.[17,18] Susceptible cattle are abundant at the northern and southern limits where periodic summer or autumn epizootics of ephemeral fever terminate. Thus, the limitations of spread were clearly imposed by lack of insect vectors at these extremes.[6,15,16]

In particular, in African countries more information is available than in other countries. Ephemeral fever was first recognized in Kenya in 1913—1914. The disease has occurred there in the form of irregular epizootics separated by periods when no cases or only sporadic cases were observed. The large epizootics of more recent years (1967—1968, 1972—1973, and 1977) were associated with higher than average rainfall; these years were also when Rift Valley fever occurred in epizootic form.[19] There is a dearth of information on the epizootics which have occurred for many years in Nigeria[20] and South Africa where the disease occurs frequently.[10,13] In South Africa, epizootics terminate at the first frost,[10] as they do in southern Australia.

The relationship between epizootics in China,[8] Taiwan,[21] Korea, Japan,[22] and countries in the same general region of eastern Asia would be interesting to study if more published information were available from the first three countries. Epizootics occurred in China in 1954, 1976, and 1983. The most recent epizootics in Taiwan occurred in 1967, 1983, and 1984.[23] Inaba[22] listed the Japanese epizootics, with morbidity figures. Epizootics occurred in each year from 1949 through 1953, 1955, 1956, 1958, and 1966. Later, epizootics occurred in 1971 and a small outbreak occurred in 1976. Recent years have been free of epizootics.[24]

Papua-New Guinea recorded a small epizootic in 1959.[16] The source of this epizootic was infected cattle, imported a few days earlier, from an area where the disease was active at the time in northern Australia. Apparently, the small and scattered cattle population of Papua-New Guinea could not provide sufficient susceptible animals to maintain the virus, as the disease died out within months, and has not recurred. Ample serology has been carried out in sentinel herds and general surveys in Papua-New Guinea confirm this.

C. Social and Economic Impact

Ephemeral fever produces its main impact[6,13] by disrupting the husbandry of beef cattle; by sharply curtailing the lactation of dairy cows; by rendering draught cattle temporarily incapable of work during a season when their labor is important; and by rendering bulls temporarily infertile.[25] The effects on international trade are substantial, but the loss is hard to quantify.[26] Average mortality is low but selective in that the best conditioned or the most productive animals, or those that continue to be worked, are the worst affected and most likely to die.[27] Owners fear the disease more than simple mortality figures suggest. In focal areas mortality can exceed 10%. The total number of animals affected in epizootics is usually unknown, but can be tens of millions in very severe epizootics. During fever, milk formation almost ceases and the milk quality is poor.[27] Lactation usually resumes on recovery, but the average loss of milk production in one study of a natural epizootic was 12% of the balance of lactation in dairy cows.[28] Under experimental conditions, the short-term milk loss was 50%.[29] In Kenya, dramatic reductions of milk production have occurred in major epizootics.[19]

The loss of body weight during illness is apparent, but usually is not quantified. In one experiment, Herefords penned in air-conditioned quarters returned to preillness body weight on *ad libitum* high-quality feed 2 weeks after recovery and return of appetite.[30] The time required to regain preillness body weight would be much longer for those animals forced to forage for lower-quality pastures under the stresses of weather. Ephemeral fever occurs mainly in the summer or rainy season, the limited season of the year when pasture is adequate in quantity and quality for body weight gain to occur, before the inadequate feed of the winter or dry season. The loss, therefore, is more serious than implied by simple consideration of a few days disability.

The impact of the disease on draught animals is known to be severe in China and Taiwan, but only in Indonesia has an attempt been made to cost this loss.[31] These workers stated that the loss of labor of the ox for 30 days was the equivalent of one half the sale value of a young animal or one seventh the cost of an adult.

II. THE VIRUS

A. Antigenic Relationships and Structure

Van der Westhuizen's[10] description of BEF virus as having the physical structure of a rhabdovirus by electron microscopy was amply confirmed.[32-34] The classification was established biochemically by Della-Porta and Brown.[35] The classification was in doubt previously as Tanaka et al.[36] considered the virus to have double-stranded RNA. The antigenic relationships with other rhabdoviruses have been more elusive, and for many years BEF virus could not be firmly related to other viruses. Tesh et al.[37] suggested an uncertain relationship to rabies virus by complement fixation (CF). However, it is becoming apparent that BEF virus is part of a complex of viruses of which Kimberley,[38] Berrimah,[39] and Adelaide River[40] viruses are closely related by immunofluorescence, but are distinct viruses by neutralization tests (N) using hyperimmune sera produced in rabbits. Fukuoka virus[41] is not now in BEF serogroups.[95] However, heterotypic antibodies are produced in cattle following sequential natural or experimental infection with certain of these viruses.[42] For instance, infection of cattle with Kimberley virus causes the development of N antibodies against that virus, plus a lower level of antibodies against BEF virus. After a later infection with BEF virus, an anamnestic response to both viruses is followed by the development of a very low level of N antibodies against Adelaide River virus.[42,43] The heterotypic BEF antibody produced in response to natural or experimental Kimberley virus infections of cattle does not appear to prevent typical ephemeral fever on subsequent exposure to BEF virus infection. The protective capacity of other combinations of BEF-related viruses is yet to be defined. The actual situation is more complex, because it is already apparent that recently isolated rhabdoviruses which are less well characterized than BEF virus are also antigenically related to the BEF serogroup, though not closely related to BEF virus.[44] This means that antibody in cattle which will neutralize BEF virus in vitro cannot be guaranteed to have been generated by that virus, or to be protective. More specific tests are required for definitive single test serology. The present N tests are valuable, though imprecise, tests for diagnosis and epizootiological studies, provided the cross reactivity is taken into account.

B. Host Range

The known vertebrate hosts of BEF group viruses are all cloven-hoofed herbivores. They include cattle (genus *Bos*), water buffaloes *(Bubalus bubalis)*,[16] Cape buffaloes *(Syncercus kaffer)*,[45] deer of at least two species *(Cervus elaphus, Axis axis)*[41] in Australia, and herbivores of various African game species.[45] Clinical signs have been reported only in cattle and water buffaloes following natural or experimental infection.[47,48] Young[49] found Australian water buffaloes resistant to experimental infection. Red deer *(C. elaphus)* were susceptible to experimental challenge with Australian BEF virus, experiencing a fever of 1 to 2 days duration.[50] Sheep have been infected experimentally without showing clinical signs,[51] but no evidence of natural infection of sheep has been found. Suckling mice, rats, and hamsters are susceptible to intracerebral (i.c.) inoculation,[7,10,52-54] but the report by Murphy et al.[34] that BEF virus can be transmitted by intraperitoneal (i.p.) inoculation has not been confirmed. Viremias have not been demonstrated in mice, rats, hamsters, or rabbits.

BEF virus has been isolated from blood by the i.v. inoculation of chicken embryos.[55] The virus either killed the embryo or, if the chicken hatched, it was abnormal. Material from chick embryos was inoculated into suckling mice to confirm the presence of BEF virus.

C. Strain Variation

In herds which frequently experience disease, farmers report that the severity of the disease varies from outbreak to outbreak. However, there is a wide variation of the severity of clinical disease within a single epizootic. Variation in virus virulence may occur, but is not yet proven by clinical or molecular criteria. Antigenic differences do occur between BEF strains isolated from cattle with clinical disease in different epizootics by in vitro comparisons as judged by cross-N tests using mouse ascitic fluids.[56] Cross protection of cattle has occurred between strains from two different epizootics in Australia in an instance where sequential experimental infections have been carried out with strains of widely different origins within the same country.[57] Monoclonal antibodies have not been used to study strain variation so far.

D. Methods for Assay

The most thorough study of ephemeral fever in cattle used susceptible cattle as the assay system.[9] This method was partially supplanted by the i.c. inoculation of suckling mice.[10] However, the method was found to be about 25% efficient in isolating BEF virus from leukocytes from 243 clinical cases of ephemeral fever.[16] Cattle are regarded as the most sensitive species for the demonstration of the presence of BEF virus. The newer methods have not been assessed objectively on field material, although they have been very effective in isolating BEF virus from leukocytes in experimental cattle.[58] Subsequently, various tissue culture systems have been developed. Interestingly, cells of bovine origin are not susceptible or are poorly susceptible,[57,59] and the significance of the homologous host tissue being poorly susceptible is unknown. BEF virus has been adapted successfully to grow in hamster kidney (BHK-21)[7,52,57] and hamster lung, in VERO cells, and also in *Aedes albopictus* (a mosquito cell line C6/36).[60] The success of primary isolation of BEF virus from the blood of naturally infected cattle in mice or cell cultures is quite erratic. The most promising method at present is to use *Ae. albopictus* for primary inoculation from leukocytes, then after 15 days, sub-inoculate cells and supernatant into BHK-21 cells to demonstrate cytopathic effect, or into VERO cells for demonstration of virus presence by immunofluorescence. The method has been very successful with experimentally infected cattle, but has not been evaluated with field material.[58]

There have been very few successful isolations of BEF virus from naturally infected insects. The first isolation, by Davies and Walker[61] from *Culicoides,* was achieved by the i.c. infection of mice, as were the second[62] and third[63] which were from mosquitoes. The most recent isolations were made in the *Ae. albopictus*/BHK-21 tissue culture combination from *Ae. aegypti* which had been inoculated intrathoracically with homogenates of insects collected in an outbreak area.[50]

IIII. DISEASE ASSOCIATIONS

A. Humans

During ephemeral fever epizootics, some farmers have reported similar signs of disease in themselves or their families. Whenever investigations were made in Australia, there was no serological evidence that BEF virus infection occurred in these people. Also, no antibody was detected in a small survey of 50 people in nearby enzootic areas.[44] There is no evidence that laboratory infection occurred when BEF virus had been inadvertently injected into a finger, thus BEF virus must be considered to be of low pathogenicity for humans. Because BEF virus does not occur in the U.S. it has been assigned a high biocontainment rating there, well above what is necessary to protect laboratory workers.

B. Domestic Animals

The species of domestic animals which suffer clinical disease are cattle and water buffalo.

Not all workers are convinced that water buffaloes are susceptible as they may resist experimental infection.[49] However, enough natural cases have been described to be sure that water buffaloes[47,64-66] do suffer illness. There is no objective evidence that any particular breed of cattle is more resistant to the disease.

The first clinical signs of natural ephemeral fever are a mild malaise in nonlactating cattle, or a severe decline of milk production in lactating cows. However, in this phase, the body temperature is elevated (40 to 42°C). This febrile phase of 12 to 24 hr is followed by a second febrile phase with the rapid development of the characteristic muscle stiffness, lameness, severe depression, ocular and nasal discharge, salivation, anorexia, and rumenal stasis. The animal may be recumbent, but able to rise if forced, or else be paralyzed in sternal or lateral recumbency. Muscular fibrillation, patchy subcutaneous (s.c.) edema, blood-flecked feces, and loss of swallowing reflex are variable signs. Dry or moist rales may be detected in the lungs occasionally followed by severe emphysema. The severe clinical signs may persist for 1 or 2 days, then recovery usually begins and proceeds quite rapidly. An animal apparently near death can be grazing normally a few hours later.

Continuous observation of experimentally infected cattle has shown that the fever is usually biphasic (sometimes polyphasic), with each phase lasting approximately 1 day. Clinical signs in the first febrile stage may be very mild in contrast to the severe signs of the second or later phases. It is the rapid resolution of the apparently severe disease which invited the name "ephemeral fever". A small percentage of cattle develop severe pneumonia or emphysema, suffer long-term paralysis, or die.[67]

The pathology of the disease has been described as a serofibrinous polyserositis. The inflammation of the endothelium of the small blood vessels[9,68,69] leads to increased vascular permeability and the effusion of fibrin-rich fluid into joints, the pericardial, pleural, and peritoneal cavities, and lung alveoli. Neural signs are superimposed. These are characteristic of hypocalcemia in the bovine, namely muscle tremor, rumenal stasis, and temporary or long-term paralysis of limbs together with loss of reflexes. The changes observed in an animal dying of natural disease depend on the stage of disease at which it dies. If death occurs early, a fibrinous exudate is found in the serous cavities of the chest, abdomen, and joints. Fibrin plaques may be found on virtually all articular surfaces. There are no lesions in the brain. The tissue damage does not appear severe enough to account for death. If death occurs late in the course of the disease, the effects of dehydration may be apparent, as many sick cattle exhibit aversion to water. Abortion can occur if cows experience disease very late in pregnancy. However, the virus does not appear to cross the placenta or affect fertility of the female.[70-72]

C. Wildlife

No disease has been described in wildlife. All the evidence of BEF infection of wildlife depends on serology.

D. Applicable Diagnostic Procedures

A neutrophilia with the appearance of immature neutrophils occurs during fever in natural or experimental ephemeral fever. This is not pathognomonic of the disease, but if not present, the diagnosis is suspect. The percentage of neutrophils in a smear of blood taken during fever can range from 46 to 90%, in contrast to the normal range of 10 to 24%.[73] The plasma fibrinogen level rises 2- to 4-fold, while the total serum calcium level falls approximately 15% in the later stages of illness.[18,67]

A single case of ephemeral fever in a nonepizootic period is difficult to diagnose. In the area of the world where ephemeral fever occurs, the main diseases requiring differential laboratory diagnosis are caused by hemoprotozoa, and the appropriate specimen to examine is a blood film taken from the ear or tail, which is fixed and stained appropriately to detect the hemoprotozoa relevant to that area.

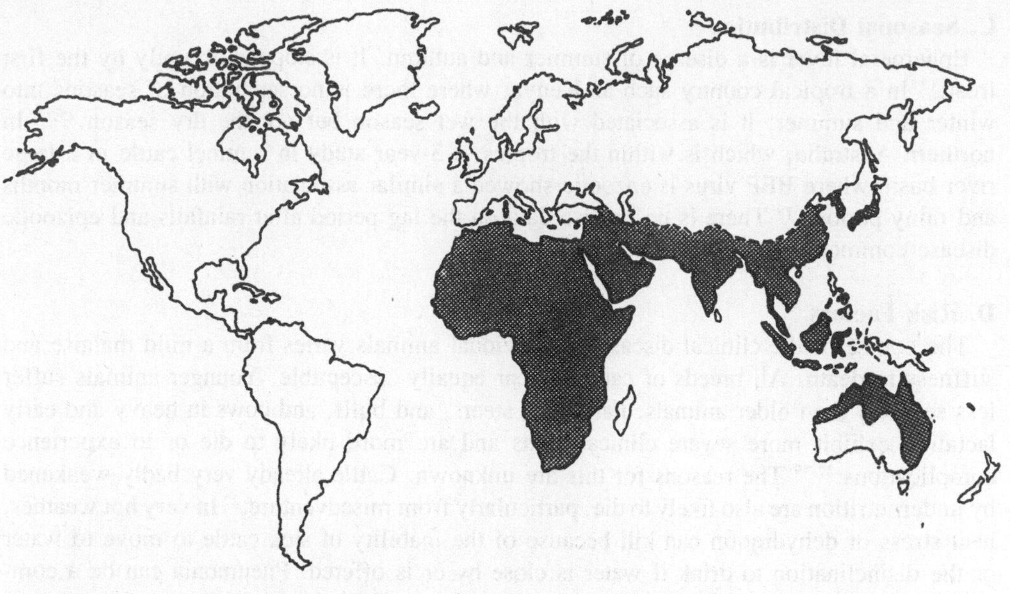

FIGURE 1. The approximate distribution of ephemeral fever in the world is shown in the shaded area. Except where there is specific information on the limits of infection within a country (Australia, Japan, China), national borders are used as boundaries. It is not implied that all countries are infected at any particular time.

BEF virus can be isolated from the leukocyte fraction of a blood sample taken during fever. A rise in neutralizing antibody titer in serum is confirmatory evidence. In immunologically naive cattle, the rise may be from negative in undiluted serum to a titer of 128 to 1024 between samples collected during illness and 2 to 3 weeks later. If cattle have been exposed previously to an antigenically related virus, a peak level may be attained within 3 to 5 days of the onset of illness.[43]

E. Adverse Effects of Virus on Vector

In nature, no ill effects are known. Mosquitoes *(Ae. aegypti* and *Cx. annulirostris)* experimentally infected with BEF virus are sensitive to CO_2,[74] an effect described by Rosen[75] for seven other rhabdoviruses and three flaviviruses and by Turell[76] for four bunyaviruses and two togaviruses.

IV. EPIZOOTIOLOGY

A. Geographic Distribution

Ephemeral fever occurs in Africa, the Middle East, Asia, and Australia, but has not been recorded in the Americas or the islands of the Pacific. All the countries in which it has appeared are contiguous or separated by relatively narrow sea barriers (less than 500 k) from each other. The full distribution was illustrated by Odend'hal,[11] but may be summarized as virtually all of Africa, and the Asian countries of Syria, Iraq, Iran, Pakistan, India, Bangladesh, Burma, China, Taiwan, and Japan and all countries to the south and southeast (Figure 1). There is a fluctuating zone of infection in the southern tip of Africa, Australia, and eastern Asia.

B. Incidence

The incidence is very variable. In explosive and well-spaced epizootics, 100% of particular herds may become sick or, in sporadic outbreaks, only 2 to 3%.

C. Seasonal Distribution

Ephemeral fever is a disease of summer and autumn. It is stopped abruptly by the first frost.[13] In a tropical country such as Kenya, where there is no separation of seasons into winter and summer, it is associated with the wet season but not the dry season.[45,77] In northern Australia, which is within the tropics, a 3-year study in sentinel cattle in a large river basin where BEF virus is enzootic showed a similar association with summer months and rainy periods.[78] There is no precise data on the lag period after rainfalls and epizootic disease commencement.[27,45]

D. Risk Factors

The course of the clinical disease in individual animals varies from a mild malaise and stiffness to death. All breeds of cattle appear equally susceptible. Younger animals suffer less severely than older animals. Fat cows, steers, and bulls, and cows in heavy and early lactation exhibit more severe clinical signs and are more likely to die or to experience complications.[27,79] The reasons for this are unknown. Cattle already very badly weakened by undernutrition are also likely to die, particularly from misadventure.[27] In very hot weather, heat stress or dehydration can kill because of the inability of sick cattle to move to water or the disinclination to drink if water is close by or is offered. Pneumonia can be a complication if chilling rain and wind occur during fever. Forced exercise is a certain way to exacerbate clinical signs and increase risk of fatality. In contrast, in areas that are enzootic or periodically epizootic, colostral antibody appears protective against infection and disease.[80] Young calves without specific colostral antibodies are susceptible to experimental infection.[81]

E. Serologic Epizootiology

In interepizootic periods in Australia, there has been an unexplained background of serological positives during periods of inactivity of clinical ephemeral fever.[82,83] It now appears that much of this antibody is heterotypically generated by antigenically related viruses. There is no doubt that during and following epizootics, the prevalence and the average titer of N antibodies rise. However, any test for antibodies which relies on group-reactive antigen such as CF, agar gel diffusion, or precipitin immunofluoresence[84] and perhaps in due course ELISA tests will not be specific. A specific test that is capable of automation is urgently needed.

V. TRANSMISSION CYCLES

A. Evidence from Field Studies

The evidence derived from the direct observations of the last 100 years is that ephemeral fever is not spread by animal-to-animal contact, feed, or fomites.[9] In insect-proof housing, no transmission occurs between infected and control cattle even if in very close contact with each other. An outbreak in a herd often commences with an index case or cases about 1 week ahead of the main wave.[15,27,80]

1. Vectors

Efforts have been made for many years to establish what species of insects transmit BEF virus.[9,27] The difficulty of isolating BEF virus from insects is one factor responsible for the poor knowledge of vectors. Intensive efforts have been made to isolate BEF virus from insects collected during epizootics[63,80] and during interepizootic phases.[85] One isolation was reported in Kenya from a mixed pool of five *Culicoides* species: *C. kingi, C. nivosus, C. bedfordi, C. pallidipennis*, and *C. cornutus*.[61] The virus was later isolated in Australia once from *C. brevitarsis*,[74] and twice from *Anopheles bancroftii*,[62,63] and from a mixed pool that included *Cx. orbostinensis, Ae. carmenti, Uranotaenia nivipes*, and *Ur. albescens*.[62] All

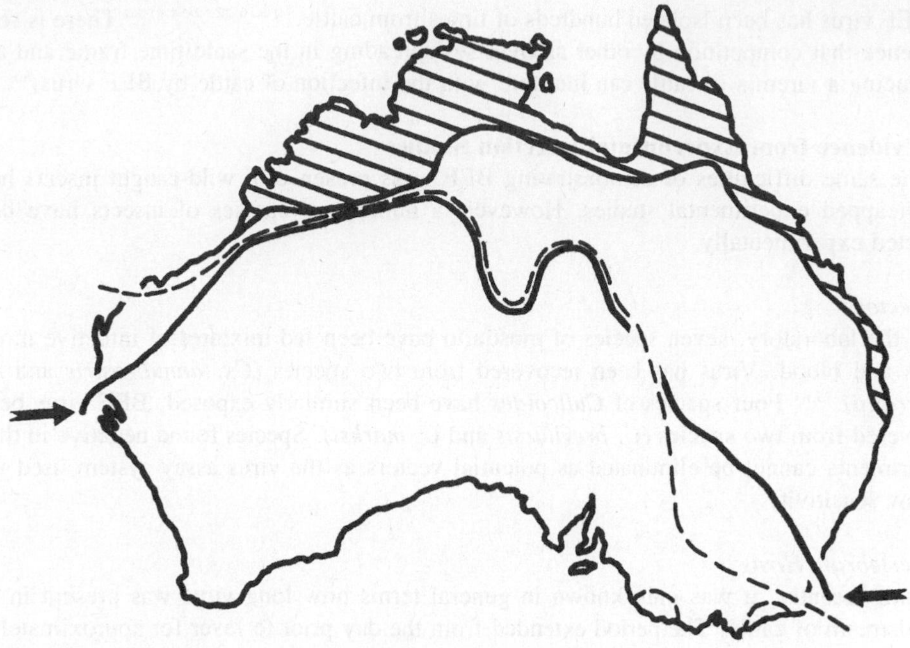

FIGURE 2. The effective limits of distribution of the insects from which BEF virus has been isolated in Australia are shown in relation to the area in which ephemeral fever has been known to occur. This area exceeds that of the occurrence of the insects. Hatched area shows the distribution of *Anopheles bancroftii*. The solid line marks the southern limit of distribution of *Culicoides brevitarsis*. The dotted line marks the extreme limit of the most extensive BEF epizootic.

these isolations were from unengorged insects. The geographical distribution of ephemeral fever in Australia exceeds that of the insect species from which it has been isolated[63] (Figure 2). However, two species of mosquito — *Cx. annulirostris* and *An. annulipes* — have distributions which match or exceed the known range of BEF in Australia; both have been collected in appreciable numbers attacking cattle during outbreaks, while *Cx. annulirostris* has been infected by feeding on blood virus mixtures through a membrane.[50] The tremendous geographic and climatic distribution of the disease in Africa and Asia also indicates that more vector species are involved in its spread than those few from which it has been isolated. The antigenically related Kimberley virus has been isolated in Australia from both mosquitoes *(Cx. annulirostris)* and *Culicoides (C. brevitarsis)*,[38] while in Japan, Fukuoka virus was also isolated from both mosquitoes *(Cx. tritaeniorhynchus)* and *Culicoides (C. punctatus)*,[41] so the involvement of very different vectors with rhabdoviruses is not unique to BEF virus.

2. *Vertebrate Hosts*

The attempts to isolate BEF virus from other vertebrate species have been few and unsuccessful. All the evidence that other herbivores are naturally infected is based on serology.[16,45,46] These herbivores (deer and African game species) live in the same areas as cattle which suffer the disease. Whether they act as reservoirs or are silently infected at the same time as the nearby cattle population is unknown. It is probable, but unproven, that the same vector species bite cattle and the wild herbivores. No antibodies to BEF virus have been found in marsupials, horses, or sheep in Australia.[86] No species of bird has been examined in sufficiently large numbers to exclude the possibility that birds can be infected. Studies are in progress on the cattle egret *(Bubulcus coromandus)*, a bird which feeds in close association with cattle, but so far the evidence does not support a role for this species in the ecology of BEF virus.[67]

BEF virus has been isolated hundreds of times from cattle.[7,10,16,22,52-54,61,80] There is some evidence that competition by other arboviruses spreading in the same time frame and also producing a viremia in cattle can interfere with the infection of cattle by BEF virus.[80]

B. Evidence from Experimental Infection Studies

The same difficulties of demonstrating BEF virus presence in wild-caught insects have handicapped experimental studies. However, a number of species of insects have been infected experimentally.

1. Vectors

In the laboratory, seven species of mosquito have been fed mixtures of infective mouse brain and blood. Virus has been recovered from two species *(Cx. annulirostris* and *An. bancroftii).*[50,87] Four species of *Culicoides* have been similarly exposed, BEF virus being recovered from two species *(C. brevitarsis* and *C. marksi).* Species found negative in these experiments cannot be eliminated as potential vectors as the virus assay system used was of low sensitivity.

2. Vertebrate Hosts

Until recently, it was only known in general terms how long virus was present in the bloodstream of cattle. The period extended from the day prior to fever for approximately 5 days.[9] Virus has been demonstrated to be present during the incubation period at a low level.[58] It is isolated regularly several hours prior to the commencement of fever and during fever which persists for 72 to 108 hr. After convalescence, BEF virus can no longer be isolated.

Many thousands of blood samples have been collected from the cattle in the sentinel herd scheme in Australia.[83] No isolations of BEF virus have been made outside the epizootic period. It can be inferred, but not proved, that the titer will be greatest at the peak of neutrophilia. Mackerras et al.[9] showed that BEF virus was in leukocytes but not in erythrocytes, and Young and Spradbrow[69] demonstrated experimentally that BEF virus was associated with neutrophils collected from fluid exudates in joints and the peritoneum. The more specific knowledge of the relationship of viremia with fever makes it possible to select a time for feeding potential vector species on cattle with confidence that BEF virus is present, or to collect blood for membrane feeding of potential vectors.

VI. ECOLOGICAL DYNAMICS

A. Climate and Weather

The disease occurs in tropical and subtropical countries and extends into some temperate areas, generally those where summer rainfall occurs. An association with recent rainfall has been noted by several observers.[15,16,19,45,78] In contrast, ephemeral fever has occurred in Egypt in the season when rain does not occur.[1,14] Murray[15] described the flooding of rivers from an area of heavy rainfall into an otherwise dry area as an important factor in a rapidly moving epizootic in inland Australia, which moved in the general direction of winds from the northwest. A rapidly changing level in a large river system could trigger the emergence of certain mosquito species. A spillover into drought-affected areas could be accounted for by wind dispersal.[15,79]

The northern and southern limits of BEF virus activity in the world are imposed by climate limitations on the vector in Japan[22] and Australia.[16] A large area of high country (over 1000 m) in eastern Australia, the New England Plateau, often escapes the epizootics of the surrounding lower country.[88] Hot or warm weather seems essential for epizootics which stop abruptly after the first frosts of winter.[13]

B. Vector Oviposition

The potential mosquito vectors, *An. bancroftii* and *Cx. annulirostris,* oviposit in ground pools. The single *Culicoides* species involved in Australia, *C. brevitarsis,* oviposits only in cattle or buffalo dung pats and rarely in dung of other herbivores such as deer. The species of *Culicoides* in the mixed pool of species from which Davies and Walker[61] isolated BEF virus breed mainly in mud or dung-mud mixtures.

C. Vector Density, Fecundity, and Longevity

There is a very wide range of intensity of epizootics even when ample susceptible vertebrate hosts are available. The mosquito species suspected of transmitting BEF virus in Australia are known to transmit other viruses at times when BEF is not spreading. Also, 15 other viruses have been isolated from *C. brevitarsis,* a probable vector species in Australia. Many of these are efficiently spread in interepizootic periods, which suggests that absence of vectors is not a reason for failure of BEF virus to spread in a particular year.

D. Biting Activity and Host Preference

The insects suspected of transmitting BEF virus feed on a range of mammal species. The species of mosquito which have been associated with BEF virus in Australia certainly bite other vertebrates besides cattle, buffaloes, and deer, but apparently do not transmit infection to them. The species of *Culicoides* from which virus has been isolated in Australia *(C. brevitarsis)* bites cattle, horses, and sheep but not man. With the exception of *C. nivipes,* the species represented in the pool from which BEF virus was isolated in Kenya[61] are known to feed on cattle.

VII. SURVEILLANCE

A. Clinical Hosts

Cattle are the only useful sentinel animals. Where surveillance is carried on between epizootics in Kenya and Australia,[16,45,83] subclinical infection has been assumed to occur between epizootics because of the development of antibodies to BEF virus in sentinel cattle. However, the discovery that heterotypic antibodies can be generated by the related Kimberley virus[43] throws doubt on these findings. Serology with these related viruses has been carried out only in Australia. but it is highly probable that BEF serogroup viruses will exist in Africa and Asia. In tropical countries, the limitations of the N test must be recognized and compensated for heterotypic virus infection by multiple parallel testing with related rhabdoviruses. Further serological surveys would be of great assistance to the understanding of the epizootiology.

B. Wild Vertebrates

Most of the serology of wild vertebrates is affected by the same problems of test specificity as with cattle when surveying for subclinical infection. However, domesticated red deer with antibodies to BEF virus were found to be free of Kimberley virus infection in an area where cattle had a high prevalence of infection.[46] Within the limits of present knowledge, serological studies of wildlife provide no useful information on the epizootiology of ephemeral fever.

C. Vectors

It is vastly easier to isolate BEF virus from cattle than from insects so there is no advantage in catching potential vectors to detect virus presence ahead of an epizootic. Also, the very slow (2 to 4 weeks) isolation methods would make this approach irrelevant.

D. Sentinels

Cattle are the only useful sentinels for monitoring subclinical or clinical disease. In the areas of Asia where water buffaloes are dominant over cattle, their value should be explored. Domestic poultry were unsatisfactory as sentinels in three successive years in which ephemeral fever occurred in northern Australia.[78]

VIII. INVESTIGATION OF EPIZOOTICS

There are several particular unresolved questions the answers to which would help in the understanding and prediction of epizootics. The place to catch the unknown vectors is at the time and the location of epizootics, in the areas where the presently suspected vectors do not occur. This would depend on a sensitive system for detecting virus in insects. The next problem needing solution is how the virus overwinters. In temperate Australia, when an epizootic is halted by winter partway through its north-south spread, it resumes its southern movement in the same approximate area the following summer,[16] but does not appear to overwinter in cattle. In tropical areas with a wet-dry seasonality, the overwintering mechanism through the dry winter season is also unknown.[78] Transovarial transmission in mosquitoes is possible. The mechanism of the movement of BEF virus across substantial water barriers from the Indonesian Islands to Australia prior to 1936, or more recently and perhaps periodically to Taiwan and Japan, remains a mystery. Transport in birds, bats, or a mass of infected insects are possibilities. More sensitive methods of detecting BEF virus would assist such an investigation.

IX. PREVENTION AND CONTROL

The economics of the disease in countries where it is periodically epizootic are heavily in favor of control. The disease is tremendously disruptive. The vectors are still substantially unknown. However, the diversity of species of mosquitoes and *Culicoides* from which BEF virus has been isolated suggests that vector control will be impractical as a means of preventing or controlling epizootics.

A. Control of Vertebrate Hosts

The only control which would seem to have any rationality are measures to prevent the movement by air of clinically ill (and therefore viremic) cattle or buffaloes to the Americas. Movement by sea would allow sufficient time for disease to display itself and recovery to occur. Recovered animals with antibodies do not seem to be a risk. Quarantine measures within a country where an epizootic is occurring have never been successful in preventing spread. There is no factor of the physical environment which can be used to affect the course of an epizootic.

B. Epizootiological Consideration of Use of Vaccines

Live or killed virus adjuvant vaccines have been developed in South Africa,[89] Japan,[90,91] and Australia.[92,93] The successful adjuvants are aluminum phosphate or hydroxide gel and Quil A. All these viruses protected vaccinated cattle under experimental conditions, but data on their value under epizootic conditions remain to be published. The loss of antigenicity as well as pathogenicity of BEF virus on serial passage in systems other than the natural host causes problems of production of an efficacious vaccine.[10,94] In order to obtain maximum value from vaccination, some prediction of epizootics is necessary. Before this can happen, the vectors in any region where epizootics occur will have to be identified and their biology understood. A persisting titer of neutralizing antibody does not develop until after a second dose of vaccine is administered 1 month after the first.[93] Annual single-dose revaccination may be necessary. The economics of vaccination balancing potential loss against cost of vaccine remain to be published.

X. FUTURE RESEARCH

The area of the world where research is particularly necessary is in the belt of tropical countries where the disease extends as epizootics to the subtropical and temperate areas. The world situation is yet to be understood and the economic effects assessed globally. More vectors remain to be identified, perhaps many species. This applies not only in countries like South Africa, Kenya, Japan, and Australia — where most of the research on ephemeral fever has been carried out — but also in other countries, particularly in Southeast Asia where little work has been done. The overwintering mechanism must be determined. Extensive studies are also necessary to determine whether species of bats or migratory birds are involved in the transfer of virus across large water barriers. The isolation and identification of antigenically related viruses need to be continued until the full spectrum of their effects on BEF virus serology of cattle and their disease potential are known. It is possible that transovarial transmission of BEF virus occurs in one or more insect species. Initial studies on this mechanism are best undertaken experimentally in the laboratory.

A. Experimental Studies

The pathogenesis of the disease still eludes complete definition, particularly the permanent paralysis which occurs occasionally. The results of treatment of the disease under field and experimental conditions are very promising. However, longer-acting and cheaper drugs are needed. A completely specific serological test capable of full automation is desirable, as is a sensitive specific test for viral antigen. An inexpensive vaccine, conveying lifelong immunity, together with a long shelf life and easy application, remains the most useful target of future research. To achieve this, both BEF virus and the disease it causes must be studied further.

REFERENCES

1. **Piot, J. B.,** Epizootic of dengue fever of cattle in Egypt. Prix Monbinne. Academic Nationale de Medicine, Paris, 1896 (CSIRO translation).
2. **Schweinfurth, G.,** *The Heart of Africa,* Frewer, E. E., Transl., Sampson Low, Marston, Low, and Searle, London, 1867, 280.
3. **Freer, G. W.,** Ephemeral fever or three-day sickness in cattle, *Vet. J.,* 66, 19, 1910.
4. **Bevan, L. E. W.,** Preliminary report on the so-called "stiff sickness" or "three-day-sickness" of cattle in Rhodesia, *J. Comp. Pathol. Ther.,* 20, 104, 1907.
5. **Bevan, L. E. W.,** Ephemeral fever, or three-day sickness of cattle, *Vet. J.,* 68, 458, 1912.
6. **Seddon, H. R.,** The spread of ephemeral fever (three-day sickness) in Australia in 1936—37, *Aust. Vet. J.,* 14, 90, 1938.
7. **Inaba, Y.,** Bovine epizootic fever, *Jpn. Agric. Res. Q.,* 3, 36, 1968.
8. **Ma Siqi,** personal communication, 1984.
9. **Mackerras, I. M., Mackerras, M. J., and Burnet, F. M.,** Experimental studies of ephemeral fever in Australian cattle, *Aust. Commonw. Sci. Ind. Res. Bull.,* 136, 1940.
10. **van der Westhuizen, B.,** Studies on bovine ephemeral fever. I. Isolation and preliminary characterization of a virus from naturally and experimentally produced cases of bovine ephemeral fever, *Onderstepoort J. Vet. Res.,* 34, 29, 1967.
11. **Odend'hal, S.,** *The Geographical Distribution of Animal Viral Diseases,* Academic Press, New York, 1983.
12. **Curasson, G.,** Three-day sickness of oxen, in *Traité de Pathologie Exotique Veterinaire et Comparee,* Vol. 1, Maladies a ultravirus, Vijot Freres, Paris, 1936, 579 (CSIRO translation).

13. **Henning, W. H.,** Ephemeral fever, three-day-sickness, drie-daesiekte, in *Animal Diseases in South Africa*, 3rd ed., Central News Agency, Johannesburg, South Africa, 1956, 1053.
14. **Rabagliati, D. S.,** Three days' fever or stiff sickness in cattle, *Vet. Rec.*, 4, 503, 1924.
15. **Murray, M. D.,** The spread of ephemeral fever of cattle during the 1967—68 epizootic in Australia, *Aust. Vet. J.*, 46, 77, 1970.
16. **St.George, T. D., Standfast, H. A., Christie, D. G., Knott, S. G., and Morgan, I. R.,** The epizootiology of bovine ephemeral fever in Australia and Papua-New Guinea, *Aust. Vet. J.*, 53, 17, 1977.
17. **Uren, M. F., St.George, T. D., and Stranger, R. S.,** Epidemiology of ephemeral fever of cattle in Australia 1975—1981, *Aust. J. Biol. Sci.*, 36, 91, 1983.
18. **Uren, M. F., St.George, T. D., Kirkland, P. D., Stranger, R. S., and Murray, M. D.,** Epidemiology of bovine ephemeral fever in Australia 1981—1985, *Aust. J. Biol. Sci.*, 39, 125, 1986.
19. **Davies, F. G.,** Ephemeral fever in Kenya, in press.
20. **Fagbami, A. H. and Ojeh, C.,** Arthropod-borne viral infections of livestock in Nigeria, *Trop. Vet.*, 1, 61, 1983.
21. **Otte, E.,** Viral diseases of cattle in Taiwan, *J. Taiwan Assoc. Anim. Husb. Vet. Med.*, 12, 1, 1968.
22. **Inaba, Y.,** Bovine ephemeral fever (three-day sickness). Stiff sickness, *Bull. Off. Int. Epiz.*, 79, 627, 1973.
23. **Chiu, S. Y. and Lu, Y. S.,** Bovine ephemeral fever in Taiwan, in Arbovirus research in Australia, St.George, T. D., Kay, B. H., and Blok, J., Eds., Commonwealth Scientific and Industrial Research Organisation and Queensland Institute of Medical Research, Brisbane, 1986.
24. **Inaba, Y.,** personal communication, 1987.
25. **Burgess, G. W. and Chenoweth, P. J.,** Mid-piece abnormalities in bovine semen following experimental and natural cases of bovine ephemeral fever, *Br. Vet. J.*, 131, 536, 1975.
26. **St.George, T. D.,** Aspects of bovine ephemeral fever influencing international trade, in *Proc. 1st Jt. Conf. Assoc. Vet. Surg. Malays. Aust. Assoc. Cattle Vet.*, Kuala Lumpur, Malaysia, 1977, 103.
27. **Anon.,** Report of a conference on three-day sickness, Circular Commun. No. 206, Department of Health, Canberra, 1937.
28. **Davis, S. S., Gibson, D. S., and Clark, R.,** The effect of bovine ephemeral fever on milk production, *Aust. Vet. J.*, 61, 128, 1984.
29. **Theodoridis, A., Giesecke, W. H., and Du Toit, I. J.,** Effects of ephemeral fever on milk production and reproduction of dairy cattle, *Onderstepoort J. Vet. Res.*, 40, 83, 1973b.
30. **Murphy, G. M.,** personal communication, 1985.
31. **Ronohardjo, P. and Rastiko, P.,** Some epidemiological aspects and economic loss of bovine ephemeral fever outbreak in Tuban and surrounding areas, East Java, Indonesia, *Penyakit Hewan*, 14(24), 25, 1982.
32. **Ito, Y., Tanaka, Y., Inaba, Y., and Omori, T.,** Electron microscopic observations of bovine epizootic fever virus, *Natl. Inst. Anim. Health Q.*, 9, 35, 1969.
33. **Holmes, I. H. and Doherty, R. L.,** Morphology and development of bovine ephemeral fever virus, *J. Virol.*, 5, 91, 1970.
34. **Murphy, F. A., Taylor, W. P., Mims, C. A., and Whitfield, S. G.,** Bovine ephemeral fever virus in cell culture and mice, *Arch. Ges. Virusforsch.*, 38, 234, 1972.
35. **Della-Porta, A. J. and Brown, F.,** The physico-chemical characterization of bovine ephemeral fever virus as a member of the family Rhabdoviridae, *J. Gen. Virol.*, 44, 99, 1979.
36. **Tanaka, Y., Inaba, Y., Ito, H., Sato, K., Omori, T., and Matumoto, M.,** Double strandedness of ribonucleic acid of bovine ephemeral fever virus, *Jpn. J. Microbiol.*, 16, 95, 1972.
37. **Tesh, R. B., Travassos da Rosa, A. P. A., and Travassos da Rosa, J. S.,** Antigenic relationships among rhabdoviruses infecting terrestrial vertebrates, *J. Gen. Virol.*, 64, 169, 1983.
38. **Cybinski, D. H. and Zakrzewski, H.,** The isolation and preliminary characterization of a rhabdovirus in Australia related to bovine ephemeral fever virus, *Vet. Microbiol.*, 8, 221, 1983.
39. **Gard, G. P., Cybinski, D. H., and St.George, T. D.,** The isolation in Australia of a new virus related to bovine ephemeral fever virus, *Aust. Vet. J.*, 60, 89, 1983.
40. **Gard, G. P., Cybinski, D. H., and Zakrzewski, H.,** The isolation of a fourth bovine ephemeral fever group virus, *Aust. Vet. J.*, 61, 332, 1984.
41. **Kaneko, N., Inaba, Y., Akashi, H., Miura, Y., Shorthose, J., and Kurashige, K.,** Isolation of a new bovine ephemeral fever virus, *Aust. Vet. J.*, 63, 29, 1986.
42. **Cybinski, D. H.,** Homologous and heterologous antibody reactions in sera from cattle naturally infected with bovine ephemeral fever group viruses, *Vet. Microbiol.*, 11, 1, 1986.
43. **St.George, T. D., Cybinski, D. H., Murphy, G. M., and Dimmock, C. K.,** Serological and biochemical factors in bovine ephemeral fever, *Aust. J. Biol. Sci.*, 37, 341, 1984.
44. **Zakrzewski, H. and Cybinski, D. H.,** personal communication, 1986.
45. **Davies, F. G., Shaw, T., and Ochieng, P.,** Observations on the epidemiology of ephemeral fever in Kenya, *J. Hyg.*, 75, 231, 1975.

46. **McKenzie, R. A., Mackenzie, A. R., Thornton, A. M., Chung, Y. S., Cybinski, D. H., and St.George, T. D.,** Serological survey of red deer, *Queensl. Dept. Prim. Ind. Proj. Rep.*, QO85006, 9, 1985.
47. **Akhtar, A. S., Ali, R., and Hussain, S.,** Clinical pathology and transmission of ephemeral fever, *Pak. Sci. Conf. Proc.*, 18/19(3), G-25, 1967.
48. **Dumag, P. U.,** Livestock diseases and parasites: prevention and its control, *Philipp. J. Anim. Ind.*, 32, 127, 1977.
49. **Young, P. L.,** Infection of water buffalo *(Bubalus bubalis)* with bovine ephemeral fever virus, *Aust. Vet. J.*, 55, 349, 1979.
50. **St.George, T. D. and Standfast, H. A.,** unpublished data, 1986.
51. **Hall, W. T., Daddow, K. N., Dimmock, C. K., St.George, T. D., and Standfast, H. A.,** The infection of merino sheep with bovine ephemeral fever virus, *Aust. Vet. J.*, 51, 344, 1975.
52. **Inaba, Y., Tanaka, Y., Sato, K., Ito, H., Omori, T., and Matumoto, M.,** Bovine epizootic fever. I. Propagation of the virus in suckling hamster, mouse and rat, and hamster kidney BHK21-W12 cell, *Jpn. J. Microbiol.*, 12, 457, 1968.
53. **Doherty, R. L., Standfast, H. A., and Clark, I. A.,** Adaptation to mice of the causative virus of ephemeral fever of cattle from an epizootic in Queensland, 1968, *Aust. J. Sci.*, 31, 365, 1968.
54. **Spradbrow, P. B. and Francis, J.,** Observations on bovine ephemeral fever and isolation of virus, *Aust. Vet. J.*, 45, 525, 1969.
55. **Elamin, M. A. C. and Spradbrow, P. B.,** Isolation and cultivation of bovine ephemeral fever virus in chickens and chicken embryos, *J. Hyg.*, 81, 1, 1978.
56. **Standfast, H. A.,** unpublished data, 1965.
57. **Snowdon, W. A.,** Bovine ephemeral fever: the reaction of cattle to different strains of ephemeral fever virus and the antigenic comparison of two strains of virus, *Aust. Vet. J.*, 46, 258, 1970.
58. **St.George, T. D. and Zakrzewski, H.,** unpublished data, 1986.
59. **Elamin, M. A. C. and Spradbrow, P. B.,** The growth of ephemeral fever virus in primary cell cultures, *Acta Virol.*, 22, 341, 1978.
60. **Hoffmann, D. and St.George, T. D.,** Growth of epizootic hemorrhagic disease, Akabane, and ephemeral fever viruses in *Aedes albopictus* cells maintained at various temperatures, *Aust. J. Biol. Sci.*, 38, 183, 1985.
61. **Davies, F. G. and Walker, A. R.,** The isolation of ephemeral fever virus from cattle and *Culicoides* midges in Kenya, *Vet. Rec.*, 95, 63, 1974.
62. **Standfast, H. A., St.George, T. D., and Dyce, A. L.,** The isolation of ephemeral fever virus from mosquitoes in Australia, *Aust. Vet. J.*, 52, 242, 1976.
63. **Standfast, H. A. and Muller, M. J.,** Vectors of bovine ephemeral fever, in *Veterinary Viral Diseases: Their Significance in South East Asia and the Western Pacific*, Della-Porta, A. J., Ed., Academic Press, North Ryde, New South Wales, 1985, 394.
64. **Thuraisingham, S.,** Reports of Veterinary Department, Malaysia, 1959/1960.
65. **Mohan, R. N.,** Diseases and parasites of buffaloes, *Vet. Bull.*, 38, 567, 1968.
66. **Malviya, H. K. and Prasad, J.,** Ephemeral fever — a clinical and epidemiological study in cross-bred cows and buffaloes, *Indian Vet. J.*, 54, 440, 1977.
67. **Uren, M. F., St.George, T. D., and Standfast, H. A.,** unpublished data, 1985.
68. **Basson, P. A., Pienaar, J. G., and van der Westhuizen, B.,** The pathology of ephemeral fever: a study of the experimental disease in cattle, *J. S. Afr. Vet. Med. Assoc.*, 40, 385, 1970.
69. **Young, P. L. and Spradbrow, P. B.,** Transmission of virus from serosal fluids and demonstration of antigen in neutrophils and mesothelial cells of cattle infected with bovine ephemeral fever virus, *Vet. Microbiol.*, 10, 199, 1985.
70. **Parsonson, I. M. and Snowdon, W. A.,** Ephemeral fever virus: excretion in the semen of infected bulls and attempts to infect female cattle by the intrauterine inoculation of virus, *Aust. Vet. J.*, 50, 329, 1974a.
71. **Parsonson, I. M. and Snowdon, W. A.,** Experimental infection of pregnant cattle with ephemeral fever virus, *Aust. Vet. J.*, 50, 335, 1974b.
72. **Tzipori, S. and Spradbrow, P. B.,** The effect of bovine ephemeral fever virus on the bovine foetus, *Aust. Vet. J.*, 51, 64, 1975.
73. **Schalm, O. W., Jain, N. C., and Carroll, E. J.,** *Veterinary Hematology*, 3rd ed., Lea & Febiger, Philadelphia, 1975.
74. **Muller, M. J.,** personal communication, 1986.
75. **Rosen, L.,** Carbon dioxide sensitivity in mosquitoes infected with sigma, vesicular stomatitis, and other rhabdoviruses, *Science*, 207, 989, 1980.
76. **Turell, M. J., Hardy, J. L., and Reeves, W. C.,** Sensitivity to carbon dioxide in mosquitoes infected with California serogroup arboviruses, *Am. J. Trop. Med. Hyg.*, 31, 389, 1982.
77. **Anon.,** The animal health position and methods of control in Malaysia, *Bull. Off. Int. Epiz.*, 93, 1231, 1981.

78. **Knott, S. G., Paull, N. I., St.George, T. D., Standfast, H. A., Cybinski, D. H., Doherty, R. L., Carley, J. G., and Filippich, C.,** The epidemiology of bovine ephemeral fever virus compared with other arboviruses in the Flinders River Basin of North Queensland, Australia, 1974 to 1977, *Queensl. Dept. Prim. Ind. Bull.*, QB83001, 1983.
79. **Morgan, I. and Murray, M. D.,** The occurrence of ephemeral fever of cattle in Victoria in 1968, *Aust. Vet. J.*, 45, 271, 1969.
80. **St.George, T. D.,** Studies on the pathogenesis of bovine ephemeral fever in sentinel cattle. I. Virology and serology, *Vet. Microbiol.*, 10, 493, 1985.
81. **Tzipori, S.,** The susceptibility of young and newbown calves to bovine ephemeral fever virus, *Aust. Vet. J.*, 51, 251, 1975.
82. **Snowdon, W. A.,** Some aspects of the epizootiology of bovine ephemeral fever in Australia, *Aust. Vet. J.*, 45, 312, 1971.
83. **St.George, T. D.,** A sentinel herd system for the study of arbovirus infections in Australia and Papua-New Guinea, *Vet. Sci. Commun.*, 4, 39, 1980.
84. **Theodoridis, A.,** Fluorescent antibody studies on ephemeral fever virus, *Onderstepoort J. Vet. Res.*, 36, 187, 1969.
85. **Standfast, H. A., Dyce, A. L., St.George, T. D., Muller, M. J., Doherty, R. L., Carley, J. G., and Filippich, C.,** Isolation of arboviruses from insects collected at Beatrice Hill, Northern Territory of Australia, 1974 and 1976, *Aust. J. Biol. Sci.*, 37, 351, 1984.
86. **Cybinski, D. H.,** personal communication, 1986.
87. **Kay, B. H., Carley, J. G., and Filippich, C.,** The multiplication of Queensland and New Guinean arboviruses in *Culex annulirostris* (Skuse) and *Aedes vigilax* (Skuse) (Diptera:Culicidae), *J. Med. Entomol.*, 12, 279, 1975.
88. **Coverdale, O. R.,** personal communication, 1985.
89. **Theodoridis, A., Boshoff, S. E. T., and Botha, M. J.,** Studies on the development of a vaccine against bovine ephemeral fever, *Onderstepoort J. Vet. Res.*, 40, 77, 1973a.
90. **Inaba, Y., Kurogi, H., Sato, K., Goto, Y., Omori, T., and Matumoto, M.,** Formalin-inactivated, aluminum phosphate gel-adsorbed vaccine of bovine ephemeral fever virus, *Arch. Ges. Virusforsch.*, 42, 42, 1973.
91. **Inaba, Y., Kurogi, H., Takahashi, A., Sato, K., Omori, T., Goto, Y., Hanaki, T., Yamamoto, M., Kishi, S., Kodama, K., Harada, K., and Matumoto, M.,** Vaccination of cattle against bovine ephemeral fever with live attenuated virus followed by killed virus, *Arch. Ges. Virusforsch.*, 44, 121, 1974.
92. **Tzipori, S. and Spradbrow, P. B.,** Studies on vaccines against bovine ephemeral fever, *Aust. Vet. J.*, 49, 183, 1973.
93. **Vanselow, B. A., Abetz, I., and Trenfield, K.,** A bovine ephemeral fever vaccine incorporating adjuvant Quil A: a comparative study using adjuvants Quil A, aluminium hydroxide gel and dextran sulphate, *Vet. Rec.*, 117, 37, 1985.
94. **Inaba, Y., Tanaka, Y., Sato, K., Ito, H., Omori, T., and Matumoto, M.,** Bovine epizootic fever. III. Loss of virus pathogenicity and immunogenicity for the calf during serial passage in various host systems, *Jpn. J. Microbiol.*, 13, 181, 1969c.
95. **Kaneko, N., Inaba, Y., Akashi, H., and Kurashige, K.,** Further studies on FUK11 virus—withdrawal of a member of bovine ephemeral fever group viruses, *Aust. Vet. J.*, 64, 289, 1987.

Chapter 18

BUNYAVIRAL FEVERS: BUNYAMWERA, ILESHA, GERMISTON, BWAMBA, AND TATAGUINE

J. P. Gonzalez and A.-J. Georges

TABLE OF CONTENTS

I. INTRODUCTION

Bwamba, Bunyamwera,Germiston, Ilesha, and Tataguine viruses, all members of the family Bunyaviridae,[1-4] are endemic in Africa and have never been recognized elsewhere. All these viruses can cause mild febrile illness, with or without rash; none is responsible for important epidemic disease or has a great social or economic impact. Interest in the study of these viruses centers on both their epidemiology and their structure. Among the five viruses, Bunyamwera, Bwamba (including Bwamba-related viruses such as Pongola), and Tataguine have been isolated most frequently.

II. HISTORICAL BACKGROUND

A. Discovery of Agents

In 1941, Smithburn et al.[5] first described the isolation of Bwamba virus. In 1946, they[6] reported the first isolation of Bunyamwera virus, which is considered the prototype of the Bunyamwera group; it was recovered in Uganda from a pool of *Aedes* mosquitoes.

In 1957, Ilesha virus was isolated in a Nigerian village of the same name from blood of a 9-year-old girl.[7] Germiston virus was recognized by Kokernot et al.[8] in 1960 in South Africa. The first isolation of Tataguine was made by Brès et al.[9] from a mixed pool of *Culex* and *Anopheles* mosquitoes collected in Senegal in 1961.

B. History of Human Cases and Geographical Spread

Bwamba virus was first recognized in the setting of a small outbreak in western Uganda; nine cases were confirmed by virus isolation.[5] Bwamba infection has been recognized by virus isolation from humans in Uganda,[5] Nigeria,[10] Cameroon,[11] Central African Republic,[11,12] Kenya, Tanzania,[5] and South Africa.[13] Identification of human infections by Bwamba group viruses was reported in Ethiopia by Ota et al. (unpublished) and, more recently (1978), in Kenya.[14]

In Bwamba County (Uganda), where Bunyamwera virus was first isolated, 2.9 to 30% of the human population was found with neutralizing antibodies.[5]

Tataguine virus is present in Senegal, where 57% of the inhabitants were found to have antibodies, and eight strains were isolated from inhabitants of Dakar suburbs.[9] It has also been found in Cameroon,[15] Nigeria,[16] the Central African Republic,[12,17] and Ethiopia.[18]

After its first isolation in Nigeria,[7] Ilesha virus was found in Cameroon, Senegal, and the Central African Republic.[11,12] Ilesha virus, but not Bunyamwera,[19] is considered to be endemic in southern Ethiopia.

None of the five viruses seems to be responsible for epidemics. They are, nevertheless, distributed widely in all of tropical Africa.

C. Social and Economic Impact

Generally responsible for only mild diseases, these viruses have a very limited recognized social and economic impact.

III. VIRUS CHARACTERISTICS

Bunyaviruses from the Western Hemisphere show serological cross reactions with those from Africa.[20] Nevertheless, it is quite easy to identify each African virus using either complement fixation (CF) or mouse protection neutralization (MPN).[21]

The genomes of Bunyaviridae consist of a single strand of RNA comprising three segments. Experimental recombination shows that genetic material can be exchanged between members of Bunyamwera group and other Bunyaviruses. This could explain the occurrence of "mos-

Table 1
ORIGIN OF VIRUS ISOLATIONS (ISOL) AND ANTIBODIES (AB) AGAINST FIVE ARBOVIRUSES (BUNYAMWERA, BWAMBA, GERMISTON, ILESHA, AND TATAGUINE)

	BUN		BWA		GER		ILE		TAT	
Origin	Isol	AB	Isol	AB	Isol	AB	Isol	AB	Isol	AB
Man	+[a]	+	+	+	+	+	+	+	+	+
Chimpanzee	ND[b]	+	ND	ND	ND	ND	ND	ND	ND	ND
Monkeys	ND	+	ND	ND	ND	ND	ND	ND	ND	ND
Domestic animals	ND	+	ND	ND	ND	ND	ND	ND	ND	ND
Rodents	ND	+	ND	ND	+	+	ND	ND	ND	ND
Birds	ND	+	ND	+	ND	ND	ND	ND	ND	ND
Donkeys	ND	ND	ND	+	ND	ND	ND	ND	ND	ND
Cattle	ND	ND	ND	ND	ND	+	ND	ND	ND	ND
Hamster (sentinel)	ND	ND	ND	ND	+	ND	ND	ND	ND	ND
Arthropods	+		+		+		+		+	

[a] Positive isolation or presence of antibodies.
[b] ND, no data.

aic'' strains of different Bunyaviruses with similar geographic and ecologic distribution, as mentioned by Iroegbu.[22] The reassortment of RNA segments can explain the diversity of members of the Bunyaviridae family, as well as the cross reactions observed using serological tests.

Germiston viral polypeptides have been studied by Ozden and Hannoun[23] and consist of three major structural polypeptides (mol wt 125 × 10^6, 27 × 10^6, 18 × 10^6) and one minor larger protein (mol wt 185 × 10^3).

A. Antigenic Relationships

Bunyamwera virus is closely related to other members of Bunyamwera group, but a prior infection by any member of this serogroup does not prevent human infection with another virus of the same group. Ukauwa virus is now considered to be a strain of Bunyamwera virus.

Bwamba virus cross reacts by neutralization test (N) with Pongola virus;[24] nevertheless, it clearly differs in a quantitative reciprocal manner. Johnson et al.[14] demonstrated that a Bwamba virus variant isolated in Kenya could present antigenic characteristics of Pongola virus.

Originally, Germiston virus was described as having significant reciprocal neutralization (adult mouse N test) with Bunyamwera; nevertheless, these two viruses are readily distinguished by hemagglutination inhibition (HI) and CF, as mentioned by Kokernot et al.[8]

Ilesha virus cross reacts to some degree with Bunyamwera by the N test;[25] by HI, the closest relationship is with Cache Valley.[26] By CF test, Ilesha virus is related more closely to Bunyamwera and Cache Valley than to other members of the Bunyamwera serogroup.[27]

B. Host Range

Table 1 summarizes the available data concerning natural host range, as evidenced by virus isolation and antibody tests. All five viruses have been isolated from humans, whereas data on other vertebrate species are scanty. Susceptibility of laboratory hosts is variable. All viruses are pathogenic for newborn mice by the intracerebral (i.c.) route; Bunyamwera and Germiston viruses are the most pathogenic, causing death in weaned mice by the peripheral route. Tataguine virus is the least pathogenic and does not produce illness in

Table 2
SUSCEPTIBILITY OF CELL CULTURE SYSTEMS FOR FIVE BUNYAVIRUSES

	BUN		BWA		GER		ILE		TAT	
Cell system	Eff[a]	Day[b]	Eff	Day	Eff	Day	Eff	Day	Eff	Day
Chick embryo	CPE	2—5	CPE	2	PLQ	3	PLQ	4—5	ND[c]	ND
VERO	PLQ	5	PLQ	5	PLQ	3	PLQ	6	PLQ	6
BHK-21	CPE	4	CPE	2	PLQ	2	ND	ND	ND	ND
LLC-MK_2	PLQ	4	PLQ	3	PLQ	4	PLQ	4	NP[d]	NP

Note: Results given can vary with the virus passage history.

[a] Eff, effect in cell culture (CPE, cytopathic effects; PLQ, plaques).
[b] Day, mean of days after isolation.
[c] ND, no data.
[d] NP, no plaque.

suckling mice inoculated i.c. or weaned mice by any route. Susceptibility of hamsters, rabbits, and other experimental hosts is described in the *International Catalogue of Arthropod-Borne Viruses*.[27]

C. Methods for Assay

All of the viruses may be assayed by i.c. inoculation of newborn mice. Cell culture systems can be used, but the cytopathic effect is not always easy to see and is sometimes absent. Susceptibilities of some cell culture systems are given in Table 2.

Techniques for virus isolation and assay using mosquito cell cultures or mosquitoes inoculated by the intrathoracic route are described below. The sensitivity and specificity of various infectivity assays are different for each virus. The N test performed with Tataguine virus in VERO cells has a high specificity and is more sensitive than the MPN test.[28]

IV. DISEASE ASSOCIATIONS

A. Humans

The most frequent symptoms are fever, headache, arthralgia. The hallmark of nearly all cases is their brief duration (4 or 5 days) and benign nature; no fatalities are recorded.

Physical examination is generally normal except, at times, for the presence of conjunctivitis or stiffness of the neck.[11,12] Convalescence is characterized by marked asthenia lasting 8 to 10 days.

Bwamba virus is responsible for a relatively severe form of generalized infection. Exanthem is nearly always present, and it is frequently associated with meningeal involvement.[74] A case of myocarditis has been reported.[11] Intestinal tract involvement, especially diarrhea, is also seen.[74]

Bunyamwera virus is generally responsible for pediatric infection. Children present with fever, headache, joint pains, and rash. Recovery occurs in less than 7 days. In some cases, visual disturbances and vertigo have been observed.[14] When infection occurs accidentally or in immunologically compromised patients, severe encephalitis can be observed.[29]

Ilesha virus infections are characteristically mild, with the primary symptom being a feeling of malaise. Fever, when present, is generally less than 39°C. A transient discrete exanthem is seen in about half of the cases.[30] Recovery occurs without sequelae, but asthenia persists for 8 to 10 days.

Tataguine virus is associated with mild disease in children and more severe symptoms in adults. Rash is present; however, fever is usually less than 39°C. Some patients complain of marked headache, gastrointestinal symptoms, and a florid, nonpruritic rash.[10]

Only two laboratory-acquired Germiston virus infections have been reported from South Africa; both were characterized by mild disease without specific symptoms and with recovery occurring in 37 hr to 3 days.[8]

B. Wild and Domestic Animals

No natural disease has been reported.

C. Diagnostic Procedures

1. Virus Isolation

Viremia is usually of short duration (24 to 48 hr); Ilesha virus, however, has been isolated 4 days after onset of disease.[12] The greatest number of successful virus isolations has been made using both i.c. and subcutaneous (s.c.) or intraperitoneal (i.p.) inoculation of suckling mice (usually 1 to 2 days of age). Material for inoculation consists of blood, serum, or plasma from humans and mammals or of arthropod pools diluted in Hanks' balanced salt solution or similar solution containing a source of protein (bovine albumin or serum). The virus can be identified at the time of harvest of a mouse brain tissue using the Chrom Elisa Technique of Lhuillier and Sarthou,[31] or it may be established by passage and identified by use of an appropriate serological test (CF or N test).

Less experience has been accumulated with the use of cell culture systems for the isolation of these viruses. Continuous cell lines, such as VERO or BHK-21 or *Aedes albopictus* (C6/36), may be useful. Inoculated C6/36 cells can be tested between days 2 and 10 for virus by indirect immunofluorescence (IFA) with specific ascitic fluids or held for as long as 2 weeks with infectious virus detectable by an appropriate technique such as mouse inoculation.

The viruses under consideration can also be isolated by intrathoracic inoculation of *Toxorhynchites* mosquitoes.[32] Antigen is demonstrated by testing mosquito head squashes by IFA or pooled mosquitoes by CF.

Detection of virus in serum by enzyme immunoassay, as described for yellow fever,[33] could represent a new approach to diagnosis, but is limited by the probable short duration and low titer of viremia in cases of most bunyaviral infections.

2. Serological Diagnosis

One must consider the serological cross reactivity within the Bunyaviridae family.[27] The "original antigenic sin" phenomenon applies to humans infected with multiple Bunyamwera group viruses. Usually, a battery of viruses and tests is required to clarify the diagnosis.[34]

Serological response to infection can be low and/or without seroconversion in the case of Tataguine virus;[12] infection can be serologically diagnosed by the N test but not the CF test.[11] Methods such as IgM antibody detection, described for other arboviruses, could represent a new approach to serodiagnosis.

V. EPIDEMIOLOGY

A. Geographic Distribution and Seroepidemiology

The distribution of these viruses determined by virus isolation has been described (in Section II.B) and is shown in Table 3. The table also shows the results of seroprevalence surveys in various countries of Africa.

No serological or virological evidence for activity of any of these agents has been found in North Africa, with the exception of Egypt, where a low prevalence (0.9%) of N test antibodies to Bunyamwera but no antibodies to Bwamba virus was reported.[35]

Table 3
DETECTION OF IMMUNITY TO BWAMBA (BWA), BUNYAMWERA (BUN), ILESHA (ILE), GERMISTON (GER), AND TATAGUINE (TAT) VIRUSES IN HUMANS: ANTIBODY PREVALENCE AND VIRUS ISOLATIONS

	BWA	BUN	ILE	GER	TAT
North Africa					
Algeria	ND[a]	0.0	0.0	ND	ND
Egypt	0.0[b]	0.9[c]	ND	ND	ND
Libya	ND	0.0	ND	ND	ND
Morroco	ND	0.0	0.0	ND	ND
Tunisia	ND	0.0	0.0	ND	ND
West Tropical Africa					
Benin	ND	3.0	ND	ND	ND
Burkina Faso	ND	17.0	ND	ND	ND
Bameroon	*[d]	8.0*	*	ND	*
CAR	*	24.0*	*	ND	*
Chad	ND	0.0	ND	ND	ND
Gambia	+[e]	+	ND	ND	ND
Ghana	ND	+	6.5*	ND	ND
Guinea	ND	*	ND	ND	ND
Guinea-Bissau	43.0	11.0	3.0	ND	ND
Ivory Coast	ND	14.0	ND	ND	ND
Liberia	ND	19.0	ND	ND	ND
Mali	ND	2.0	ND	ND	ND
Niger	ND	3.0	ND	ND	ND
Nigeria	33.0—40.0*	0.0—23.0*	27.0—45.0*	10.0	26.0—61.0*
Senegal	ND	19.0*	*	ND	*
Sierra Leone	ND	*	ND	ND	ND
Togo	ND	0.5	ND	ND	ND
East Tropical Africa					
Ethiopia	+	1.8—19.2*	*	+	ND
Somalia	ND	*	ND	ND	ND
West Equatorial Africa					
Congo	ND	7.0—25.0	ND	ND	ND
Equatorial Guinea	+	+	ND	ND	ND
Gabon	ND	8	ND	ND	ND
Rwanda	ND	0.0—10.0	ND	ND	ND
East Equatorial Africa					
Kenya	*	+	*	*	ND
Tanzania	75.0*	11.1	ND	ND	ND
Uganda	44.0* 37.0*	*	*	ND	
Southern Tropical Africa					
Angola	+	52.0	ND	28.0	ND
Madagascar	ND	0.0—6.5	ND	ND	ND
Mozambique	24.7*	24.1*	ND	ND	ND
Zimbabwe	ND	15.9	ND	2.3	ND
Austral Africa					
Botswana	3.3	53.3	ND	90.1	ND
Namibia	ND	42.1	ND	56.0	ND
SAR	0.0—80.0	0.0—45.7*	ND	1.4*	ND

[a] ND, no data.
[b] Sera tested, but no evidence for antibody.
[c] Prevalence (%) of viral antibodies.
[d] Virus isolation.
[e] Presence of viral antibodies; prevalence not defined.

In tropical parts of West Africa, the prevalence of bunyamwera antibodies has been high in many countries.[36] In Equatorial Guinea, the prevalence of antibodies to Bunyamwera and Bwamba viruses appeared to be higher than to most other arboviruses.[37] In contrast, a low prevalence of Bunyamwera antibodies was reported in Togo, Benin, Mali, Niger, and Chad.[38,39]

In Central Africa, 100% of the populations inhabiting the rainforest area of the Congo had antibodies after 10 years of age, whereas only 8.0% of persons in the same age group in Gabon were immune.[40] Considering Bunyamwera antibody prevalence in Africa as a whole, it is apparent that it is highest in the tropics, low in North Africa, and shows a sharp decline in the Republic of South Africa, being 0% in Cape Province and 45% in North Natal.[41]

In East Africa, Bunyamwera and Bwamba viruses appear to be endemic in Uganda, Tanzania, and Mozambique.[41,42] A low prevalence of bunyamwera antibodies has been found in Madagascar, with the highest percentage positive on the north coast.[43]

In South Africa, Bunyamwera and Bwamba antibody prevalences are significantly higher in the Simbu Pan area of North Natal than in other regions. Natal seems to be the southern limit for arboviruses, as they require a tropical ecology. Germiston virus seems to be active only in Austral Africa, with a specific importance in Angola and Botswana. In South Africa, Germiston antibodies are found in livestock.[36,44]

Tataguine virus has a wide geographic distribution encompassing West and Central Africa. In Nigeria, the highest prevalence of antibodies was found in the derived savannah zone (61%) followed by lowland rain forest (42%).[45]

B. Incidence

Evidence for human infections has been accumulated by many studies and is summarized in Table 3.[13,15,36,40,43,44,46,56] After chikungunya, Semliki Forest, Sindbis, yellow fever, Uganda S, West Nile, Wesselsbron, and Zika, Bunyamwera and Bwamba appear as the ninth and tenth most frequent arboviruses infecting humans in the African continent.[36] In the Central African Republic, Bwamba was the virus most often isolated from human cases of arboviral infection.[12] Tataguine virus also appears to be a very common human infection in some areas, e.g., Nigeria,[45] where multiple virus isolations have been made from febrile patients.[46]

C. Seasonal Distribution

Bunyamwera virus has been isolated in either the middle or, more frequently, at the end of the rainy season. In the Central African Republic, Ilesha and Bwamba viruses were most often isolated during the dry season. All Tataguine virus isolations were made during the dry season. Germiston strains from Kenya were isolated at the beginning of the rainy season.[47]

D. Risk Factors

Bunyamwera antibodies seem to appear earlier in age in females than in males, and the seroprevalence in females remains higher lifelong.[49,50]

VI. TRANSMISSION CYCLES

A. Evidence from Field Studies

1. Vectors

Bunyamwera virus has been isolated from mosquitoes belonging to three genera (*Aedes*, *Mansonia*, and *Culex*), but *Aedes* spp. appear to play the predominant role in transmission. Multiple strains of the virus were recovered from *Ae. circumluteolus* collected at the same time and place as a naturally infected human in South Africa.[55] The wide array of mosquito species which have yielded virus suggests that high viremia levels may occur in a variety of different vertebrate hosts.

Table 4
EVIDENCE OF VIRUS CIRCULATION IN WILD AND DOMESTIC VERTEBRATE HOSTS

		Serological evidence	
	Virus isolation	Wild animals	Domestic animals
BWA	No	*Arvicanthys niloticus* *Boedon fuliginosus* *Varanus niloticus* *Turtur ater* Monkeys	Goats, sheep, cattle
BUN	No	Chimpanzee, cattle, sheep, rodents, nonhuman primates, goats	Goats, sheep
ILE	No	No	No
GER	*Herpestes ichneumon* *Dasymys incomtus* *Rattus rattus* *Lophuromys* spp. *Arvicanthus niloticus*	*Arvicanthus niloticus* Horses	Goats, sheep, cattle
TAT	No	No	No

Little field evidence has been accumulated for Ilesha virus; the virus has been isolated from *Anopheles gambiae* in the Central African Republic and Senegal and from *Cx. thalassius* in Senegal.[57] Germiston virus has been recovered repeatedly from *Cx. rubinotus* in South Africa,[8,58] Zimbabwe,[58] Mozambique,[58] Kenya,[57] and Uganda.[59] High minimum infection rates in this species suggest that transovarial transmission of the virus may occur.

Bwamba virus has been isolated from *An. funestus* in Uganda,[27] Senegal, Nigeria,[27] Central African Republic,[57] and the Ivory Coast[57] and from *An. gambiae* and *Ae. furcifer* in Senegal.[57] Twelve strains of Bwamba virus were isolated during field studies in riverine forest in Nigeria in 1971,[60] the majority from *Ae. (Neomelanoconion)* spp. and *Ae. (N.) circumluteolus*, but also from *An. coustani* and *Ma. uniformis*. The *Aedes* were captured on human bait, suggesting that these mosquitoes, as well as anophelines, may be responsible for human infections.

Tataguine virus has been isolated principally from anopheline mosquitoes. Multiple strains have been recovered from *An. gambiae* in Cameroon,[15] Central African Republic, Senegal, and Ethiopia; from *An. funestus* in Nigeria, Central African Republic,[27,57] and Ethiopia; and from *An. nili* in Senegal.[27,57] An isolate has also been made from *Coquillettidia aurites* in Cameroon.[27,57] The association between virus isolations from humans and anopheline mosquitoes suggests that transmission occurs in the domestic habitat with humans serving as a viremic host.[15]

2. *Vertebrate Hosts*

No isolations of Bunyamwera, Ilesha, Bwamba, or Tataguine viruses have been made from naturally infected vertebrates. Germiston virus has been isolated on multiple occasions from rodents (Table 4): from *Dasymys incomtus* in Kenya[48] and from *Rattus rattus, Arvicanthus niloticus, Lophuromys sikapusi,* and *L. flavopunctatus* in Uganda.[59,61] An isolate was made from a mongoose (*Herpestes ichneumon*) in Kenya.[48] Multiple isolations of Germiston virus were also made from sentinel hamsters in South Africa and Mozambique.[62]

The interpretation of serological evidence for involvement of vertebrate hosts with Bunyamwera group viruses is limited somewhat by the problem of antigenic cross reactivity among members of the serogroup. Neutralizing antibodies to Bunyamwera, Bwamba, and

Germiston viruses have been found in domestic livestock (Table 4). A high seroprevalence to Germiston virus has also been found in rodents, in accord with virus isolation data implicating them as hosts. A high prevalence of antibodies to Tataguine virus in humans and absence of antibodies in wild and domestic animals[45] further support the conclusion that humans serve as hosts in the transmission cycle. Antibodies to Bwamba virus have been found in birds in South and East Africa.[27,48] A high prevalence of antibodies in monkeys was reported in Uganda.[63] In South Africa, the seroprevalence in humans to Bwamba and Bunyamwera viruses was significantly higher than in monkeys, birds, or domestic livestock.[41]

B. Evidence from Experimental Studies

1. Vectors

After intrathoracic inoculation, Bunyamwera virus has been shown to replicate in *Ae. vexans*,[64] *Ae. canadensis*,[64,65] *Ae. aegypti*,[64-66] *Ae. triseriatus*,[65,66] *Psorophora ferox*,[66] *Cx. pipiens*,[27] and *An. quadrimaculatus*.[27] Oral infection of and transmission by *Ae. aegypti* has been demonstrated for both Bunyamwera[64,65,67] and Ilesha viruses.[67]

Cx. rubinotus has been shown to become infected after feeding on Germiston virus and to transmit virus to hamsters.[27] Bwamba virus has been shown to infect *Ae. aegypti*, *An. quadrimaculatus*, and *Cx. pipiens* after intrathoracic inoculation,[27] but susceptibility to oral feeding has not been investigated. *Cx. pipiens* fed on virus-soaked pledgets become infected with but are incapable of transmitting Tataguine virus.[68]

2. Vertebrate Hosts

Wild African rodents (*Arvicanthus abyssinicus* and *Cricetomys gambianus*) experimentally infected with Bunyamwera virus develop viremias sufficient to infect mosquitoes, suggesting that rodents could play a role in natural transmission cycles.[69] Similar results have been reported for *Tadarida* bats.[70] Monkeys develop viremia, fever, and inapparent or mild illness.[71] Germiston virus behaves similarly in experimental animals, including rodents (*Arvicanthus niloticus*[59] and *Tatera brantsi*[72]) and monkeys.[71] Little or no useful information is available regarding experimental infections with Bwamba or Ilesha viruses.

C. Summary

Except for Germiston virus, for which convincing evidence of a rodent-*Cx. rubinotus* cycle is available, the natural history of the viruses under consideration remains obscure. All viruses are mosquito-borne; for Ilesha, Bwamba, and Tataguine viruses, anophelines appear to be the principal vectors, especially endo- and anthropophilic species (*An. gambiae* and *funestus*). A human-*Anopheles* cycle is suggested for Tataguine and possibly Bwamba viruses, but a role for wild vertebrate hosts cannot be excluded.

VII. ECOLOGICAL DYNAMICS

A. Macro- and Microenvironment

Bunyamwera virus serological studies show virus activity in association with rivers and riverine forests.[73] HI antibody prevalence in Nigeria was higher in forest and savannah zones than in swampy areas.[36] Virus activity in the Central African Republic is associated with gallery forests in the moist savannah zone.[49,50]

Ilesha antibody prevalence in the endemic area of Nigeria appears to be higher in the savannah than in the plateau and rain forest, and the virus seems to be more active in rural than in urban communities.[73] Germiston virus isolations have been associated with irrigated areas, where cattle have a high antibody prevalence. Tataguine antibodies have been found in highest prevalence in the derived savannah vegetational zone.[45]

VIII. PREVENTION AND CONTROL

Vaccines are not available. Because of their relatively low pathogenicity and lack of economic impact, specific efforts have not been developed to reduce the incidence of these diseases.

REFERENCES

1. **Murphy, F. A., Whitfield, S. D., Coleman, P. H., Calisher, C. H., Rabin, E. R., Melnick, J. L., Edwards, M. R., and Whitney, E.,** California group arboviruses: electron microscopic studies, *Exp. Mol. Pathol.*, 9, 44, 1968.
2. **Murphy, F. A., Harrison, A. K., and Tzianabos, T.,** Electron microscopic observations of mouse brain infected with Bunyamwera serologic group arboviruses, *J. Virol.*, 2, 1315, 1968.
3. **Murphy, F. A., Harrison, A. K., and Whitfield, S. G.,** Bunyaviridae: morphologic and morphogenetic similarities of Bunyamwera serologic supergroup viruses and several other arthropod-borne viruses, *Intervirology*, 1, 297, 1973.
4. **Porterfield, J. S., Casals, J., Chumakov, M. P., Gaidamovich, S., Hannoun, C. L., Holmes, I. H., Horzinek, M. C., Mussgay, M., Oker-Blom, N., and Russell, P. K.,** Bunyaviruses and Bunyaviridae, *Intervirology*, 6, 13, 1976.
5. **Smithburn, K. C., Mahaffy, A. F., and Paul, J. H.,** Bwamba fever and its causative virus, *Am. J. Trop. Med. Hyg.*, 21, 75, 1941.
6. **Smithburn, K. C., Haddow, A. J., and Mahaffy, A. F.,** A neurotropic virus isolated from *Aedes* mosquitoes caught in the Semliki forest, *Am. J. Trop. Med. Hyg.*, 26, 189, 1946.
7. West African Council for Medical Research, Yaba Annual Report, Yaba, Lagos, Nigeria, 1957, 123.
8. **Kokernot, R. H., Smithburn, K. C., Paterson, H. E., et al.,** Isolation of Germiston virus, a hitherto unknown agent, from Culicine mosquitoes, and a report of infection in two laboratory workers, *Am. J. Trop. Med. Hyg.*, 9, 62, 1960.
9. **Brès, P., Williams, M. C., and Chambon, L.,** Isolement au Senegal d'un nouveau prototype d'arbovirus, la souche "Tataguine" (IPD/A 252), *Ann. Inst. Pasteur*, 111, 585, 1966.
10. **Moore, D. L., Causey, O. R., Carey, D. E., et al.,** Arthropod-borne viral infections of man in Nigeria, 1964—1970, *Ann. Trop. Med. Parasitol.*, 69, 49, 1975.
11. **Digoutte, J. P., Salaun, J. J., Robin, Y., Brès, P., and Cagnard, V. J. M.,** Les arboviroses mineures en Afrique Central et Occidentale, *Med. Trop. (Marseilles)*, 40, 523, 1980.
12. **Georges, A. J., Saluzzo, J. F., Gonzalez, J. P., and Dussarat, G. V.,** Arboviroses en Centrafrique: incidence et aspects diagnostiques chez l'homme, *Med. Trop. (Marseilles)*, 40, 561, 1980.
13. **Woodall, J. P.,** Human infections with arboviruses of the Bunyamwera group, in *Arboviruses of the California Complex and the Bunyamwera Group*, Bardos, V., et al., Eds., Publishing House of the Slovak Academy of Sciences, Bratislava, 1969, 317.
14. **Johnson, B. K., Chanas, A. C., Squires, E. J., Shockley, P., Simpson, D. I., and Smith, D. H.,** The isolation of a Bwamba virus variant from man in western Kenya, *J. Med. Virol.*, 2, 15, 1978.
15. **Salaun, J. J., Rickenbach, A., Brès, P., Germain, M., Eouzan, J. P., and Ferrara, L.,** Isolement au Cameroun de trois souches de virus Tataguine, *Bull. Soc. Pathol. Exot.*, 61, 557, 1968.
16. **MacNamara, F. N., Horn, D. W., and Porterfield, J. S.,** Yellow fever and other arthropod-borne viruses, *Trans. R. Soc. Trop. Med. Hyg.*, 53, 202, 1959.
17. **Digoutte, J. P., Brès, P., Nguyen-Trung-Wong, and Durand, B.,** Isolement du virus Tataguine a partir de deux cas de fievre exanthematique, *Bull. Soc. Pathol. Exot.*, 62, 72, 1969.
18. **Ota, W. K., Watkins, H. M. S., Neri, P., Schmidt, M. L., and Schmidt, J. R.,** Arbovirus recoveries from mosquitoes collected in Gambela, Illuhabor Province, Ethiopia, 1970, *J. Med. Entomol.*, 13, 173, 1976.
19. **Rhodain, F., Hannoun, C., and Metselaar, D.,** Enquete epidemiologique et serologique sur les arboviroses dans la basse vallee de l'Omo (Ethiopie meridionale), *Bull. WHO*, 47, 295, 1972.
20. **Whitman, L. and Shope, R. E.,** The California complex of arthropod-borne viruses and its relationship to the Bunyamwera group through Guaroa virus, *Am. J. Trop. Med. Hyg.*, 11, 691, 1962.
21. **Saluzzo, J. F., Germain, H., Huard, M., Robin, Y., Gonzalez, J. P., Herve, J. P., Georges, A. J., Hjeme, E., and Digoutte, J. P.,** Le virus Bozo (Arb7343): un nouvel arbovirus du groupe Bunyamwera isole en Republique Centrafricaine. Sa transmission experimentale par *Aedes aegypti*, *Ann. Virol. (Inst. Pasteur)*, 134E, 221, 1983.

22. **Iroegbu, C. U. and Pringle, C. R.,** Genetic interactions among viruses of the Bunyamwera complex, *J. Virol.*, 37, 383, 1981.
23. **Ozden, S. and Hannoun, C. L.,** Biochemical and genetic characteristics of Germiston virus, *Virology*, 103, 232, 1980.
24. **Tomori, O. and Fabiyi, A.,** Differentiation of Bwamba and Pongola viruses by agar-gel diffusion and immunoelectrophoretic techniques, *Am. J. Trop. Med. Hyg.*, 25, 289, 1976.
25. **Hunt, A. R. and Calisher, C. H.,** Relationships of Bunyamwera group viruses by neutralization, *Am. J. Trop. Med. Hyg.*, 28, 740, 1979.
26. **Okuno, T.,** Immunological studies relating two recently isolated viruses, Germiston virus from South Africa and Ilesha virus from West Africa, to the Bunyamwera group, *Am. J. Trop. Med. Hyg.*, 10, 223, 1961.
27. **Karabatsos, N., Ed.,** *International Catalogue of Arthropod-Borne Viruses and Certain Other Viruses of Vertebrates*, 3rd ed., American Society of Tropical Medicine and Hygiene, Washington, D.C., 1985.
28. **Fagbami, A. H.,** Growth, plaque assay and immunofluorescent studies on Tataguine virus in cell culture, *Cytobios*, 26, 37, 1979.
29. **Southam, C. M. and Moore, A. E.,** West Nile, Ilheus, and Bunyamwera virus infections in man, *Am. J. Trop. Med. Hyg.*, 31, 724, 1951.
30. **Brottes, H. and Salaun, J. J.,** Isolement au Cameroun d'une souche d'arbovirus a partir d'une fievre exanthematique, *Arch. Inst. Pasteur Tunis*, 1, 77, 1965.
31. **Lhuillier, M. and Sarthou, T. L.,** Chrom Elisa: a new technique for rapid identification of arboviruses, *Ann. Virol. (Inst. Pasteur)*, 134E, 349, 1983.
32. **Gonzalez, J. P., Saluzzo, J. F., and Herve, J. P.,** Interet de la technique d'inoculation intrathoracique a *Aedes aegypti* dans l'isolement et le reisolement des arbovirus, *Ann. Virol. (Inst. Pasteur)*, 132E, 519, 1981.
33. **Monath, T. P. and Nystrom, R. R.,** Detection of yellow fever virus in serum by enzyme immunoassay, *Am. J. Trop. Med. Hyg.*, 33, 151, 1984.
34. **Peters, C. J. and LeDuc, J. W.,** Bunyaviruses, Phleboviruses, and related viruses, in *Textbook of Human Virology*, Belshe, R. B., Ed., PSF Publishing, Littleton, Mass., 1984.
35. **Smithburn, K. C., Taylor, R. M., Rizk, F., and Kader, A.,** *Am. J. Trop. Med. Hyg.*, 3, 9, 1954.
36. **Brès, P.,** Recent data from serological survey on the prevalence of arbovirus infections in Africa, with special reference to yellow fever, *Bull. WHO*, 43, 223, 1970.
37. **Pinto, M. R.,** Survey for antibodies to arboviruses in the sera of children in Portuguese Guinea, *Bull. WHO*, 37, 101, 1967.
38. **Brès, P., Carrie, J., Desbois, A., Lartigue, J. J., and Mace, G.,** Les arbovirus en Haute Volta, *Ann. Inst. Pasteur*, 108, 341, 1965.
39. **Robin, Y., Brès, P., Lartigue, J. J., Gidel, R., Lefevre, M., Athawet, B., and Hery, G.,** Arboviroses en Afrique de l'Ouest, *Bull. Soc. Pathol. Exot.*, 61, 833, 1968.
40. **Chambon, L., Brès, P., Chippaux, C. L., et al.,** Role des arbovirus dans l'etiologie des fievres exanthematiques en Afrique Centrale, *Mel. Afr. Noire*, 185, 1969.
41. **Kokernot, R. H., Smithburn, K. C., and Weinbren, M. P.,** Neutralizing antibodies to arthropod-borne virus in human being and animals in the Union of South Africa, *J. Immunol.*, 77, 313, 1956.
42. **Kokernot, R. H., Smithburn, K. C., Gandara, A. F., McIntosh, B. M., and Heymann, C. S.,** Provas de neutralizao com soros de individuos residentes en Mozambique contra determinados virus isoladas em Africa transmiditos par arthropodes, *An. Inst. Med. Trop.*, 17, 201, 1960.
43. **Sureau, P.,** Arbovoses a Madagascar, *Arch. Inst. Pasteur Madagascar*, 38, 27, 1965.
44. **Kokernot, R. H., Szlamp, E. L., Levitt, J., and DeMeillon, B.,** Survey for antibodies against arthropod-borne viruses in the sera of indigenous residents of Caprivi strip and Bechuanaland Protectorate, *Trans. R. Soc. Trop. Med. Hyg.*, 59, 553, 1965.
45. **Fagbami, A. H., Monath, T. P., Tomori, O., Lee, V. H., and Fabiyi, A.,** Studies on Tataguine infection in Nigeria, *Trop. Geogr. Med.*, 24, 298, 1972.
46. **Fagbami, A. H. and Tomori, O.,** Tataguine virus isolations from humans in Nigeria, 1971—1975, *Trans. R. Soc. Trop. Med. Hyg.*, 75, 788, 1981.
47. **Johnson, B. K., Shockley, P., Chanas, A. C., et al.,** Arbovirus isolations from mosquitoes: Kano Plain, Kenya, *Trans. R. Soc. Trop. Med. Hyg.*, 71, 518, 1977.
48. **Johnson, B. K., Chanas, A. C., Shockley, P., et al.,** Arbovirus isolations from, and serological studies on, wild and domestic vertebrates from Kano Plain, Kenya, *Trans. R. Soc. Trop. Med. Hyg.*, 71, 512, 1977.
49. **Gonzalez, J. P., Saluzzo, J. F., Herve, J. P., and Geoffroy, B.,** Enquetes serologiques sur l'incidence des arbovirus chez l'homme en milieu forestier et periforestier de la region de la Basse Lobaye (Empire Centrafricain), *Bull. Soc. Pathol. Exot.*, 72, 416, 1979.
50. **Saluzzo, J. F., Gonzalez, J. P., Herve, J. P., and Georges, A. J.,** Enquetes serologiques sur la prevalence de certains arbovirus dans la population humaine du Sud-Est de la Republique Centrafricaine en 1979, *Bull. Soc. Pathol. Exot.*, 74, 490, 1981.

51. **Rickenbach, A., Germain, M., Eouzan, J. P., and Poirier, A.,** Recherches sur l'epidemiologie des arboviroses dans une region forestiere du Sud-Cameroun, *Bull. Soc. Pathol. Exot.*, 61, 266, 1968.
52. **Boorman, J. P. T. and Draper, C. C.,** Isolation of arboviruses in the Lagos area of Nigeria, and a survey of antibodies to them in man and animals, *Trans. R. Soc. Trop. Med. Hyg.*, 62, 269, 1968.
53. **McIntosh, B. M., Serafini, E. T., Dickinson, D. B., et al.,** Antibodies against certain arboviruses in sera from human beings resident in the coastal areas of Southern Natal and Eastern Cape Provinces of South Africa, *S. Afr. J. Med. Sci.*, 77, 1962.
54. **Courtois, G., Osterrieth, P., and Blanes-Ridaura, G.,** *Ann. Soc. Belge Med. Trop.*, 40, 29, 1960.
55. **Kokernot, R. H., Smithburn, K. C., DeMeillon, B., and Paterson, H. E.,** Isolation of Bunyamwera virus from a naturally infected human being and further isolation from *Aedes (Banksinella) circumluteolus* Theo., *Am. J. Trop. Med. Hyg.*, 7, 579, 1958.
56. **Kokernot, R. H., Heymann, C. S., Muspratt, J., et al.,** Studies on arthropod-borne viruses of Tongaland. V. Isolation of Bunyamwera and Rift Valley fever viruses from mosquitoes, *S. Afr. J. Med. Sci.*, 22, 71, 1957.
57. Dakar, Senegal, Annual Report, Institut Pasteur, Paris, 1985.
58. **McIntosh, B. M., Jupp, P. G., Santos, I. S. L., and Meenehan, G. M.,** *Culex (Eumelanomyia) rubinotus* Theobold as vectors of Banzi, Germiston, and Witwatersrand viruses. I. Isolation of virus from wild populations of *Cx. ribinotus*, *J. Med. Entomol.*, 12, 637, 1976.
59. **Monath, T. P., Henderson, B. E., and Kirya, G. B.,** Characterization of viruses (Witwatersrand and Germiston) isolated from mosquitoes and rodents collected near Lunyo Forest, Uganda, in 1978, *Arch. Ges. Virusforsch.*, 38, 125, 1972.
60. **Lee, V. H., Monath, T. P., Tomori, O., Fagbami, A., and Wilson, D. C.,** Arbovirus studies in Nupeko Forest, a possible natural focus of yellow fever in Nigeria. II. Entomological investigations and viruses isolated, *Trans. R. Soc. Trop. Med. Hyg.*, 68, 39, 1974.
61. **Henderson, B. E., McCrae, A. W. R., Kirya, G. B., Ssenkubuge, Y., and Sempala, S. D. K.,** Arbovirus epizootics involving man, mosquitoes, and vertebrates at Lunyo, Uganda, 1968, *Ann. Trop. Med. Parasitol.*, 66, 343, 1972.
62. Annual Report, South African Institute of Medical Research, Johannesburg, 1972.
63. **Dick, G. W. A.,** Epidemiologic notes on some viruses isolated in Uganda, *Trans. R. Soc. Trop. Med. Hyg.*, 47, 13, 1953.
64. **Peers, R. R.,** Bunyamwera virus replication in mosquitoes, *Can. J. Microbiol.*, 18, 741, 1972.
65. **Ogunbi, O.,** Ukauwa virus proliferation in mosquitoes, *Can. J. Microbiol.*, 14, 125, 1968.
66. **Hayes, C. G.,** A comparison of suckling mouse and mosquito susceptibility to infection by the Bunyamwera group arboviruses, *Mosq. News*, 32, 172, 1972.
67. **Boorman, J. P. T.,** Studies on the growth and transmission of two viruses of the Bunyamwera group in *Aedes aegypti* Linn., *Trans. R. Soc. Trop. Med. Hyg.*, 60, 332, 1966.
68. **Fagbami, A. H.,** Studies on transmission of Tataguine virus by *Culex (pipiens) fatigans* mosquitoes, *Afr. J. Med. Sci.*, 8, 31, 1979.
69. **Simpson, D. I. H.,** Experimental Bunyamwera virus infection in two species of African rats, *Trans. R. Soc. Trop. Med. Hyg.*, 59, 198, 1965.
70. Annual Report, East African Virus Research Institute, 1964, 5 and 46.
71. **Schwartz, A. and Allen, W. P.,** Experimental infection of monkeys with Bunyamwera and Germiston viruses, *Infect. Immun.*, 2, 762, 1970.
72. **McIntosh, B. M.,** Susceptibility of some African wild rodents to infection with various arthropod-borne viruses, *Trans. R. Soc. Trop. Med. Hyg.*, 55, 63, 1961.
73. **Fagbami, A. H. and Fabiyi, A.,** A survey for Ilesha (Bunyamwera group) virus antibodies in sera from domestic animals and humans in three ecological zones of Nigeria, *Virologia*, 26, 27, 1975.
74. **Gonzalez, J. P. and Georges, A. J.,** unpublished data.

Chapter 19

CALIFORNIA GROUP VIRUS DISEASE

Paul R. Grimstad

TABLE OF CONTENTS

I. INTRODUCTION

A number of excellent reviews of the California (CAL) group have been published including those by Henderson and Coleman,[1] Parkin et al.,[2] and LeDuc.[3] The Proceedings of the 1982 International Symposium on California Serogroup Viruses, edited by Calisher and Thompson,[4] provided a detailed review of La Crosse (LAC) virus in particular. Since 1945, many hundreds of reports have been published globally on CAL group viruses. Many hundreds more have been published on the bionomics and control of various mosquito species principally, or perhaps only incidentally, involved in CAL group virus cycles in nature. Space will permit reference to only a limited number of key or representative individual papers for any one of the 15 viruses and specific discussion sections presented in this review. In light of earlier reviews and the vast CAL group literature, this chapter presents an updated summary of the epidemiologically relevant information currently available on each of the CAL group virus diseases.

Many North American studies of the CAL group prior to the early 1970s are somewhat confusing in that the etiological agents discussed were not clearly identified. Workers often used a single serotype (e.g., BSF-283) in serological studies. In others, the isolates were not typed beyond their placement in the CAL group. As a result of more recent studies, however, one can retrospectively identify the probable serotype of virus from earlier studies through association with specific vertebrates, arthropods, or geographic areas. Cautious use of earlier investigations lends much to our understanding of several major agents when viewed in the light of recent reports where thorough typing and/or complete serological analyses were done. Some authors consistently refrained from referring to the serotype under investigation and wrote only of California, California encephalitis, or CAL virus, creating everlasting confusion in the literature. The reader should be aware that from a serological standpoint, only minor differences occur throughout the CAL group. In addition, the serotypes differ only slightly in molecular characteristics[5] and are indistinguishable from one another by electron microscopy.[6] However, each serotype or variety is associated with a unique epidemiological cycle, often quite unrelated to other closely related serotypes. Since each serotype is distinct from all the others in this most relevant of terms, this review presents each subtype or variety as a distinct virus.

Official records list all clinical cases as California encephalitis. This terminology may be acceptable in clinical circles since treatment for severe viral infections would be similar for CAL group neuropathogens; from an epidemiological standpoint, however, this is insufficient. Given the increased awareness of the epidemiological differences among CAL group viruses and recognition that a number of serotypes cause clinically recognized disease, "failure to determine the etiologic agent is a disservice to the general population and, given the availability of techniques (for typing and) for detecting antibody (to CAL group viruses), to the scientific community."[7]

II. HISTORICAL BACKGROUND

A. Discovery of Agents and Vectors

In 1943 in Kern County, California, a virus isolation made from *Aedes melanimon* mosquitoes (then called *Aedes dorsalis*) by Hammon et al.[8] established the CAL serogroup. Two additional isolations were made the following year from *Ae. melanimon* and *Culex tarsalis* mosquitoes in the same area.[8] The 1944 *Ae. melanimon* isolate, designated BFS-283 and named California encephalitis (CE) virus (Table 1), became the prototype virus of the CAL serogroup. Several years later, in 1948, a second virus was isolated from *Ae. trivittatus* mosquitoes near Bismark, North Dakota, by Eklund and colleagues.[8] This isolate, named trivittatus (TVT) virus, was shown to be serologically related to, but distinct from, CE virus. Six years later, in 1955, Melao (MEL) virus was isolated at the Trinidad (Port of Spain) Regional Virus Laboratory from *Ae. scapularis* mosquitoes collected in northeast Trinidad.[9] In the subsequent 11 years, 12 new viruses serologically related to, but distinct from, BFS-283 were isolated and characterized by numerous workers (Table 1).[8-19]

Hammon and Reeves[20] had implicated CE virus as the causal agent in three human cases of encephalitis in 1945 in the San Joaquin Valley of California. During the early to mid-1960s, workers in Europe associated numerous cases of "influenza-like" illness and a rare occurrence of encephalitis with Tahyna (TAH) virus infections.[21,22] However, it was in North America that the California group gained wide public health recognition beginning in the early 1960s. The death of a 4-year-old Minnesota girl in a La Crosse, Wisc., hospital in 1960 from severe encephalitis led to the isolation in 1964 of LAC virus from autopsy brain tissue by Thompson and associates.[15] Retrospective serological analyses by Thompson and Evans[23] of sera collected from patients in Wisconsin between 1960 and 1964 with central nervous system (CNS) illness and sera from other febrile patients demonstrated the past occurrence of CAL group virus infections in the upper Midwest. [Interestingly, during their early study of LAC virus, Thompson and Evans discovered what have become the earliest documented human infections of Jamestown Canyon (JC) virus. Of special note was their 1965 report of three young men who experienced mild febrile illnesses and subsequently developed neutralizing antibody to CE virus. Both snowshoe hare (SSH) and CE viruses were used in the initial serological tests; later epidemiological studies and retesting of the young men's sera with JC virus in neutralization (N) tests suggested that these young men had experienced infection with JC virus.[24]] Meanwhile, in Ohio, evidence was concurrently accumulating which indicated the public health importance of LAC virus.[1] In the decade following the isolation of LAC virus, numerous workers contributed to provide a detailed description of the natural cycle of that virus, evidence of its epidemiological importance to the human population, and an understanding of the bionomics of the primary vector, *Ae. triseriatus*.

While the isolation of a new virus from arthropod tissues is exciting, its status as an arbovirus is based on the discovery of a vector and subsequent field and laboratory studies

Table 1
HISTORICAL RECORD OF CALIFORNIA SEROGROUP VIRUS ISOLATIONS[a]

Virus name and abbreviation	Prototype strain	Year source collected/year virus isolated	Source	Locality of isolation	Ref.
California encephalitis (CE)	BSF-283	1943/1943	*Aedes melanimon*	California	8
Trivittatus (TVT)	933	1948/1948	*Aedes trivittatus*	North Dakota	8
Melao (MEL)	TRVL-9375	1955/1955	*Aedes scapularis*	Trinidad	9
Guaroa (GRO)	352111	1956/1956	Human female	Colombia	10
San Angelo (SA)	20230	1958/1958	*Anopheles p. pseudopunctipennis*	Texas	11
Tahyna (TAH)	92	1958/1958	*Aedes caspius*	Czechoslovakia	12
Snowshoe hare (SSH)	Original	1959/1959	*Lepus americanus*	Montana	13
Lumbo (LUM)	AR 1881	1959/1959	*Aedes pembaensis*	Mozambique	14
La Crosse (LAC)	LAX-177	1960/1964	Human female	Wisconsin	15
South River (SR)	NJO-94F	1960/1960	*Anopheles crucians*	New Jersey	16
Jamestown Canyon (JC)	61V-2235	1961/1962	*Culiseta inornata*	Colorado	11
Jerry Slough (JS)	BFS 4474	1963/1963	*Culiseta inornata*	California	11
Inkoo (INK)	KN 3641	1964/1964	*Aedes communis/punctor*	Finland	17
Keystone (KEY)	B64-5587.05	1964/1964	*Aedes atlanticus-tormentor*	Florida	18
Serra do Navio (SDN)	BeAr103645	1966/1966	*Aedes fulvus*	Brazil	19

[a] Lumbo is now considered to be a strain of Tahyna virus; SIRACA considers Jerry Slough to be a strain of Jamestown Canyon virus.[19] The remaining 13 viruses are registered in the *International Catalogue of Arboviruses* (1985).[19]

designed to prove vector status. Rice and Pratt[25] outlined four basic criteria for proving the vector status of a particular arthropod species: (1) isolation of the disease-producing agent from wild-caught specimens, (2) demonstration of its ability to become infected by feeding upon vertebrate hosts, (3) demonstration of its ability to transmit by bite, and (4) collection of field evidence that confirms the association of the infected arthropod with the vertebrate population in which the infection is occurring. It is now well recognized that marked variation in vector competence occurs among geographic strains, or even regional populations, of the same vector species and it is necessary to speak of vector populations.[26] Discovery of a primary vector and implicating it in the natural cycle of a number of CAL group viruses [Guaroa (GRO), Inkoo (INK), MEL, San Angelo (SA), Serra do Navio (SDN)] has not been accomplished as yet. With others [CE, JC, Keystone (KEY), LAC, SSH, TAH, TVT], the primary vector(s) have been identified and all are mosquitoes, primarily of the genus *Aedes*. In addition, what is emerging for the CAL group in general is an intimate virus-vector relationship — transovarial (vertical) transmission — which is primarily responsible for maintaining these latter seven viruses in nature.[27-29]

Turell and LeDuc,[30] in a comprehensive table, summarized CAL group isolations from hematophagous arthropods, noting the principal vector(s) and additional species from which isolates were made. The primary update of their list would be the inclusion of *Ae. stimulans* as a principal vector of JC virus.[28] While the arthropod vectors of GRO, INK, MEL, SA, SDN, and SR viruses remain unknown, the mosquito species listed in Table 1 from which each virus was first isolated have generally been considered potential vector species. For a number of CAL group viruses, there is an extensive record of isolations from numerous arthropod species in addition to those recognized as the primary vectors.[2,30] These additional species probably have little or no relevance to the transmission of the various CAL group viruses in nature. For example, of 12 mosquito and 2 tabanid species yielding LAC virus isolates,[30] only *Ae. triseriatus*,[31] and to a lesser extent *Ae. canadensis*,[32] have been shown to be of importance in the natural cycle. TVT virus has been isolated from 14 species of mosquitoes;[30] only *Ae. trivittatus*, and perhaps *Ae. infirmatus* in Florida, are the principal vectors. During periods of virus transmission to primary vertebrate hosts, it is likely that other mosquito species might attempt to feed on these hosts. Wright and DeFoliart[33] have documented the catholic host perferences of woodland Aedine mosquitoes in Wisconsin; ingestion of a small amount of viremic blood or contamination of mouthparts in unsuccessful feeding attempts on a variety of hosts can occur. If these "contaminated" individuals are collected perhaps within 1 to 2 days, virus isolation is possible. These additional isolation records for the CAL group have made identification of the primary vectors more difficult. The possibility that some of these other species may be of importance in horizontal amplification must be evaluated, however.

B. History of Epidemics

Between 1963 and 1984, 1611 cases of CNS illness, serologically confirmed as CE, were reported to the Centers for Disease Control (CDC) (Table 2);[34-37] the vast majority of these cases undoubtedly resulted from infection with LAC virus.[34] Numerous other probable or suspect cases remain unreported, existing in the records of the various state health departments. Many of these cases may represent infection with CAL group viruses other than LAC. The first recognized outbreak of CE occurred in southeastern Indiana in 1964.[1] In retrospect, this outbreak was almost certainly due to LAC virus infections.[38] In 1975, 174 cases of CE were reported to the CDC. This was the highest number of cases annually reported to date (Table 2) and probably reflected the increased awareness in the medical community of mosquitoborne encephalitis that year as a result of the nationwide St. Louis encephalitis (SLE) epidemic,[39] rather than an outbreak/epidemic situation. Indeed, Illinois and Ohio, two of the northcentral states that had the highest number of SLE cases in 1975,

Table 2
RECORD OF REPORTED CALIFORNIA ENCEPHALITIS CASES IN THE U.S., 1963 TO 1985[a]

State reporting	No. of cases reported by year																							Total
	1963	1964	1965	1966	1967	1968	1969	1970	1971	1972	1973	1974	1975	1976	1977	1978	1979	1980	1981	1982	1983	1984	1985[b]	
Arkansas													1			1	1				1			4
Florida	1																							1
Georgia																	1			9				10
Illinois				1		2				2	3		42	6	15	9	21	6	5	8	8	2		130
Indiana		12	7	2	2	1					4	1	7	2	5	1	4	5	1	6	4	15	12	91
Iowa			6		11	3	6	3			3	2	8	5	2	2	13	7	4	7		5		87
Kentucky						1	1																	2
Louisiana					2	9	1																	12
Maryland																				1				1
Michigan								1					1			11			1					14
Minnesota			1	12	9	7	16	16	17	9	21	10	23	4	11	24	33	5		11	12	10		251
Mississippi					1																			1
Missouri											2	2					2	1		2	1	2		12
New Jersey												1	1			1								3
New York				1			2	2	4	2	5	1			1	3	1		5	11		2		40
North Carolina		1	3														4	1	5	6		3		23
Oklahoma																						1		1
Ohio		25	28	34	25	43	18	36	25	14	31	2	66	18	13	11	24	16	34	37	25	21	33	579
Pennsylvania								3		1													1	5
South Carolina																				1		1		2
Tennessee																1				2		1		4
Utah																		2						2
West Virginia							3	2																5
Wisconsin		4	14	14	3	5	20	26	12	17	16	11	25	12	17	46	36	30	21	24	13	11		377
Total	1	42	59	64	53	71	67	89	58	45	85	30	174	47	64	110	140	73	76	125	64	74	46	1657

[a] See References 34 to 37.
[b] Provisional data only.

both reported record high numbers of CE that year (Table 2). In 1978, 1979, and again in 1982, numbers of reported CE cases increased markedly, but for no apparent reason. However, active surveillance programs have led to increased case detection,[40] especially in Ohio.[41]

In 1980, Fauvel and colleagues[42] reported several cases of CNS illness in Ontario residents and implicated SSH virus as the etiological agent. That same year, Grimstad and associates[7] implicated JC virus as the causal agent of severe encephalitis in an 8-year-old southwestern Michigan girl. Deibel and co-workers[43] noted 11 additional cases of JC virus encephalitis, meningitis, or febrile illness in adult New York and Ontario residents; a number of additional clinical cases have been confirmed serologically in North American residents.[44,45] Evidence is accumulating that TVT virus may also cause CNS illness,[46] and INK virus has been associated with mild febrile illness in Finland,[19] as has GRO in Brazil.[19] With the increasing interest in the CAL group, most or all agents may eventually be linked to human illness, since antibodies to KEY and SDN viruses have been reported in human serological surveys or in clinical case workups. Expanded serological surveys in the eastern Atlantic states and Texas might detect persons with prior SR or SA virus infections, respectively.

C. Social and Economic Impact

Social impact is marked in those diseases where urban epidemics or widespread rural infections produce severe disease and elevate public awareness to the problem. With CAL group viruses, human disease occurs endemically,[34] often in rural or suburban populations, and is generally mild or inapparent.[47] Thus, the social impact of the CAL group has been limited for the most part with little recognized impact on human populations in general. The notable exception has been LAC virus with its predilection for causing moderate to severe clinical illnesses in young children with subsequent sequelae in many individuals.[48,49] Four urban localities in North America consistently have been associated with increased numbers of LAC virus encephalitis cases: La Crosse, Wisc.;[4] Peoria, Ill.;[50] Ft. Wayne, Ind.;[38] and Gambier, Ohio.[41,51] In Illinois, 61% of the LAC virus encephalitis cases reported in that state have come from Peoria County.[50] Balfour et al.[52] have described the focal occurrence of LAC virus infections in residents of a southeastern Minnesota community. Almost 82% (1314/1611) of officially reported CE cases have come from six northcentral states — Ohio, Wisconsin, Minnesota, Illinois, Indiana, and Iowa (Table 2).

The economic impact of the CAL group has also been slight except where the individual and family are concerned. In the case of LAC virus encephalitis cases, the cost can be substantial. Clark et al.[50] have noted that hospitalization costs for cases in Illinois (estimated in 1981 dollars) ranged from $3967 to $5750; the 5-year (1976 to 1980) cost of care for LAC virus encephalitis cases in Illinois alone ranged from approximately $242,000 to $350,000. The long-term effects of sequelae to LAC virus infections can profoundly affect children with a social and economic impact that is considerable.[48] Gundersen[53] has estimated that the long-term care of an individual suffering recurrent seizures (as a sequela to a LAC virus infection) may exceed $450,000 over the reduced lifetime of the individual.

Despite the known risks of LAC virus infection in children and the knowledge that certain sites are natural foci, public officials in some midwestern areas remain unconcerned. In at least one midwestern focus, school officials still conduct nature walks for children through a known focus with little obvious concern, despite records of LAC virus encephalitis cases from that locality, numerous LAC virus isolations from the site, and warnings from state public health officials. In contrast, the community-wide concern shown by the residents of La Crosse County, Wis., has led to the virtual elimination of LAC virus encephalitis cases in that county since 1981.[54]

Table 3
ANTIGENIC RELATIONSHIPS AMONG CALIFORNIA SEROGROUP MEMBERS[a]

Group	Complex	Virus	Subtype	Variety	Variants/ strains
California	Guaroa	Guaroa			
	California encephalitis	Trivittatus			
		Melao	Jamestown Canyon	Jamestown Canyon	Jerry Slough[b]
				South River[b]	
			Keystone		
			Melao		
			Serra do Navio		
		California encephalitis	California encephalitis		
			Inkoo		
			La Crosse	La Crosse	A, B, C
				Snowshoe hare	
			San Angelo		
			Tahyna	Tahyna	
				Lumbo[b]	

[a] Relationships based on Calisher[56] and Karabatsos;[19] Table 3 based on Calisher.[18]
[b] Jerry Slough is considered by SIRACA to be a strain of Jamestown Canyon;[19] South River is an unregistered virus closely related to Jamestown Canyon and may represent a geographic variety; Lumbo is also an unregistered virus and represents the African variety of Tahyna virus.

III. EPIDEMIOLOGICALLY RELEVANT CHARACTERISTICS OF THE VIRUS

A. Antigenic Relationships

The California serogroup is but one of 16 serogroups within the genus *Bunyavirus* of the family Bunyaviridae.[19,55] Two antigenic complexes, CE and GRO, are presently recognized (Table 3);[19,56] their antigenic relationships have been summarized by Bishop and Shope[55] and Calisher.[56] Three viruses, CE, MEL, and TVT, comprise the California complex. In turn, CE virus and MEL virus have been serologically separated into seven and five subtypes/varieties, respectively. Subtypes of GRO and TVT viruses have not been identified as yet. Three recognized variants of LAC, designated A, B, and C by Klimas et al.,[57] show specific geographic distributions; the latter may have resulted from a variation in the LAC virus-mosquito vector relationships. Jerry Slough (JS) is a western U.S. strain of JC associated with *Culiseta inornata* mosquitoes;[19] it will likely be shown to be one of many variants of the geographically diverse JC subtype. South River (SR), an unregistered eastern U.S. variety of JC,[3] may also represent a geographic variant of JC virus.

B. Vertebrate Host Range

Numerous vertebrate species have been found with antibodies to CAL group agents, or have yielded viral isolates. Not all species found with antibody to each virus are of epidemiological significance, however. A number of species with detectable N antibody either develop no detectable viremia,[58-61] or develop a low-titered viremia of limited duration insufficient to infect any arthropod vector. These animals are important from the standpoint of their being excellent natural sentinels. Table 4 summarizes the primary vertebrate host(s)

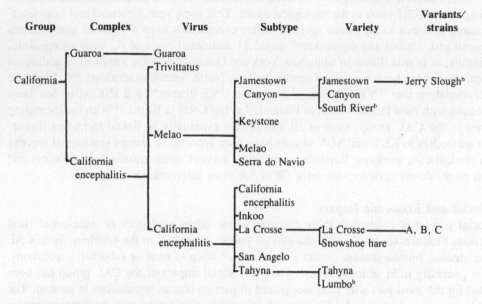

Table 4
MAJOR VERTEBRATE HOSTS[a] OF CALIFORNIA GROUP VIRUSES[3,4,19]

Virus	Principal vertebrate host(s)	Secondary vertebrate hosts and *sentinels*
California encephalitis	California ground squirrel	Cottontail and jackrabbits, *domestic animals?*
Guaroa	Unknown	Unknown
Inkoo	Reindeer	Moose, *cattle*
Jamestown Canyon	White-tailed deer	Antelope, elk, moose, mule deer, *domestic animals*
Jerry Slough	Unknown	Unknown
Keystone	Gray squirrels, cottontail rabbits	Cotton rats, *white-tailed deer, goats, horses*
La Crosse	Eastern chipmunk, gray squirrels	Red foxes, *domestic rabbits*
Melao	Unknown	Unknown
San Angelo	Unknown	Unknown
Serra do Navio	Unknown	Unknown
Snowshoe hare	Snowshoe hares, *Citellus* ground squirrel	Small rodents, *moose, deer, horses*
Tahyna	European hares, rabbits	European hedgehog, *swine*
Trivittatus	Cottontail rabbits	*Domestic rabbits*

[a] Humans are assumed to be a dead-end host for these viruses. See the *International Catalogue of Arboviruses* (1985)[19] and Parkin et al.[2] for additional serological records. Secondary vertebrate hosts probably play a limited role in virus amplification and/or dissemination. Sentinel animals include those showing high antibody prevalence rates but may now develop viremias; these hosts presumably do not contribute to virus amplification and dissemination in the natural cycles.

and major secondary hosts known or suspected to be involved in the natural cycle of each virus. The significance of these host species is discussed below in relation to transmission cycles. Of interest is the lack of any avian involvement in CAL group virus cycles.[19]

C. Strain Variation

Numerous molecular studies have utilized CAL group viruses, primarily LAC, SSH, and TAH. These studies have been summarized by Beaty and Trent[62] in their chapter, "Virus Variation and Evolution: Mechanisms and Epidemiologic Significance" in Volume I.[62] Of major epidemiological importance is the ability of CAL group viruses (LAC, SSH, TAH) to reassort within a mosquito host,[55,63-65] both in the laboratory and apparently in the field.[57] This may lead to viruses with altered levels of virulence,[65] and potentially may be the mechanism for generation of new CAL group genotypes in nature.[57,63]

Strain variation in levels of N, hemagglutination-inhibition (HI), and complement-fixation (CF) antibodies induced, plaque size, virulence, thermal stability, and antigenic relationships suggest that TAH virus is quite heterogeneous, not only between strains isolated in geographically distinct European regions, but also in a single Czechoslovakian region.[66,67] Malkova and Reddy[68] noted distinct changes in titer and plaque size, but not virulence or thermostability on serial passage in VERO cells. Plaque variant strains have also been noted for a number of CAL group viruses. The effect of passage on plaque morphology and cytopathic activity has been investigated by a number of workers for JC,[69] LAC,[70,71] and TAH[72,73] viruses. Temperature-sensitive (ts) mutants have been described for LAC,[74] LUM,[75] and SSH[74,76] viruses. Attenuation of virulence in strains of LAC,[77-79] SSH,[79] and TAH[80] has also been documented. Miller[77] noted the inability of one plaque-purified LAC variant to disseminate from midgut epithelial cells of *Ae. triseriatus*, and thus it was not transmitted to suckling mice. While strain variation may in part be due to laboratory artifacts and growth in various mammalian cell lines, the variation potentially resulting from continued passage through various vector species in nature needs to be assessed thoroughly. Transmission of virus by species "A" may result primarily in inapparent infections, while transmission by species "B" may result in a higher ratio of clinical cases. This mechanism may be occurring in the transmission of JC virus.

D. Methods of Viral Assay

Numerous species of experimental animals, including suckling mice, hamsters, guinea pigs, rabbits, and others, have been used for CAL group isolation and pathology studies.[19] However, the suckling mouse has been the animal of choice for isolation. VERO cells have been used most widely for isolation and virus typing.[19] BHK, Hep-2, duck embryo, chick embryo, HeLa, Singh's *Ae. aegypti*, C6/36 *Ae. albopictus*, and LLC-MK2 cells have also been used by a number of workers. Typing of viral isolates has generally been by CF, followed by HI and N tests.[81] Recently, Artsob et al.[82] described an efficient enzyme immunoassay (ELISA) typing system for CAL group viruses. Detection of virus in host tissues, primarily arthropod, has been accomplished by immunofluorescence;[65,83,84] these studies have sought to delineate the sequence of virus infection in mosquito tissues.

IV. DISEASE ASSOCIATIONS

A. Humans

As was noted earlier, CAL group viruses frequently produce mild or inapparent infections in humans resulting in inapparent to apparent ratios that are higher than those seen with other more pathogenic mosquitoborne arboviruses such as SLE.[26,39] Several estimates of these ratios are available for LAC virus and range from 322:1 in Ohio to 1571:1 in Indiana.[38] These ratios represent each of the states 0- to 16-year age populations. In contrast, Monath et al.[85] noted an inapparent to apparent ratio of 26:1 for the 0- to 15-year age group in a geographically small but intense LAC virus focus in rural Minnesota.

Four separate generalized syndromes characterize CAL group clinical infections:[19] (1) mild to moderate febrile illness without apparent CNS involvement (e.g., GRO, INK); (2) febrile illness with marked respiratory involvement and rare CNS involvement (e.g., TAH); (3) febrile illness with frequent CNS involvement (e.g., CE, LAC, SSH, TVT); and (4) febrile illness with both frequent CNS involvement and frequent respiratory system involvement (e.g., JC). Clinical illness caused by KEY, LUM, MEL, SA, SDN, or SR viruses has not been documented yet and infection with any of these may perhaps result in only a mild febrile illness at most.

The first generalized syndrome — mild to moderate febrile illness without CNS involvement — is probably representative of most human CAL group infections even with the neuropathogenic serotypes. In a Minnesota LAC virus case, Monath et al.[85] reported that the child experienced only a fever, sore throat, nausea, and vomiting; Thompson and Gundersen[24] noted that one child with an apparent TVT virus infection experienced a "brief mild flu-like illness".

The second syndrome — febrile illness with marked respiratory involvement and rare CNS involvement — is characteristic of TAH virus in Europe. Bardos et al.[86] described two cases in young boys. The first presented with a sudden onset of fever, sore throat, and a generalized malaise; he later developed pharyngitis, a mild cough, and conjunctivitis. The second youth experienced a febrile onset followed by sore throat, other influenza-like symptoms, and a suggestive meningeal involvement. Bardos and Sluka[87] described the case histories of 25 patients in which the symptoms included fever, headache, nausea, anorexia, and cough. In reporting the first isolation of TAH virus from the blood of a patient, Simkova and Sluka[88] reported that the patient had a 4-day history of headache, anorexia, and generalized weakness. Sluka[89] described 6 clinical syndromes associated with TAH virus infection: (1) brief flu-like febrile illness; (2) viral pneumonia and pleuritis; (3) acute arthritis; (4) abdominal pain; (5) fever and pharyngitis; and (6) acute CNS illness.[89] The latter was seen in approximately 3% of 58 cases.[89]

The third generalized syndrome — febrile illness with frequent CNS involvement — is typical of several CAL group viruses, especially LAC virus. Excellent summaries of clinical LAC virus encephalitis cases have been presented by Balfour et al.,[90] Chun et al.,[91] Cramblett et al.,[92] Gundersen and Brown,[48] Henderson and Coleman,[1] Hilty et al.,[93] Thompson and Inhorn,[94] and Young.[95] Grimstad et al.[7] summarized the clinical laboratory findings associated with LAC viral infections. The common symptoms of LAC virus infection include fever, headache, nausea and vomiting, nuchal rigidity, lethargy, seizures, and, occasionally, coma. Of considerable relevance in LAC virus infections is the occurrence of seizures; Gundersen and Brown[48] reported that approximately 42% of patients had seizures and stated that "La Crosse virus in the upper Midwest is highly epileptogenic". Parents of LAC virus encephalitis patients stated frequently that they found the child convulsing, but that the child had been well just hours before and this was the first sign of illness they noted.[96] Fauvel et al.[42] described the first cases of SSH virus encephalitis in Quebec residents in 1980. These patients presented with headache, fever, and vomiting, with CNS involvement. Embil et al.[97] noted that an 11-year-old New Brunswick (Canada) boy experienced fever, vomiting, lethargy, and seizures as a result of an SSH virus infection. Mahdy et al.[98] detailed three cases of SSH virus infection with varying symptomology: one patient had an influenza-like illness initially thought to be Q fever; the second patient was diagnosed as having viral meningitis; and the third patient, a 14-year-old female, was comatose and died 5 days after onset. Her case was diagnosed initially as Reye's syndrome. The importance of this case cannot be overlooked with regard to the need to consider CAL group viruses in differential diagnoses involving CNS syndromes. While TVT virus has been considered to be of minimal public health importance, several cases with CNS involvement have been documented. In 1965, Quick et al.[99] detailed a severe encephalitis case with seizures in a 12-year-old Florida female which occurred in 1963 (Table 2). This was only the fourth case of CAL group infection documented in North America. While her illness was described only as CE, subsequent studies in Florida implicate TVT virus as the etiological agent.[100,101] Recent serological analyses of CNS case sera from Illinois patients revealed three persons with apparent TVT virus infections.[46] One hospitalized case presented with a sudden onset of unilateral headache, vomiting, and dehydration — no spinal tap or clinical laboratory workup was apparently performed. The other two patients presented with symptoms diagnosed as aseptic meningitis by their doctors. In all three cases, the serology detected significant specific neutralizing antibody to TVT virus.

The fourth generalized syndrome — febrile illness with frequent CNS and respiratory system involvement — is typical of the clinical picture emerging for JC virus infections.[26] A summary of 39 serologically documented clinical cases indicates that meningitis is more commonly diagnosed (23 cases) compared to encephalitis (10 cases); other CNS involvement (e.g., "Cerebral dysfunction") was reported for 2 cases and the remainder had no specific reference to CNS involvement.[7,43-45,102] In addition, in 7 of 15 cases (of the 39) where additional symptoms were provided by the attending physician, respiratory involvement was marked (e.g., sore throat, pneumonia, upper respiratory tract infections, nasal congestion, cough). Interestingly, 7 of 15 cases experienced a prodromal syndrome that included respiratory symptoms of sore throat, cough, "influenza-like illness", and fever for 2 days to 2 weeks. Of the 39 cases, 2 cases as well as 1 earlier case presented with rash. Thompson and Evans[23] noted that 1 Wisconsin youth had a rash for 3 days, Grimstad et al.[45] described the case of an Ohio youth who had a 3-week history of a herpes-like genital rash, and a New York patient presented with erythema multiforme.[102] The spectrum of clinical symptoms presented by JC virus is perhaps the broadest of any of the CAL group members.

Kalfayan[103] has provided the pathology reports of two fatal LAC virus encephalitis cases autopsied at the La Crosse Lutheran Hospital. The brain of each child at autopsy was "diffusely congested and edematous with flattening of the convolutions". The histopathology was similar to that seen with meningoencephalitis caused by other viral neuropathogens. Inflammatory lesions were seen throughout most of the cortical gray matter, midbrain, and pons. Nerve cells were seldom seen in these lesions, and vasculitis was observed with lymphocytic cuffing. Kalfayan[103] suggested that the "absence of inflammatory lesions in the cerebellum, medulla, and spinal cord may prove to be a distinguishing feature of La Crosse virus encephalitis". In addition, he suggested that "the acute vasculitis of cerebral capillaries and small veins plays a key role in the pathogenesis of La Crosse virus encephalitis".

Of major importance to the patient with LAC virus encephalitis is the frequent occurrence of sequelae and of recurrent seizures (epilepsy) in particular.[104] Chun[49] noted that most focal neurological findings resolve completely; only rare cases continue for as long as 8 years. While 86 to 100% of cases show abnormal electroencephalographic (EEG) findings during the acute period, according to Chun, 33% of subjects still had EEG abnormalities 1 to 8 years later. On the whole, LAC virus encephalitis victims function normally with regard to academic performance. However, there are children who suffer permanent destructive lesions that lower IQ and adversely affect school performance.[49] Perhaps the most serious sequela is seizure. Grabow et al.[105] noted that of children with seizures in the acute period, approximately half had recurrent seizures. Deering[104] noted that 20% of Wisconsin and Minnesota children in his experience developed recurrent epileptic seizures. Deering also noted that the mean time to the initiation of recurrent seizures in these children was 45 months, ranging from a few days to 135 months; in one case, seizures occurred 8 years after the acute period. Thus, the cost to a patient, the family, and ultimately to society can be considerable when the lost potential of these children is considered.

B. Domestic Animals

As was noted earlier, numerous vertebrates, including domestic species, have been found with antibody to CAL group viruses, including some avian species. Generally, only small (e.g., squirrels, rabbits) or large mammals (e.g., white-tailed deer) are part of the natural virus cycles (Table 4). Natural infection of domestic animals apparently occurs with eight CAL group viruses: CE, INK, JC, KEY, LAC, SSH, TAH, and TVT.[19] Antibody production has been induced in burros with GRO virus.[61] Many CAL group viruses do not produce detectable viremias on infection of domestic animals, and detectable clinical illness apparently does not occur in these vertebrates. Thus, domestic animals play little if any active role in CAL group natural cycles in general.

C. Wildlife

Natural infection of numerous wildlife species with CAL group viruses has been documented including CE, INK, JC, KEY, LAC, SSH, TAH, and TVT.[19] Despite marked levels of viremia noted in many species, clinical illness does not occur in wildlife infected with these viruses. These hosts serve in the amplification and/or dissemination of various CAL group viruses and are believed to be important to the continued long-term maintenance of these viruses by supplementing the role of transovarial transmission in the natural cycles. Of interest is the apparent complete lack of involvement of wild avian species in CAL group transmission cycles despite the finding of antibody in a number of species.[19]

D. Applicable Diagnostic Procedures

Diagnosis of clinical infection with CAL group viruses traditionally had been made following a fourfold or greater rise or fall in CF, HI, and/or neutralizing antibodies.[48,106] In the U.S., the CDC has provided state health departments with standardized antigens for use in CF and HI tests. In 1981, Calisher and Bailey[107] reported that only 33 states tested for CAL group antibody in clinical sera (23 states used the CF test and 10 used the HI test in screening for CAL group infections; all used the LAC virus antigen, and none routinely used N tests). Lindsey et al.[108] had speculated earlier that use of only LAC virus antigen might preclude diagnosis of infections caused by viruses other than LAC. Calisher and Bailey[107] emphasized the ineffectiveness of the CF test for LAC virus case detection and urged use of the N test for confirmation of HI test results and subtype identification. Recent studies using sera from persons with JC virus infections have shown clearly that less than 1.5% of JC virus infections will be detected and classified as CE if LAC virus is used in CF and/or HI tests.[7,45] Use of LAC virus in N tests would permit detection of approximately 5% of JC virus infections; however, use of JC virus in N tests detects virtually all JC infections and apparently all LAC virus infections.[45] The CF test with LAC virus antigen is no longer recommended for routine diagnostic workups in the U.S.;[56] HI tests, however, are quite reliable in detecting LAC virus infections.[107] In Indiana, the annual rate of LAC case detection has doubled since HI tests were substituted for CF. This is not surprising since Calisher and Bailey[107] had shown that only 50% of seroconversions subsequent to LAC virus infections were detected when the CF test was used alone. The lack of interest in diagnosing CAL group infections shown by numerous state diagnostic laboratories, and even less interest in identifying the etiological agents, contributes to "a rather disconcerting picture"[56] with regard to case detection.

In severe CNS infections with a differential diagnosis of herpes encephalitis, it is essential to determine as soon as possible after onset if herpes or an arbovirus has caused the infection, because chemical therapy is available for the former. Several investigators have used immunoglobulin M (IgM) antibody capture enzyme immunoassay (MAC-ELISA) procedures to detect early infection in LAC virus encephalitis cases with encouraging success.[109-111] Development and widespread utilization of a MAC-ELISA for rapid diagnosis of all CAL group infections is a high priority. Given the emergence of JC as an important neuropathogen and its wide geographic distribution throughout temperate North America, sera from all suspect cases must be assayed initially for CAL group antibodies. It will remain an essential step to eventually determine the etiological agent, probably through routine N testing in cell culture.[81] Until these procedures are followed throughout the U.S., CAL group virus infections will remain vastly underreported.[26,38,56]

E. Adverse Effects of the Viruses on Their Vectors

Several studies utilizing LAC virus have evaluated the adverse effects of virus infection on the natural vector *Ae. triseriatus*. Using orally infected female *Ae. triseriatus*, Grimstad et al.[112] determined that viral infection significantly increased the frequency of probing in a population of *Ae. triseriatus* and also significantly decreased the amount of engorgement. They had shown earlier that geographic strains of *Ae. triseriatus* obtained from regions of the U.S. where LAC virus was enzootic had significantly reduced vector competence compared to strains from the nonenzootic regions.[113] Their work suggested that the net effect of reduced engorgement over time would be selection for more resistant strains of *Ae. triseriatus* in the enzootic regions, as the laboratory studies had shown. In a subsequent study, Patrican and DeFoliart[114] evaluated the effect of infections acquired transovarially on strains of *Ae. triseriatus*. They concluded that no adverse effects occurred with regard to "duration of larval period, sex ratio, hatching success, time to ovarian maturation, fecundity, or adult survival through the second oviposition". This is in contrast to a report from Miller that orally infected *Ae. triseriatus* had a reduced fecundity.[3] Similar deleterious effects have not been seen with other CAL group viruses in their natural vectors. To be epidemiologically relevant, future studies must use both the virus and the vector that naturally interact in the ecosystem.

V. EPIDEMIOLOGY

A. Geographic Distribution

The CAL group is distributed primarily throughout the Western Hemisphere, particularly in North America. Only INK, TAH, and LUM viruses have been isolated on other continents. Recently, Gu et al.[115] reported detecting antibodies to SSH virus in 3.9% (5/126) of residents of suburban Shanghai, suggesting that this virus may have the broadest global geographic distribution. The geographic distribution of all viruses listed in Table 1 (with the exception of JS virus) has been illustrated recently by Calisher.[56] Briefly, CE virus is distributed throughout parts of the western U.S. from California and Utah and east to Texas.[19,56] It has been isolated recently in Manitoba, Canada, however.[116] TVT virus has a more northern distribution — from the Dakotas east to New York; it has also been reported from Utah, Texas, Mississippi, Georgia, and Florida.[19,56] MEL virus is confined to the states of Para and Sao Paulo in Brazil, as well as Panama and Trinidad.[19,56] GRO virus also ranges through Brazil's Para State as well as Colombia and Panama[56] however, antibodies have been detected in residents of Argentina and Peru.[19] SA virus appears to be confined to the southwestern states of Arizona, Colorado, New Mexico, and Texas.[56] TAH virus has a broad distribution throughout Europe and into the southern regions of the U.S.S.R. (i.e., Czechoslovakia, France, Germany, Italy, Lithuania, and east to the Azerbaijan and Tajik Soviet Socialist Republics).[19] It has also been isolated as far north as Norway.[19] Its African variety, LUM virus, has been isolated only in Kenya and Mozambique.[56] LAC virus is distributed throughout the midwestern U.S. from Minnesota and Texas east to New York and Georgia.[56] While isolations of LAC virus have been made from only 13 states,[56] clinical cases have been reported from residents of perhaps 24 states (Table 2). SR virus has a very limited distribution in the eastern states of Pennsylvania, New Jersey, and New York.[56] When both isolation records and results of serological surveys for specific N antibody are considered, JC virus has the widest North American distribution of all the CAL group viruses. Its range encompasses Alaska,[56] most of Canada including Newfoundland and the Maritime Provinces,[117] and in the lower 48 states from California and Utah east through Texas, the Dakotas,[118] and most states east of the Mississippi.[26,45] Grimstad[46] recently has detected N antibody specific for JC virus in deer from 13 eastern and southeastern states from Massachusetts to Florida, west to Arkansas, Louisiana, and inland to Tennessee. JS virus has been isolated only from California, and the known range of INK virus is confined to Finland at present.[19] KEY virus is widely distributed along the eastern and Gulf coasts of the U.S., from New York and Massachusetts south to Florida and west to Texas, and inland to Tennessee.[19,56] Finally, SDN virus has been isolated only in Para State, Brazil.[56] Collectively in North America, CAL group viruses, or N antibody specific for certain subtypes/varieties, have been identified in virtually every Canadian province and American state north of Mexico with the exception of the Pacific Northwest, the three northernmost New England states, and a few states in between. Undoubtedly, expanded serological surveys or virus isolation attempts would detect the presence of CAL group viruses in these geographic locations as well.

B. Incidence

While numerous retrospective serological surveys have shown widespread past infection of the human population in North America and Europe with CAL group viruses, little is known of the annual incidence of infections since large-scale, long-term prospective studies have yet to be conducted. Several limited studies have shown a low annual incidence. Thompson and Evans[23] detected seroconversion in a 1963 survey of conservation workers in northern Wisconsin. In a second study, Thompson and Gundersen[24] collected sera from 233 persons in 1972 and 1973 in the La Crosse, Wisc., area and again collected sera from some of these earlier volunteers as well as new recruits five seasons later in 1977. Four persons were found with antibody in 1977: two with antibody to LAC and two with antibody to JC viruses. However, only one of the four persons acquired infection in the La Crosse area. Rowley et al.[119] surveyed the grade school population in Allamakee County, Iowa, in the summers of 1976 and 1977 for antibody to LAC and TVT viruses. They noted an annual incidence of 2.2/100 individuals in 1976 and 3.3/100 individuals in 1977 for LAC and TVT viruses combined. One other small prospective study currently in progress has revealed an annual incidence of infection of 1.0% with JC virus of permanent residents of a Cape Cod-area island. Interestingly, the year following the removal of the deer herd from the island, the annual incidence increased to 8%.[120] Sera will be collected at that site for several additional years to determine the incidence trend resulting from the removal of the primary natural vertebrate host of JC virus on the island.

C. Seasonal Distribution

As might be expected, cases of CAL group infection have had onset only during the period of active mosquito transmission. However, specific seasonal distribution patterns are evident in several diseases. In Europe, Aspock et al.[121] reported that TAH virus circulated only from mid-June through mid-September, however, that period could be much shorter in years when the populations of the primary vector were low. Bardos et al.[122] indicated that isolations of TAH virus from the blood of infected children were most likely during the main period of mosquito activity — August and September — in Czechoslovakia. Canadian workers have noted a variation in the seasonal transmission of SSH virus. Of ten human cases reported by Artsob,[117] four had onset in June, three had onset in July, and three had onset in August. Mahdy et al.[98] additionally reported the May onset of three cases in residents of southern Ontario; elsewhere in Ontario, cases have had onset in June and July.[123] Bellonick et al.[124] noted that SSH virus circulated before mid-July in Quebec; however, on Newfoundland, Mokry et al.[125] reported the greatest virus activity later in the summer. McLean[126] has made extensive collections of northern mosquitoes for years and isolated numerous strains of SSH virus from early spring into the summer, primarily in the Yukon. LAC virus isolations frequently have been made shortly after *Ae. triseriatus* adults have emerged in the spring,[127,128] as well as from transovarially infected larvae.[129] Clark et al.[50] noted that the earliest date of onset of a LAC virus encephalitis case in Illinois was May 30, 1979. However, in all areas, human infections peak in July through September,[34] as shown in Figure 1. No LAC virus encephalitis cases have had onset in the December through April period. In Ohio, Berry et al.[41] have noted a seasonal distribution very similar to that shown in Figure 1, with a peak in late August to early September. In contrast to the single peak of LAC virus case distribution, the seasonal occurrence of JC virus encephalitis cases shows a bimodal peak (Figure 1) which may reflect the seasonal activity of two separate vectors (or groups of vectors).

D. Risk Factors

Virtually nothing is known of risk factors for CAL group virus infections with regard to race, other genetic factors, nutrition, immunological background, and so forth. Clark et al.[50] noted that all LAC virus encephalitis cases reported in Illinois were in Caucasian children; however, Kappus et al.[130] have documented the occurrence of LAC cases on a North Carolina Indian reservation. Data are available for age, sex, and, to some extent, occupation. Since CAL group virus infections occur endemically in the human population, antibody prevalence increases with the age of the population.[38] Clinical illness resulting from infection with LAC virus is seen most often in the 1- to 12-year age group;[34,41] the mean age of children with clinical cases in Ohio is 8.8 years.[41] With JC virus clinical infections, the mean age is 26.6 years (mean of 39 cases) and ranges from 8 to 87 years.[43] TAH virus infections are commonly seen in children.[22] There is insufficient data available to determine the age at risk for other CAL group viruses.

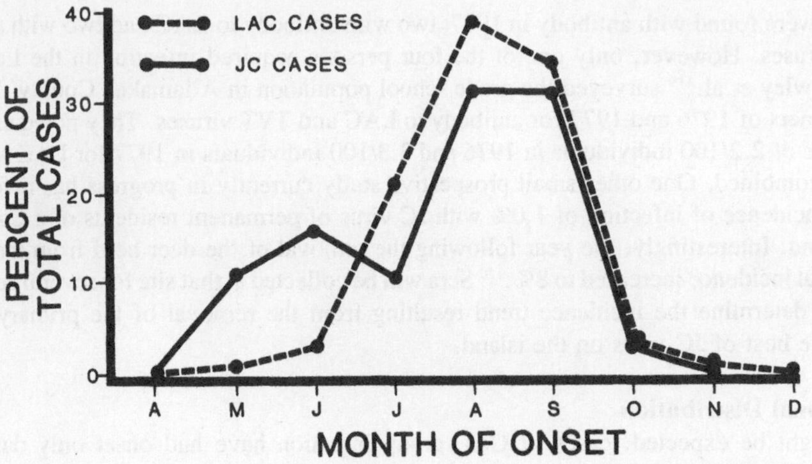

FIGURE 1. Percent of total La Crosse (LAC) virus encephalitis cases reported to the Centers for Disease Control between 1963 and 1981 by month of onset (N = 806),[34] compared with the percent of total Jamestown Canyon (JC) virus encephalitis cases documented to date by month of onset (N = 39).[7,43-45,102] A number of additional JC central nervous system infections are being investigated in New York, however, all have onset in the August to September period.[102]

With both LAC and JC, more males than females have been reported with clinical cases. Kappus et al.[34] noted that 64% of LAC virus encephalitis cases reported through 1981 were in males. The sex ratio of 39 documented JC virus cases to date shows a nonsignificant 1.2:1 male:female ratio;[7,43-45,102] however, Grimstad et al.,[45] in a serological survey using premarital bloods where a 1:1 sex ratio occurred in the sample population, reported a male:female ratio of 1.8:1 ($p < 0.01$) for those with N antibody to JC virus. Danielova and Marhoul[131] noted 62% males vs. 55% females infected by TAH virus. There is insufficient data on other viruses to suggest a sex ratio difference.

Persons working outdoors would be expected to exhibit a higher prevalence of antibody to CAL group viruses. An early survey by Thompson and Evans[23] found that 34% of conservation workers in Wisconsin had antibody to CAL group viruses. Zarnke et al.[132] noted the high prevalence of antibody (54% to JC virus and 42% to SSH virus) in Alaskan Eskimo reindeer herders. Grimstad et al.[45] reported more than twice the antibody prevalence rate to JC virus in residents of Michigan's Upper Peninsula (42%) vs. residents of the Lower Peninsula (19.8%) which was attributed to the higher level of outdoor activity enjoyed by the residents of the Upper Peninsula. However, studies in LAC virus foci in Minnesota (noted below) demonstrate the considerable variation present in residents of even single communities, such as Zumbra Ridge,[52] and mere residence in a focal area may result in a greater risk than for the surrounding population. Focality of CAL group viruses undoubtedly accounts for much of the variation in antibody prevalence and disease occurrence documented to date.

E. Serologic Epidemiology

Seropositive rates for CAL group viruses of <1 to >90% have been reported for residents of focal areas. Henderson and Coleman[1] listed numerous serological studies conducted between 1945 and 1968 in North America. In those early serosurveys, a variety of test antigens (often a single antigen, e.g., BSF-283) and assay systems (HI, CF, N) were used and it is difficult, if not impossible, to tell which etiological agent(s) was the cause of past human infection in each case. Generally, antibody prevalence rates <20% in select populations were seen in these early studies. Some surveys probably did not detect past LAC virus infections (e.g., survey of Seminole Indians in Florida with a 33 to 40% antibody prevalence, probably to KEY/TVT viruses),[1] however, other surveys in the north central and northeastern states probably detected primarily LAC virus infections.

Human infection with TAH virus in central Europe commonly occurs but at various rates. Bardos[133] noted N antibody prevalence rates of 2 to 30% in Czechoslovakian residents; the higher antibody prevalence rates were associated with increased populations of *Ae. vexans*, the primary vector of TAH virus in that country. In another survey, Bardos et al.[134] surveyed residents of four villages in an enzootic area and reported that persons 1 year or less in age had a 22% antibody prevalence and those under 5 years had a 50% prevalence; those approximately 15 years old had an 85% prevalence and older persons approached 96%. Heinz et al.[135] recorded antibody prevalence rates from 6.5 to 21%. In general, antibody prevalence to TAH virus is highest in central European residents;[136] residents of Albania, Italy, and Yugoslavia had lower antibody prevalence rates.

The first serological survey for CAL group virus antibody was conducted by Hammon and Reeves[20] who noted an 11% prevalence in 188 convalescent sera from patients admitted to the Kern County (California) General Hospital. Eleven years later, Gresikova et al.[137] reported that 37% (44/118) of workers in rural agricultural areas of Kern County, Calif., had CE virus antibody. However, a 1969 survey failed to find any rural or urban Kern County children with antibody.[138]

Subsequent surveys in North America focused on LAC virus encephalitis, particularly in the upper Midwest. Three studies are of particular interest. Monath et al.[85] surveyed the rural and urban populations of Winona, Minn., and found that 27% (158/590) of the rural population had N antibody to LAC virus. This rate dropped to 15.3% (118/770) among the urban residents. Thompson and Gundersen[24] serologically analyzed 233 sera from residents of La Crosse, Wis. They reported that 36 sera (15.4%) had antibody to CAL group viruses, including 15 (6.4% of total sample population) with specific antibody to LAC virus. Balfour et al.[52] sampled 79 residents of Zumbra Ridge, Minn. (southwest of Minneapolis), for N and HI antibody to LAC virus. Overall, 29% (23/79) of sera from Zumbra Ridge residents had CAL group antibody, primarily to LAC virus. Balfour and associates found that 31.4% (11/35) of residents of the northwest sector of Zumbra Ridge had antibody to LAC virus compared to only 6.8% (3/44) of the residents of the remaining Zumbra Ridge sectors. In addition, they surveyed 533 area schoolchildren and found that 27 (5.1%) had HI antibody to LAC virus. Their study clearly documented the extreme focality of LAC virus — probably not unlike other CAL group viruses. The largest serosurvey for CAL group viruses was conducted in Indiana in 1978—1979.[38] Of more than 10,000 residents of that state tested for N antibody, 2.3% (239/10,194) had LAC virus antibody. Grimstad et al.[38] also used a computer graphics program to map the distribution of persons with antibody to LAC virus and found a close correlation with location of reported cases. Residents of the southeastern area of Indiana, where the first outbreak was reported in 1964,[1,34] had antibody prevalence rates exceeding 10%.[38]

Numerous serosurveys for LAC virus have uncovered past infections with JC virus. Thompson and Gundersen[24] found 5.2% (12/233) of sera from La Crosse residents with antibody to JC virus. In Zumbra Ridge, Balfour et al.[52] detected 4 of the 79 residents (5.1%) with antibody to JC virus. Monath et al.[85] also reported the same rate (5.1%) in Winona residents. In Alaska, Zarnke et al.[132] found 54% of 121 Eskimo reindeer herders with antibody; theirs is the highest rate of antibody to this virus reported to date for a small select population. In surveys for antibody to SSH virus in Canada, Mokry et al.[125] found 6.5% (43/660) of Newfoundland residents with antibody to JC virus. Deibel et al.[43] reported that 3.5 to 10% of New York residents less than 16 years of age had antibody to JC virus; this rate rose to 8 to 12.9% in those more than 16 years of age. They further documented antibody prevalence in southwest Wisconsin residents: 2.5% in those children less than 15 years of age and 10% in those 15 or older compared to a 5% antibody prevalence in those under 15 and 10% in those 15 or older in the northern two thirds of Wisconsin where JC virus predominates. Watts et al.[139] tested the sera of residents of the Delmarva Peninsula and found 12% (3/25) with N antibody to JC virus. The statewide Indiana survey conducted by Grimstad et al.[38] also detected persons with N antibody to JC virus, but at rates lower than those found for LAC virus. The 0.6% statewide estimate for antibody to JC virus derived from that study has now been determined to have been too conservative since only LAC virus was used to initially screen sera for CAL group antibodies.[38] More than 3500 of the original 10,200 sera have been rescreened and a statewide antibody prevalence estimate of 2 to 3% is more accurate; residents of northern Indiana counties have antibody prevalence rates ranging from 3 to 15%.[46] A second statewide survey, that of the Michigan population, utilized 780 premarital sera.[45] There, Grimstad and associates reported that 19.8% (100/505) of the sera from Lower Peninsula residents and 41.8% (116/275) of the sera from Upper Peninsula residents had specific N antibody to JC virus. The statewide average was 27.7% (216/780). Of importance is their finding that none of the sera had specific N antibody to LAC virus. Earlier, 14 cases of CE had been reported to the CDC by the Michigan Department of Public Health (Table 2)[34] and were presumed to be LAC virus infections. The counties of residence of those 14 persons were in areas of Michigan where Grimstad et al.[45] found residents had a high prevalence of antibody to JC virus, but no persons with LAC virus N antibody. Those workers suggest that the 14 reported cases probably represent the few sera that reacted in the HI tests conducted by the Michigan diagnostic laboratory using LAC virus and thus represent approximately 1.3% of the estimated 1075 JC virus encephalitis cases that might have been reported out of Michigan if a N test with JC virus had been used in place of the HI with LAC virus. As in the Indiana survey,[38] computer-generated maps were used to show the geographic distribution of persons with antibody to JC virus. A close correlation was seen when comparing the distribution of persons with antibody to JC virus and the distribution of the deer population in the Lower Peninsula.[38]

VI. TRANSMISSION CYCLES

Basic transmission cycles have been elucidated for seven CAL group viruses: CE, JC, KEY, LAC, SSH, TAH, and TVT. Each of these seven cycles apparently is maintained (overwinters) principally, or to an important extent, in the field by transovarial (vertical) transmission. A number of recent reviews have provided excellent summaries of these CAL group transmission cycles.[3,30,31] Space permits only a brief outline of the basic epidemiological concepts associated with each cycle which the most recent literature has provided.

California encephalitis virus — Repeated isolation of CE virus from *Ae. melanimon* in California,[138] from *Ae. dorsalis* in Utah,[140] and the demonstrated ability of both species to vertically transmit CE virus[141] suggest that these are the primary and secondary vectors in the western U.S. CE virus also has been isolated from other *Aedes* as well as from species of *Anopheles, Culex, Culiseta*, and *Psorophora*.[30,116] These latter species are probably of no importance in the natural cycle and represent chance "contamination" in the field as discussed previously. The two Aedine vectors readily feed on a number of small vertebrates (Table 4); however, the importance, or necessity, of these hosts in the amplification and/or dissemination of CE virus in the field is unknown.

Jamestown Canyon virus — In the western U.S., JC virus has been associated exclusively with *Cs. inornata* mosquitoes; Turell and LeDuc[30] suggested earlier that this was the primary vector of JC virus. Elsewhere in the U.S., this species appears to be much less important, however, compared to the *Ae. communis* group and *Ae. stimulans* mosquitoes. In New York, JC virus was isolated almost every year from 1972 to 1980, most often from the *Ae. communis* group.[142] This mosquito group also had the lowest minimum field infection rate (MFIR) of

13 species in the New York study: 1:1274. These workers also noted "a dramatic increase in prevalence of JC virus from 1978 to 1980, presumably as a result of unseasonably mild winters that fostered a population explosion in the white-tailed deer population".[142] In northern Indiana where the *Ae. communis* group is rare, *Ae. stimulans* has been shown to be the primary vector of JC virus through laboratory transmission trials, through field isolation studies (which provided evidence of vertical transmission of JC virus by *Ae. stimulans*), and through vector-host relationship studies.[143] A MFIR of 1:755 for JC virus in *Ae. stimulans* was noted in those Indiana studies. In Ohio, Berry et al.[29] isolated JC virus from eggs of *Ae. triseriatus* collected in the field, and their subsequent studies have yielded additional isolations of JC virus from adult *Ae. triseriatus* reared from field-collected eggs; these eggs were collected from the field as late as November.[144] Their work suggests that *Ae. triseriatus* may be important in the maintenance of JC virus in Ohio. In laboratory trials with geographic strains of *Ae. triseriatus* ingesting as much as 5.0 $\log_{10}$/0.025 mℓ of JC virus in defibrinated rabbit blood, no infected mosquitoes have been found.[46] The results of these two studies suggest the possibility that marked variation exists in the susceptibility of *Ae. triseriatus* for JC virus and that specific populations may be susceptible, while others are completely refractory to infection. Nasci[145,146] has shown the high prevalence of deer blood feeding in midwestern populations of *Ae. triseriatus*; similar results have been shown in Wisconsin by Burkot and DeFoliart.[147] In addition, there is growing circumstantial evidence that Anophelines may be involved in the transmission and maintenance of JC virus. Grimstad et al.[148] have detected seroconversion to JC virus in penned deer in the early spring at a time when no potential vectors other than *Anopheles punctipennis* were present. Isolations of JC virus from *An. punctipennis* and *An. quadrimaculatus* in Ohio show MFIR of 1:128 to 1:733 and 1:274, respectively.[144] These isolates have come primarily in the late summer (September) presumably as these species are entering overwintering sites. The bimodal onset (Figure 1) of JC virus cases may represent early season transmission by *Ae. stimulans*, the *Ae. communis* group, and perhaps other early spring *Aedes* mosquitoes, and late season transmission by *Ae. triseriatus* and/or *An. punctipennis* or *An. quadrimaculatus*. A number of studies have shown clearly that the white-tailed deer is the primary vertebrate host for JC virus.[139,149-152] Other regional hosts probably include moose and elk,[153,154] mule deer,[118] and antelope.[155] Issel[156] suggested that maternal antibody in the colostrum protected newborn fawns from a primary infection for their first summer of life.[156] This hypothesized protection in a natural focus of JC virus has been substantiated by Grimstad et al.[148] using a penned deer herd in north central Michigan. Given its vast geographic distribution, the variety of mosquito species that probably serve as regional vectors, and the number of large wild ungulates that apparently serve as hosts, JC virus is the most epidemiologically complex of the CAL group viruses.

Keystone virus — The numerous elegant studies conducted by LeDuc, Watts, and associates have contributed most of what is known about the natural cycle of KEY virus. The quantitative model of KEY transmission developed by Fine and LeDuc,[157] provides an excellent basis for similar modeling of other CAL group virus cycles. On the eastern seaboard of the U.S., *Ae. atlanticus*, a floodwater mosquito, is the primary vector of KEY virus.[3] Numerous isolates have come from this species, and seroconversion in primary vertebrate hosts and sentinels (Table 4) have been detected coincident with the emergence of *Ae. atlanticus* adults. Field evidence suggests that vertically infected *Ae. atlanticus* adults transmit KEY virus with the initial bloodfeeding.[3] In Florida, a separate cycle (transmission of KEY virus by *Ae. infirmatus* to cotton rats and cottontail rabbits) has been proposed.[3] However, *Ae. atlanticus* also ranges throughout this region,[158] and its presence may greatly diminish the role of *Ae. infirmatus* in the southern transmission of KEY virus. *Ae. atlanticus* readily feeds on white-tailed deer, and KEY virus-infected deer develop N antibody without a viremia.[151] Watts et al.[152] have shown that marked cross protection occurs as the result of sequential JC-KEY or KEY-JC virus infections. On the Delmarva Peninsula, both JC and

KEY viruses circulate; the overall effect of this cross protection on the circulation of these two serotypes, particularly of JC virus in deer, poses an interesting question.

La Crosse virus — The basic transmission cycle and maintenance mechanism is well documented for LAC virus and is better understood than any of the other CAL group viruses.[31] LAC virus is transmitted both horizontally to vertebrate hosts (Table 4)[60,159-161] and vertically by *Ae. triseriatus*.[128,162,163] This mosquito is a container-breeder whose range encompasses the eastern half of the U.S.[158] The eastern variant ("C") of LAC virus[57] is also readily transmitted by *Ae. canadensis*,[32] presumably from eastern Ohio through New York. Frequent passage of LAC virus in the field through this latter Aedine mosquito may have led to the evolution of this variant. Gray squirrels and eastern chipmunks are important blood meal sources for *Ae. triseriatus*[145-147] and serve as the primary amplifying hosts for LAC virus (Table 4) in the natural cycle[159,164,165] despite their defensive (antimosquito) behavior that impacts on the feeding success of older *Ae. triseriatus*.[166,167] Nasci[145] has shown through blood meal analyses that *Ae. triseriatus* will feed readily on canine hosts.[145] This is of considerable importance since Yuill and students[60] have demonstrated the potential role of red foxes in disseminating LAC virus in the upper Midwest. Infection of foxes can occur directly by ingestion of viremic hosts (e.g., chipmunks), as well as by mosquito transmission.[60] Recent studies have shown that *Ae. triseriatus* also utilizes white-tailed deer as a major source of blood, and in some areas deer constitute the principal source of blood.[145-147] However, these vertebrates develop only a low-titered transient viremia, and probably play no role in amplification and dissemination of LAC virus. LAC virus is vertically transmitted in the eggs of *Ae. triseriatus*.[168] Once established in a focal woodlot, LAC virus can persist for years and annually be isolated from the same treehole.[127] A second mechanism, that of venereal transmission, has been shown uniquely in this cycle. Vertically infected male *Ae. triseriatus* can transmit LAC virus in semen to uninfected females;[169] the females infected in this manner can orally and vertically transmit LAC virus in the laboratory[168] and, presumably, in the field. The importance of vertical transmission to progeny is well established; however, the role of venereal transmission in the natural cycle remains to be ascertained. Finally, recent studies have confirmed that a sibling species, *Ae. hendersoni*, can transmit LAC virus in the laboratory and presumably in the field.[170] This species is quite abundant in LAC virus foci throughout the upper Midwest at population levels far beyond what has previously been believed.[171] Difficulty in separating *Ae. triseriatus* from *Ae. hendersoni*, especially in the adult stage,[172] precludes accurate separation without the use of electrophoresis on every single individual mosquito.[146] The distinct probability exists that some past isolations of LAC virus from what were identified as *Ae. triseriatus* may actually have been *Ae. hendersoni*; transmission in the field to vertebrate hosts by *Ae. hendersoni* may also occur.[146]

Snowshoe hare virus — The harsh environment of many parts of Canada and the limited summer season requires that indigenous arbovirus transmission be very efficient, and this is true for SSH virus. LeDuc[3] reviewed the evidence for three possible overwintering mechanisms for SSH virus which when combined would provide an efficient means of assuring virus survival. In Canada, larval univoltine *Aedes* species, primarily *Ae. implicatus*, and other *Aedes (canadensis, communis, hexodontus, nigripes)* have yielded SSH virus isolates.[126,173,174] These isolations suggest that vertical transmission by one or more of these species assures annual transmission and is the primary maintenance mechanism. Other studies have provided a second mechanism involving *Cs. inornata* — overwintering in adult mosquitoes. This mosquito is capable of long-term maintenance (138 to 194 days) of SSH virus at temperatures of 0°C in the laboratory, and it will also readily transmit SSH virus in the laboratory.[126] McLean et al.[175] reported spring isolations of SSH virus from *Cs. inornata* which document the early season initiation of the cycle in snowshoe hares and perhaps *Citellus* ground squirrels. Laboratory studies suggest that *Ae. communis* may be the primary

Aedine vector.[175] Finally, the possible persistence of SSH virus in small hibernating arctic mammals such as voles, lemmings, and other species has been hypothesized as a third overwintering mechanism.[3]

Tahyna virus — In Europe, TAH virus appears to also have multiple overwintering mechanisms. The primary vector, *Ae. vexans*, has been shown to both horizontally and vertically transmit TAH virus.[176-179] Horizontal transmission is additionally accomplished by *Ae. cinereus*,[180] *Ae. caspius*,[181] and perhaps other *Aedes* species. Transmission to the primary vertebrate hosts, European hares and rabbits, as well as to other small- to moderate-sized vertebrates, is readily accomplished in the field.[182-184] TAH virus has been recovered from larval *Culiseta annulata*.[185] This species has also been shown in the laboratory to maintain TAH virus for extended periods in a simulated hibernation;[186] it can readily transmit to primates.[187] However, *Cs. annulata* is relatively refractory to infection in the laboratory,[188] and may become infected in the field only after feeding on lagomorphs or other vertebrates circulating high titers of virus. Other workers have suggested that TAH virus may overwinter in *Culex modestus* adults.[189] In a very interesting study, European hedgehogs were infected with TAH virus, then placed into hibernation for up to 140 days; subsequent to their awakening, some of the hedgehogs developed a viremia sufficient to infect feeding mosquitoes.[190]

Trivittatus virus — The natural cycle of TVT virus is one of the simplest in the CAL group. *Ae. trivittatus* mosquitoes transmit TVT virus vertically to their progeny,[191,192] and horizontally to cottontail rabbits throughout the summer in the north central U.S.[193-195] This mosquito readily feeds on a variety of hosts,[195] including deer,[149] however, only cottontail rabbits are apparently involved in the natural cycle.[119] *Ae. trivittatus* is absent in Florida where TVT virus is known;[158] however, *Ae. infirmatus* has yielded numerous isolates of TVT virus there and appears to be the primary vector in Florida.[196] *Ae. infirmatus* also readily feeds on rabbits.[3] In the southern foci, LeDuc[3] suggests that warm temperatures may permit year-round transmission of TVT virus.[3]

Little is known of the natural transmission cycles of the other viruses. INK virus may be similar to JC virus in that Aedine mosquitoes apparently transmit this serotype to large ungulates (reindeer and moose) in Finland.[19,56] While antibody to SA virus has been found in white-tailed deer, subcutaneous (s.c.) inoculation of deer with virus failed to cause a viremia.[151] Alternative hosts have not been identified.

VII. ECOLOGICAL DYNAMICS

The vast number of components interacting in any disease cycle include both internal and external interactions as they relate to the virus, vector, and host; Grimstad[26] has related these interactions to the concept of the ecological "niche" as a way to better understand arbovirus disease transmission.[26] The following topics relate primarily to the external interactions operating in select CAL group virus cycles.

A. Macro- and Microenvironments

Most of the mosquito species vectoring CAL group viruses breed in ground pools that may be transient to semipermanent (e.g., *Aedes, Culiseta*) or in permanent waters (e.g., *Anopheles*).[197] Within these aquatic microenvironments or habitats, all nutrients needed to reach the adult stage must be obtained. In many vernal pools and permanent water habitats, nutrients may not be a limiting factor; in others, nutrients may have a considerable impact on larval development. In the container (treehole, tire) microenvironment where *Ae. triseriatus* larvae develop, nutrients are most often very limited, with nutrient input coming primarily during the autumn leaf drop.[198] The end result is that adults of reduced size emerge from these containers,[199,200] having an increased vectorial capacity for LAC virus.[201] Emerging adults enter a macroenvironment where temperature, and to a lesser extent relative humidity, can largely determine longevity. This is true not only for CAL group vectors, but also for vectors of most other arboviruses as well. The environmental temperature will also determine the length of the extrinsic incubation period (EIP: the period of time from when a vector ingests virus in a blood meal until it is able to orally transmit) for orally infected mosquitoes.[202] This EIP is approximately 18 to 21 days at 21°C for *Ae. triseriatus* — essentially the mean ambient temperature these adults experience in the upper midwestern forest habitats.[31]

B. Climate and Weather

Climate and weather are important considerations because they limit in part the geographic distribution of both vertebrate hosts and arthropod vectors in the CAL group virus cycles. The cold temperature climate of winter in Europe and North America necessitates efficient overwintering mechanisms. The CAL group viruses have solved this problem through vertical transmission in the diapausing eggs of the vectors. *Aedes* mosquitoes, with their drought-resistant eggs that can readily survive several dry seasons when larval habitats are non-existent,[197] provide the primary survival mechanism for CAL group viruses.

C. Vector Density, Fecundity, Oviposition, Longevity, and Body Size

Density of CAL group vectors is related in part to the type of breeding site, but is ultimately dependent on genetically regulated fecundity, spontaneous movement, and oviposition.[26] Most CAL group *Aedes* vectors breed in temporary aquatic habitats that often have virtually disappeared by the time of oviposition. There, ovipositing females seek out the moist substrate of these drying habitats which are often under the vegetative canopy, and lay eggs among loose layers of detritus.[197,203] Species breeding in snow-melt and/or floodwater pools often reach enormous population densities and are of medical importance as pest species in addition to vectoring CAL group viruses. For example, if *Ae. dorsalis* has more than one summer generation, populations will build to very large numbers by mid to late summer.[204] *Ae. stimulans* develop in huge numbers in many parts of their range; in southern Ontario, it is the worst pest species in early spring before *Ae. vexans* emerges there.[204] *Ae. vexans* is considered to be the principal pest species in the U.S.[203] Its relative abundance is influenced by spring floods in the flood plains and summer rains that flood other sites; variation in local flooding produces considerable variation in *Ae. vexans* density.[203] In Europe, TAH virus transmission is assured by the summer-long occurrence of its primary vector, *Ae. vexans*, of which several generations each year are normal.[203] In contrast, *Ae. triseriatus*, which breeds in treeholes, tires, and other water-impounding containers, is generally found in small discrete populations. In a 10.1-ha Indiana woodlot, Sinsko and Craig[205] determined that approximately 1200 females were present in late August. The oviposition of this species results in a vertical distribution of eggs; most are placed in basal-level treeholes, however, a portion of the population's eggs are laid in canopy-level treeholes.[206,207]

From an evolutionary standpoint, a relatively high vertical LAC virus transmission rate (0.5% based on field isolation rates) reported for *Ae. triseriatus* populations in Wisconsin,[31] is probably essential given the low density at which this vector exists. In contrast, stabilized infections would persist at much lower levels in high-density field populations (e.g., a peak field infection rate of 0.081% for *Ae. melanimon*,[208] based on field isolation rates in California).

Fecundity in *Ae. triseriatus* (and other *Aedes* species) is affected by host blood sources and the duration of the gonotropic cycle;[26,209] in turn, the duration of the gonotropic cycle is determined in part by temperature and host blood sources.[26,209] Fecundity is also affected by larval nutrition since nutritionally deprived larvae give rise to smaller adults with a reduced fecundity. For example, Patrican and DeFoliart[210] found that nutritionally stressed

(SMALL) *Ae. triseriatus* oviposited significantly fewer eggs (mean of 48.9) compared to a mean of 73.3 eggs for LARGE females not nutritionally stressed in the larval stages.

Longevity is dependent in part on environmental conditions, especially temperature, and perhaps on larval nutrition. While the effect of larval nutrition on longevity does not show up in *Ae. triseriatus* through the second oviposition in the laboratory,[114] McCombs[199] has shown in the laboratory that adult survival rates decline with declining adult size, thus significantly affecting the population biology of *Ae. triseriatus*.[200] Daily survivorship rates of *Ae. triseriatus* are approximately 0.87,[205] although a number of other studies have shown slight variations on this figure.

In addition to affecting fecundity, body size of *Ae. triseriatus* (as influenced by larval nutrition) specifically affects vectorial capacity and vector competence in general. For example, Grimstad and Haramis[201] showed a significant difference in rates of LAC virus transmission to suckling mice by orally infected SMALL females (82% transmission) compared to LARGE females (52% transmission). Patrican and DeFoliart[210] reported similar results for vertically infected *Ae. triseriatus* (68% for SMALL vs. 53% for LARGE).

D. Host Preferences and Biting Activity

The broad host preference of woodland *Aedes* was alluded to earlier. In Wisconsin, Wright and DeFoliart[33] observed that "most of the *Aedes* mosquitoes appeared to attack the most available host in an area".[33] Their studies showed that *Ae. canadensis* fed on a variety of hosts from leopard frogs to white-tailed deer. *Ae. trivittatus* was the most opportunistic feeder, in both Wisconsin[33] and Indiana,[211] while *Ae. abserratus-punctor* (a possible vector of INK virus) had a limited host range — red foxes, raccoons, and white-tailed deer for the most part.[33] DeFoliart[31] has pointed out that feeding on nonamplifier species is "wasted" with regard to the perpetuation of CAL group viruses by these opportunistic feeders. Use of precipitin tests for blood meal identification has been a key to establishing the host-vector relationships for the CAL group.[212] A new technique even permits studies of the age of hosts using the binding of estradiol to mosquito blood meals.[213]

The biting activity of CAL group vectors on human hosts ranges from aggressive and persistent to shy reluctance. For example, *Ae. stimulans* exhibits maximum activity during the first hour after sunset, but it will feed during the day in shady wooded areas.[197] *Ae. trivittatus* is an aggressive feeder day or night;[197] the same is true of *Ae. vexans*, and both are thus important pest species in North America and the latter also in parts of Europe.[203] In contrast, *Ae. triseriatus* is easily disturbed and is much less aggressive. However, field observations suggest that tire-bred populations of *Ae. triseriatus* are more aggressive in feeding on human hosts than their treehole-bred counterparts.[214] This is important because of the high frequency of tire breeding associated with LAC virus encephalitis cases in Ohio (discussed below with regard to control).

E. Vertebrate Host Density and Immunological Background

In Michigan, the neutralizing antibody prevalence to JC virus in the human population is closely correlated with the population density of the white-tailed deer population;[45] the same is true in the northern third of Indiana.[46] The 1961 to 1963 serosurvey of Wisconsin wildlife workers conducted by Thompson et al.[215] determined that the geographic distribution of wildlife workers with CAL group antibodies fit the distribution of deer with similar antibody — now determined to have been JC virus antibody. This area of Wisconsin also has a higher deer density compared to the southwestern area where LAC virus predominates. The field study of Gauld et al.[165] on the dynamics of chipmunk populations in which LAC virus was circulating suggests that the level of virus is related both to the density of the chipmunk and the *Ae. triseriatus* populations. Their studies also demonstrated the importance of nonimmune hosts in the horizontal transmission of LAC virus. The spring cohort of juvenile chipmunks lost maternal antibody and became susceptible to a primary LAC virus infection during the summer amplification of that virus. The late summer cohort lost maternal antibody by the late fall and provided a susceptible host cohort the following spring as vertically infected *Ae. triseriatus* adults emerged and began feeding.[165] Presence of these susceptible cohorts may be important from another standpoint. Thompson showed that female *Ae. triseriatus* fed on immune chipmunks, then venereally infected, had lower oral transmission rates compared to those females fed on nonimmune chipmunks.[216] As noted earlier, the protection afforded newborn white-tailed deer fawns by the maternal antibody in colostrum assures a large susceptible cohort of 1-year-old deer the following spring as JC virus-infected *Aedes*, and perhaps overwintering *Anopheles*, emerge and amplification begins.[148,156] Yuill[60] noted that in areas of Wisconsin where there is a high trapping pressure exerted on the red fox population, 80% of the free-ranging animals are juveniles which then constitute a population highly susceptible to LAC virus. These rapid turnovers provide for vertebrate populations whose immunological background permits substantial horizontal transmission and virus amplification in any of the CAL group cycles.

F. Movements and Migrations of Vectors and Hosts

The variable movement of vectors can in part determine the density and distribution of foci and establish the size of the vertebrate population at risk. *Ae. stimulans* females tend to stay near larval sites with minimal dispersion.[197] In one northern Indiana JC virus focus, Boromisa and Grimstad[28] isolated JC virus from *Ae. stimulans* in only 1 of 5 study sites within the 2000-ha Kingsbury State Fish and Wildlife Area which were sampled weekly in the summers of 1982 to 1984. Their results and the limited movement of this vector both suggest that where JC virus is transmitted primarily by *Ae. stimulans*, JC virus foci might be very unevenly distributed; the potential role of white-tailed deer in disseminating JC virus is of considerable interest in those areas. *Ae. trivittatus* also tends to stay by larval sites, however, a few females migrate short distances (up to 2.5 km).[197] *Ae. triseriatus* females tend to stay within woodlots. Their movements between woodlots are quite limited, however; interwoodlot movement in the upper Midwest appears to be along fence rows more than across open areas.[217] Other studies have shown that considerable *Ae. triseriatus* movement occurs within woodlots. Fluorescent dust-marked females released at one point in a 10.1-ha woodlot in northern Indiana were found to have dispersed evenly 24 hr later.[205] Of the CAL group vectors, *Ae. vexans* is perhaps the best known for its migrations. In summarizing the bionomics of that species, Horsfall[197] presented vivid descriptions of vast clouds of *Ae. vexans* approaching Chicago in 1949 and 1951. He noted that *Ae. vexans* females commonly disperse more than 20 km from their breeding sites, at altitudes as high as 1500 m, and some disperse up to 48 km. One marked female was recovered 22.4 km from its release point 24 hr later. Migratory flights also occur over open water. Thus, dispersion of TAH virus in Europe could readily occur via mosquito flight in contrast to the probable dispersion of other CAL group viruses by one or more of their vertebrate hosts. *Ae. dorsalis* also migrates considerable distances, sometimes more than 30 km.[197]

Most vertebrate hosts in CAL group cycles show limited movements. Gauld et al.[165] indicated that chipmunks are territorial and tend to remain within their limited (0.3 to 0.5 ha) territories. Most lagomorphs involved in the CAL group cycles also have rather limited dispersion. White-tailed deer, despite their size, generally range over a few square miles in North America,[218] although some bucks have been known to move more than 80 km in one season. On the other hand, red foxes move over considerable areas. Dispersion over 3 to 10 km within 24 hr and up to 65 km in 3 days apparently occurs.[60]

G. Human Element in CAL Group Disease Ecology

The emerging problem of controlling *Ae. triseriatus* breeding in man-made containers as discussed below is a direct result of our throw-away society and disregard for the environment. Suburbanization of many wooded areas of the eastern and central U.S. has brought children directly into contact with LAC virus foci literally in their own backyards. In Peoria, La Crosse, Ft. Wayne, and Zumbra Ridge (Section II.C), most LAC virus infections have occurred in suburban children. Agricultural practices and political pressure from hunting interests on state departments of natural resources have led to an explosion of the white-tailed deer population throughout much of the U.S., bringing with the expanding herd an apparent increase in human JC virus infections in the upper Midwest and perhaps elsewhere.[26,45] Increased outdoor activity as noted earlier also contributes globally to an increased probability of contact with CAL group virus-infected vectors.

H. Vector Competence

The seven previous subjects all relate to the concept of vector competence with regards to the CAL group. Vector competence in itself is a very broad concept that cannot be summarized even in a few pages. Recent reviews relating to vectors of the CAL group and other arboviruses have explored this concept from different points of view.[26,202,219,220] In presenting other sections of this review, vector competence studies have been used repeatedly as examples. At the risk of repetition, a number of these bear emphasis in this section. The numerous studies by Wisconsin workers from the late 1960s through the present which demonstrate that *Ae. triseriatus* is the primary vector of LAC virus and the existence of vertical transmission of an arbovirus by a mosquito comprise the most comprehensive investigation of a CAL group virus. Most of the examples of vector competence investigations relate to the LAC virus-*Ae. triseriatus* interaction. Other studies showing the vectorial capacity of *Ae. canadensis* for the LAC virus "C" variant,[32] the variation in vectorial capacity by geographic strains of *Ae. triseriatus*,[113] the deleterious effect of LAC virus infections on this vector,[112] the significant effect of larval nutrition on adult vectorial capacity,[201] the vector-host relationships detected through blood meal analyses of field-collected *Ae. triseriatus*,[145,146] and the demonstration of barriers to infection and transmission in *Ae. triseriatus* and its sibling species, *Ae. hendersoni*,[170] all contributed to a better understanding of the LAC virus cycle. Studies by numerous other workers on the vectorial capacity of *Ae. vexans* for TAH virus,[177] *Ae. atlanticus* for KEY virus,[3] *Ae. stimulans* for JC virus,[28] *Ae. dorsalis* and *Ae. melanimon* for CE virus,[141] *Ae. trivittatus* for TVT virus,[191,192] and *Ae. communis* and *Cs. inornata* for SSH virus[126] all have contributed to an understanding of these CAL group virus cycles. Despite these advances, much remains to be investigated. A complete elucidation of the three main components of vector competence[26,221] — vector exposure to the pathogen, susceptibility to the pathogen and multiplication of the pathogen in the vector, and transmission of the pathogen — and the factors associated with each of these three main components are essential to the understanding of any CAL group cycle and serve as a model for future investigations.

VIII. SURVEILLANCE: RATIONALE, METHODS, AND RELATIONSHIPS TO CONTROL

A. Clinical Hosts

Surveillance for CAL group virus infections of humans has been less than enthusiastic in the U.S., given the results of Calisher and Bailey's[107] survey. The emergence of JC virus as a neuropathogen with a continent-wide distribution, and the recognition that CAL group viruses other than just LAC in North America cause CNS illness, may increase awareness and lead to increased surveillance. The anticipated availability of an ELISA in the near future for detection of potentially all CAL group infections will do much to solve the diagnostic problem. Undoubtedly, N tests still will be necessary to type infections, however. Serological surveillance studies cited earlier have done much to increase awareness of past human infection in North America and Europe and have exposed the tremendous size of the "submerged iceberg" of inapparent and undiagnosed infections.

B. Wild and Domestic Vertebrates

Surveillance of select wild and domestic vertebrate populations (Table 4) is very useful for determining the range of serotypes. Use of domestic animals as sentinels in particular is a relatively easy means of surveillance and they have been used extensively in surveillance of other virus diseases.[222] In many areas of the U.S., it is mandatory for deer hunters to bring their kill to check stations so that wildlife biologists might obtain biological data used in herd management. At these times it is easy to remove blood pooled in the body cavity for antibody surveillance; this procedure has been used successfully in CAL group studies.[149,223]

C. Vectors

Since all of the known vectors of CAL group viruses are mosquitoes, and *Aedes* spp. in particular, surveillance of their populations is routine. Service[224] has provided an excellent summary of numerous techniques for studying mosquito bionomics and for monitoring in surveillance programs.[224] Snow-melt and floodwater species tend to emerge synchronously in mass populations in contrast to the staggered emergence of container-breeding mosquitoes in temperate and tropical areas.[197] The literature is replete with references to mosquito surveillance; however, new procedures are constantly being developed. Issues of the *Journal of the American Mosquito Control Association (Mosquito News)* are perhaps the best general source of information on CAL group vector surveillance and control.

D. Sentinels

Table 4 lists the vertebrates that have most often been found with N antibody to the respective viruses.[3,4,19] However, viremia has not been detected in all species in laboratory studies as noted earlier. Various species of domestic animals are important and useful sentinels. With JC virus, antibody prevalence rates in Michigan cattle and horses range from 30 to 60%.[46] Godsey[59] has also found seropositive cattle, horses, goats, swine, and dogs in Wisconsin with antibody prevalence to JC virus ranging from 10 to 29%. However, his experimental studies have shown that these domestic vertebrates do not develop viremia on experimental inoculation.[59] Canadian workers have also documented the sentinel role of domestic animals in SSH virus surveillance.[125,225,226] It must be emphasized that records of CAL group virus antibodies in any animal species are of limited value unless workers perform N tests to determine the specific etiological agents and not simply report a "California virus" infection.

IX. INVESTIGATION OF EPIDEMICS

As noted earlier, CAL group human infections occur endemically throughout their range and rarely are there sufficient cases to constitute an outbreak in any specific area. Exceptions include TAH virus focal areas in Europe, and North American foci of JC and LAC viruses. A major question relating to the latter virus is why there are specific foci of LAC virus in La Crosse, Peoria, Fort Wayne, Gambier, and Zumbra Ridge (Section II.C) and apparently no foci in adjacent suburban areas? It is imperative that the level of undetected CAL group infections caused by JC, TVT, and perhaps KEY in the U.S., and JC and SSH viruses in Canada be investigated. Retrospective surveys to detect N antibody and computer mapping of focal areas with subsequent prospective studies will establish the relative public health importance of these agents, especially with regard to the occurrence of mild clinical "flu-like" illnesses that undoubtedly can contribute to time lost from the workplace. The failure to determine the etiological agent of approximately 80% of primary encephalitis cases reported every year in the U.S. (see annual summaries of *Morbidity Mortality Weekly Report*, USDHHS/CDC, Atlanta) potentially results in many CAL group virus infections going undiagnosed. The same could be said for aseptic meningitis cases. Use of premarital bloods, residual sera from hospital or clinic submissions, and blood from other volunteer programs in Indiana and Michigan led to the statewide mapping of LAC and JC virus foci.[38,45] These data have proved useful to state health officials in long-range planning. The same type of serological surveys and mapping could be accomplished easily anywhere in Europe or North America and at a reasonable cost.

X. PREVENTION AND CONTROL

A. Vector Control

Numerous effective procedures are available for control of snow-melt and floodwater mosquitoes including the CAL group vectors *Ae. communis, Ae. dorsalis, Ae. melanimon, Ae. stimulans, Ae. trivittatus, Ae. vexans,* and other species. These effective procedures are based on judicious use of larvicides and adulticides; mosquito control workers must be ever mindful of the rapid development of resistance that has historically occurred in vector populations. Resistance to the new third-generation insecticides may be inevitable if the substances are misused. While control of vernal pool or permanent water-breeding mosquitoes is relatively successful, a major problem exists with control of container breeders such as *Ae. triseriatus*.[227] The relatively recent movement of *Ae. triseriatus* into tire habitats in the north central states poses a unique control problem.[171,228] Some of the enormous tire dumps in the Midwest containing more than 1 to 5 million used steel-belted tires — all potential breeding containers for *Ae. triseriatus* — challenge the best control techniques. In La Crosse, Wisc., a countywide public education program and response led to a great reduction in the *Ae. triseriatus* population and a most impressive reduction in LAC virus encephalitis cases there.[54] Case reduction in surrounding counties (where no control efforts directed at source reduction were made) has not occurred.[96] As with CAL group snow-melt and floodwater vector species, source reduction for container-breeding mosquitoes is a most effective and environmentally acceptable means of control. Possible biological control of *Ae. triseriatus* through the use of *Toxorhynchites* as larval predators is an alternative under consideration.[227]

B. Control of Vertebrate Hosts

Vertebrate population manipulation is probably not a viable alternative for control of any CAL group virus disease. The removal of deer in an island situation led to a marked increase in the incidence of human infection with JC virus as discussed previously.[120] Deer probably act as a "sponge" for JC virus (and perhaps for KEY and LAC viruses), and other vertebrates (including dead-end domestic animals) possibly function in a similar manner with other CAL group viruses, thus keeping human infection rates lower than they might be otherwise. Anyone who has aspirated host-seeking mosquitoes intent on engorging on a sentinel horse, for example, has probably been amazed at the lack of mosquitoes attempting to feed on the collector, as this author has been in eastern equine encephalomyelitis virus enzootic areas of Indiana and Michigan. The potential for enhancing contact between serotypes with possible reassortants that might result from vertebrate population reduction in any circumstance must be considered.

C. Environmental Modification

Environmental modification as it relates to source reduction for control of specific CAL group snow-melt and floodwater vector mosquitoes may be of use in focal areas. It may be preferential to intense insecticide use from an environmental (and public relations) standpoint as well. Sanitation — removal of water-impounding junk, especially tires, and other artificial containers — is an effective means of source reduction for LAC virus control that is environmentally acceptable and very effective as demonstrated by the La Crosse case.[54] Berry[228] stated that in Ohio, the trash littering the environment that provided *Ae. triseriatus* breeding sites, increased that mosquitoes' populations by a factor of 7.7. His reference to Pogo, the cartoon character of the same name, saying, "We have met the enemy and he is us", is appropriate with regard to this environmentally related breeding problem with *Ae. triseriatus* populations. In Ohio, Peterson has found approximately 11.5% of the *Ae. triseriatus* in treeholes compared to 88.5% in man-made water-holding trash and containers in her epidemiological investigations of LAC virus encephalitis cases.[144,228]

D. Epidemiological Consideration of the Use of Vaccines

Prophylactic use of vaccines for CAL group disease control is not currently justified given the high inapparent to apparent ratios and the very low mortality rate (<1.0%). If future reassortants prove to be highly pathogenic, limited use of vaccines in focal areas might be the last resort for disease control. However, effective vector control programs in those areas coupled with public education programs remain the best means of disease prevention. Use of personal protection in focal areas is readily provided by insect repellants and protective clothing; these measures need to be encouraged whenever possible for disease prevention.

XI. FUTURE RESEARCH

In summarizing the recommendations generated by participants at the International Symposium on California Serogroup Viruses (November 12 to 13, 1982, Cleveland), Reeves[229] presented a long list of unsolved problems and unanswered questions. Subsequent studies have made inroads in answering some questions, however, much remains to be investigated. Reeves[229] stated that "Much of the research that was recommended [in the Symposium] will be of an applied nature and must be done in the field rather than in a laboratory environment". The current deemphasis on field studies in pursuit of molecular solutions in the laboratory will ultimately hinder, rather than promote, a clearer elucidation of the epidemiology of the natural CAL group virus cycles, unless a balance is quickly struck where field investigations are again encouraged and funded. Reeves' emphasis on field studies was echoed by DeFoliart et al.[230] in their review of changing patterns in transmission of mosquitoborne arboviruses, including CAL group members.

A. Field Studies

While much information is available on the basic field bionomics of CAL group vectors,[197,203,204] field investigations relating directly to the involvement of vectors in natural transmission cycles need to be expanded. Seasonal variation in vector populations, improved methods of population measurement, elucidation of vector-host relationships with an increased emphasis on the role of the host (e.g., antimosquito behavior), identification of all primary vectors and determination of the role of secondary vectors in CAL group cycles (e.g., *Ae. hendersoni* for LAC virus), identification of all sentinel hosts (particularly domestic animals to aid in surveillance), and expanded prospective human serological and clinical studies are but a few of the many investigations sorely needed.

B. Experimental Studies

Expanded vector competence studies using reassortants with regional virus strains are needed to show the vast variation in susceptibility and vectorial capacity. All the components of vector competence,[26,202] to say nothing of virus and host competence, have yet to be investigated completely for a single CAL group vector. This is especially true in the area of basic genetic variability and genetic regulation of susceptibility to CAL group viruses. Use of significant numbers of mosquitoes in vector competence studies must be emphasized so that statistically significant differences might be seen and accurate conclusions drawn. The effect of larval nutrition and other extrinsic factors on vector competence, the development of a standard membrane-feeding procedure or artificial viremia in animals, along with rigorous use of a control ''white mosquito'' strain in all laboratory transmission trials,[26] are also just a few of the pressing areas of investigation for CAL group virus studies. Molecular studies certainly have their place; the study of the role of the middle-sized RNA segment of LAC virus in dissemination from *Ae. triseriatus* midgut cells by Beaty et al.[231] provided very critical information of epidemiological value. However, molecular studies must be judiciously used as a tool to look at the virus-vector-host relationships, rather than as an end in themselves; too often a mosquito that has been in colony for generations and is easy to rear is injected with a virus that never infects the experimental vector in nature, and the resulting study is of questionable academic interest only. Given the limited resources disease research is facing in the immediate future, fiscal conservation mandates that studies be epidemiologically relevant.

ACKNOWLEDGMENTS

The author would like to dedicate this review to Dr. Gene R. DeFoliart, a leading pioneer in LAC virus-vector research, a mentor who introduced me to the exciting world of mosquitoes and arboviruses and a friend. Thanks are extended to Dr. George B. Craig, Jr., for helpful discussions during the preparation of this manuscript. The long hours of patient editing by R. D. Grimstad and her continued support and encouragement are deeply appreciated.

The author was supported in part by NIH grant AI 19679 during the writing of this manuscript. Research support for the unpublished data (of P. R. Grimstad) cited herein was supported in part by AI 19679 and a service contract from the Indiana State Board of Health. The literature review for this manuscript was completed in December 1985.

REFERENCES

1. **Henderson, B. E. and Coleman, P. H.,** The growing importance of California arboviruses in the etiology of human disease, in *Progress in Medical Virology,* Vol. 13, Melnick, J. L., Ed., S. Karger, Basel, 1971, 404.
2. **Parkin, W. E., Hammon, W. McD., and Sather, G. E.,** Review of current epidemiological literature on viruses of the California arbovirus group, *Am. J. Trop. Med. Hyg.,* 21, 964, 1972.
3. **LeDuc, J. W.,** The ecology of California group viruses, *J. Med. Entomol.,* 16, 1, 1979.
4. **Calisher, C. H. and Thompson, W. H., Eds.,** *California Serogroup Viruses,* Alan R. Liss, New York, 1983.
5. **Bishop, D. H. L., Fuller, F. J., and Akashi, H.,** Coding assignments of the RNA genome segments of California serogroup viruses, in *California Serogroup Viruses,* Calisher, C. H. and Thompson, W. H., Eds., Alan R. Liss, New York, 1983, 107.
6. **Murphy, F. A., Whitfield, S. G., Coleman, P. H., Calisher, C. H., Rabin, E. R., Jenson, A. B., Melnick, J. L., Edwards, M. R., and Whitney, E.,** California group arboviruses: electron microscopic studies, *Exp. Mol. Pathol.,* 9, 44, 1968.
7. **Grimstad, P. R., Shabino, C. L., Calisher, C. H., and Waldman, R. J.,** A case of encephalitis in a human associated with a serologic rise to Jamestown Canyon virus, *Am. J. Trop. Med. Hyg.,* 31, 1238, 1982.
8. **Hammon, W. McD., Reeves, W. C., and Sather, G. E.,** California encephalitis virus, a newly described agent. II. Isolations and attempts to identify and characterize the agent, *J. Immunol.,* 69, 493, 1952.

9. **Spence, L., Anderson, C. R., Aitken, T. H. G., and Downs, W. G.,** Melao virus, a new agent isolated from Trinidadian mosquitoes, *Am. J. Trop. Med. Hyg.*, 11, 687, 1962.
10. **Groot, H., Oya, A., Bernal, C., and Barreto-Reyes, P.,** Guaroa virus, a new agent isolated in Colombia, South America, *Am. J. Trop. Med. Hyg.*, 8, 604, 1959.
11. **Hammon, W. McD. and Sather, G. E.,** History and recent reappearance of viruses of the California encephalitis group, *Am. J. Trop. Med. Hyg.*, 15, 199, 1966.
12. **Bardos, V. and Danielova, V.,** The Tahyna virus — a virus isolated from mosquitoes in Czechoslovakia, *J. Hyg. Epidemiol. Microbiol. Immunol.*, 3, 264, 1959.
13. **Burgdorfer, W., Newhouse, V. F., and Thomas, L. A.,** Isolation of California encephalitis virus from the blood of a snowshoe hare *(Lepus americanus)* in western Montana, *Am. J. Hyg.*, 73, 344, 1961.
14. **Kokernot, R. H., McIntosh, B. M., Worth, C. B., DeMorais, T., and Weinbren, M. P.,** Isolation of viruses from mosquitoes collected at Lumbo, Mozambique, *Am. J. Trop. Med. Hyg.*, 11, 678, 1962.
15. **Thompson, W. H., Kalfayan, B., and Anslow, R. O.,** Isolation of California encephalitis group virus from a fatal human case, *Am. J. Epidemiol.*, 81, 245, 1965.
16. **Murphy, F. A. and Coleman, P. H.,** California group arboviruses: immunodiffusion studies, *J. Immunol.*, 99, 276, 1967.
17. **Brummer-Korvenkontio, M., Saikku, P., Korhonen, P., Ulmanen, I., Reunala, T., and Karvonen, J.,** Arboviruses in Finland. IV. Isolation and characterization of Inkoo virus, a Finnish representative of the California group, *Am. J. Trop. Med. Hyg.*, 22, 404, 1973.
18. **Bond, J. O., Hammon, W. McD., Lewis, A. L., Sather, G. E., and Taylor, D. J.,** California group arboviruses in Florida and a report of a new strain, Keystone virus, *Public Health Rep.*, 81, 607, 1966.
19. **Karabatsos, N., Ed.,** *International Catalog of Arboviruses Including Certain Other Viruses of Vertebrates*, 3rd ed., American Society of Tropical Medicine and Hygiene, San Antonio, Tex., 1985.
20. **Hammon, W. McD. and Reeves, W. C.,** California encephalitis virus, a newly described agent. I. Evidence of natural infection in man and other animals, *Calif. Med.*, 77, 303, 1952.
21. **Bardos, V. et al.,** *Arboviruses of the California Complex and Bunyamwera Group*, Slovak Academy of Science, Bratislava, 1969.
22. **Bardos, V.,** Acute infections induced by Tahyna virus: evaluation of the years 1959 to 1976, *Cas. Lek. Cesk.*, 116, 995, 1977.
23. **Thompson, W. H. and Evans, A. S.,** California encephalitis virus studies in Wisconsin, *Am. J. Epidemiol.*, 81, 230, 1965.
24. **Thompson, W. H. and Gundersen, C. B.,** La Crosse encephalitis: occurrence of disease and control in a suburban area, in *California Serogroup Viruses*, Calisher, C. H. and Thompson, W. H., Eds., Alan R. Liss, New York, 1983, 225.
25. **Rice, P. L. and Pratt, H. D.,** *Epidemiology and Control of Vector-Borne Diseases*, USDHEW/PHS Training Manual, Atlanta, 1974.
26. **Grimstad, P. R.,** Mosquitoes and the incidence of encephalitis, in *Advances in Virus Research*, Vol. 28, Lauffer, M. A. and Maramorosch, K., Eds., Academic Press, New York, 1983, 357.
27. **Tesh, R. B.,** Transovarial transmission of arboviruses in their invertebrate vectors, in *Current Topics in Vector Research*, Vol. 2, Harris, K. F., Ed., Praeger, New York, 1984, 57.
28. **Boromisa, R. D. and Grimstad, P. R.,** Virus-vector-host relationships of *Aedes stimulans* and Jamestown Canyon virus in a northern Indiana enzootic focus, *Am. J. Trop. Med. Hyg.*, 35, 1285, 1986.
29. **Berry, R. L., LaLonde-Weigert, B. J., Calisher, C. H., Parsons, M. A., and Bear, G. T.,** Evidence for transovarial transmission of Jamestown Canyon virus in Ohio, *Mosq. News*, 37, 494, 1977.
30. **Turell, M. J. and LeDuc, J. W.,** The role of mosquitoes in the natural history of California serogroup viruses, in *California Serogroup Viruses*, Calisher, C. H. and Thompson, W. H., Eds., Alan R. Liss, New York, 1983, 43.
31. **DeFoliart, G. R.,** *Aedes triseriatus:* vector biology in relationship to the persistence of La Crosse virus in endemic foci, in *California Serogroup Viruses*, Calisher, C. H. and Thompson, W. H., Eds., Alan R. Liss, New York, 1983, 89.
32. **Berry, R. L., Parsons, M. A., LaLonde-Weigert, B. J., Lebio, J., Stegmiller, H., and Bear, G. T.,** *Aedes canadensis*, a vector of La Crosse virus in Ohio (California serogroup), *J. Am. Mosq. Control Assoc.*, 2, 73, 1986.
33. **Wright, R. E. and DeFoliart, G. R.,** Association of Wisconsin mosquitoes and woodland vertebrate hosts, *Ann. Entomol. Soc. Am.*, 63, 777, 1970.
34. **Kappus, K. D., Monath, T. P., Kaminski, R. M., and Calisher, C. H.,** Reported encephalitis associated with California serogroup virus infections in the United States, 1963 to 1981, in *California Serogroup Viruses*, Calisher, C. H. and Thompson, W. H., Eds., Alan R. Liss, New York, 1983, 31.
35. USDHHS/PHS/CDC, Annual summary 1982: reported morbidity and mortality in the United States, *Morbid. Mortal. Wkly. Rep.*, 31, 1983.
36. USDHHS/PHS/CDC, Human arboviral encephalitis — United States, 1983, *Morbid. Mortal. Wkly. Rep.*, 33, 339, 1984.

37. USDDH/PHS/CDC, Arboviral infections of the central nervous system — United States, 1984, *Morbid. Mortal. Wkly. Rep.*, 34, 283, 1985.
38. **Grimstad, P. R., Barrett, C. L., Humphrey, R. L., and Sinsko, M. J.,** Serologic evidence for widespread infection with La Crosse and St. Louis encephalitis viruses in the Indiana human population, *Am. J. Epidemiol.*, 119, 913, 1984.
39. **Monath, T. P., Ed.,** *St. Louis Encephalitis*, American Public Health Association, Washington, D.C., 1980.
40. **Hurwitz, E. S., Schell, W., Nelson, D., Washburn, J., and LaVenture, M.,** Surveillance for California encephalitis group virus illness in Wisconsin and Minnesota, 1978, *Am. J. Trop. Med. Hyg.*, 32, 595, 1983.
41. **Berry, R. L., Parsons, M. A., Restifo, R. A., Peterson, E. D., Gordon, S. W., Reed, M. R., Calisher, C. H., and Halpin, T. J.,** California serogroup virus infections in Ohio: an 18-year retrospective survey, in *California Serogroup Viruses*, Calisher, C. H. and Thompson, W. H., Eds., Alan R. Liss, New York, 1983, 215.
42. **Fauvel, M., Artsob, H., Calisher, C. H., Davignon, L., Chagnon, A., Skvorc-Ranko, R., and Bellonick, S.,** California group virus encephalitis in three children from Quebec: clinical and serologic findings, *Can. Med. Assoc. J.*, 122, 60, 1980.
43. **Deibel, R., Srihongse, S., Grayson, M. A., Grimstad, P. R., Mahdy, M. S., Artsob, H., and Calisher, C. H.,** Jamestown Canyon virus: the etiologic agent of an emerging human disease?, in *California Serogroup Viruses*, Calisher, C. H. and Thompson, W. H., Eds., Alan R. Liss, New York, 1983, 313.
44. **Srihongse, S., Grayson, M. A., and Deibel, R.,** California serogroup viruses in New York State: the role of subtypes in human infections, *Am. J. Trop. Med. Hyg.*, 33, 1218, 1984.
45. **Grimstad, P. R., Calisher, C. H., Harroff, R. N., and Wentworth, B. B.,** Jamestown Canyon virus (California serogroup) is the etiologic agent of widespread infection in Michigan humans, *Am. J. Trop. Med. Hyg.*, 35, 376, 1986.
46. **Grimstad, P. R.,** unpublished data, 1985.
47. **Monath, T. P.,** Arthropod-borne encephalitides in the Americas, *Bull. WHO*, 53, 513, 1979.
48. **Gundersen, C. B. and Brown, K. L.,** Clinical aspects of La Crosse encephalitis: preliminary report, in *California Serogroup Viruses*, Calisher, C. H. and Thompson, W. H., Eds., Alan R. Liss, New York, 1983, 169.
49. **Chun, R. W. M.,** Clinical aspects of La Crosse encephalitis: neurological and psychological sequelae, in *California Serogroup Viruses*, Calisher, C. H. and Thompson, W. H., Eds., Alan R. Liss, New York, 1983, 193.
50. **Clark, G. G., Pretula, H. L., Lankop, C. W., Martin, R. J., and Calisher, C. H.,** Occurrence of La Crosse (California serogroup) encephalitis viral infections in Illinois, *Am. J. Trop. Med. Hyg.*, 32, 838, 1983.
51. **Berry, R. L., Parsons, M. A., LaLonde, B. J., Stegmiller, H. W., Lebio, J., Jalil, M., and Masterson, R. A.,** Studies on the epidemiology of California encephalitis in an endemic area in Ohio in 1971, *Am. J. Trop. Med. Hyg.*, 24, 992, 1975.
52. **Balfour, H. H., Jr., Edelman, C. K., Bauer, H., and Siem, R. A.,** California arbovirus (La Crosse) infections. III. Epidemiology of California encephalitis in Minnesota, *J. Infect. Dis.*, 133, 293, 1976.
53. **Gundersen, C. B.,** La Crosse encephalitis: a cost analysis, oral presentation, Int. Symp. Calif. Serogroup Viruses, Cleveland, 1982.
54. **Parry, J. E.,** Control of *Aedes triseriatus* in La Crosse, Wisconsin, in *California Serogroup Viruses*, Calisher, C. H. and Thompson, W. H., Eds., Alan R. Liss, New York, 1983, 355.
55. **Bishop, D. H. L. and Shope, R. E.,** Bunyaviridae, in *Comprehensive Virology*, Vol. 14, Fraenkel-Conrat, H. and Wagner, R. R., Eds., Plenum Press, New York, 1979, 1.
56. **Calisher, C. H.,** Taxonomy, classification, and geographic distribution of California serogroup bunyaviruses, in *California Serogroup Viruses*, Calisher, C. H. and Thompson, W. H., Eds., Alan R. Liss, New York, 1983, 1.
57. **Klimas, R. A., Thompson, W. H., Calisher, C. H., Clark, G. G., Grimstad, P. R., and Bishop, D. H. L.,** Genotypic varieties of La Crosse virus isolated from different geographic regions of the continental United States and evidence for a naturally occurring intertypic recombinant La Crosse virus, *Am. J. Epidemiol.*, 114, 112, 1981.
58. **Issel, C. J., Trainer, D. O., and Thompson, W. H.,** Experimental studies with white-tailed deer and four California group arboviruses (La Crosse, trivittatus, snowshoe hare, and Jamestown Canyon), *Am. J. Trop. Med. Hyg.*, 21, 979, 1972.
59. **Godsey, M. L.,** California Serogroup Virus Infections in Wisconsin Domestic Animals, M.S. thesis, University of Wisconsin, Madison, 1986.
60. **Yuill, T. M.,** The role of mammals in the maintenance and dissemination of La Crosse virus, in *California Serogroup Viruses*, Calisher, C. H. and Thompson, W. H., Eds., Alan R. Liss, New York, 1983, 77.

61. **March, R. W. and Hetrick, F. M.,** Studies on Guaroa virus. III. Experimental infection of burros, *Am. J. Trop. Med. Hyg.*, 16, 200, 1967.
62. **Beaty, B. J. and Trent, D. W.,** Arbovirus evolution: mechanisms and epidemiologic significance, in *Arboviruses,* Vol. 5, Monath, T. P., Ed., CRC Press, Boca Raton, Fla., 1988.
63. **Beaty, B. J., Fuller, F., and Bishop, D. H. L.,** Bunyavirus gene structure — function, relationships, and potential for RNA segment reassortment in the vector: La Crosse and snowshoe hare reassortant viruses in mosquitoes, in *California Serogroup Viruses,* Calisher, C. H. and Thompson, W. H., Eds., Alan R. Liss, New York, 1983, 119.
64. **Beaty, B. J., Sundin, D. R., Chandler, L. J., and Bishop, D. H. L.,** Evolution of Bunyaviruses by gene reassortment in dually infected mosquitoes *(Aedes triseriatus), Science,* 230, 548, 1985.
65. **Tignor, G. H., Burrage, T. G., Smith, A. L., and Shope, R. E.,** California serogroup gene structure-function relationships: virulence and tissue tropisms, in *California Serogroup Viruses,* Calisher, C. H. and Thompson, W. H., Eds., Alan R. Liss, New York, 1983, 129.
66. **Bardos, V. and Pesko, J.,** Biological and antigenic variants among Tahyna virus strains isolated in Czechoslovakia, *Arch. Virol.*, 68, 65, 1981.
67. **Malkova, D.,** Comparative study of two variants of Tahyna virus, *Acta Virol.*, 18, 407, 1974.
68. **Malkova, D. and Reddy, G. N.,** Influence of early passages on the character of freshly isolated strains of Tahyna virus, *Acta Virol.*, 19, 333, 1975.
69. **Issel, C. J., Pantuwatana, S., Yuill, T. M., and Hanson, R. P.,** Selection for plaque variants of two California group arboviruses (Jamestown Canyon and La Crosse) by passage in natural vertebrate hosts, *Acta Virol.*, 19, 318, 1975.
70. **Seymour, C. and Yuill, T. M.,** Studies on selection of La Crosse virus variants by natural vertebrate hosts and vector mosquitoes, *Acta Virol.*, 25, 87, 1981.
71. **Ksiazek, T. G. and Yuill, T. M.,** Selection of La Crosse virus variants by sentinal squirrels *(Sciuris carolinensis)* and chipmunks *(Tamias striatus), Acta Virol.*, 21, 119, 1977.
72. **Sefcovicova, L.,** Effect of various modes of passaging on the cytopathic activity of Tahyna virus, *Acta Virol.*, 9, 502, 1965.
73. **Danielova, V.,** The property changes of two Tahyna virus strains by influence of serial passages in various environments, *Zentralbl. Bakteriol. I. Orig. Ser. A,* 229, 323, 1974.
74. **Beaty, B. J., Rozhon, E. J., Gensemer, P., and Bishop, D. H. L.,** Formation of reassortant bunyaviruses in dually infected mosquitoes, *Virology,* 111, 662, 1981.
75. **Ozden, S. and Hannoun, C.,** Isolation and preliminary characterization of temperature-sensitive mutants of Lumbo virus, *Virology,* 84, 210, 1978.
76. **Vezza, A. C., Repik, P. M., Cash, P., and Bishop, D. H. L.,** In vivo transcription and protein synthesis capabilities of bunyaviruses: wild-type snowshoe hare virus and its temperature-sensitive group I, group II, and group I-II mutants, *J. Virol.*, 31, 326, 1979.
77. **Miller, B. R.,** A variant of La Crosse virus attenuated for *Aedes triseriatus* mosquitoes, *Am. J. Trop. Med. Hyg.*, 32, 1422, 1983.
78. **Gonzalez-Scarano, F., Jannsen, R. S., Najjar, J. A., Pobjecky, N., and Nathanson, N.,** An avirulent G1 glycoprotein variant of La Crosse bunyavirus with defective fusion function, *J. Virol.*, 54, 757, 1985.
79. **Rozhon, E. J., Gensemer, P., Shope, R. E., and Bishop, D. H. L.,** Attenuation of virulence of a bunyavirus involving a large RNA defect and isolation of La Crosse-snowshoe hare virus-La Crosse virus and La Crosse virus-snowshoe hare virus-snowshoe hare virus reassortants, *Virology,* 111, 125, 1981.
80. **Janssen, R., Gonzalez-Scarano, F., and Nathanson, N.,** Mechanisms of bunyavirus virulence: comparative pathogenesis of a virulent strain of La Crosse and an avirulent strain of Tahyna virus, *Lab. Invest.*, 50, 447, 1984.
81. **Calisher, C. H., Monath, T. P., Karabatsos, N., and Trent, D. W.,** Arbovirus subtyping. Applications to epidemiologic studies, availability of reagents, and testing services, *Am. J. Epidemiol.*, 114, 619, 1981.
82. **Artsob, H., Spence, L. P., and Th'ng, C.,** Enzyme-linked immunosorbent assay typing of California serogroup viruses isolated in Canada, *J. Clin. Microbiol.*, 20, 276, 1984.
83. **Beaty, B. J. and Thompson, W. H.,** Delineation of La Crosse virus in developmental stages of transovarially infected *Aedes triseriatus, Am. J. Trop. Med. Hyg.*, 25, 505, 1976.
84. **Beaty, B. J. and Thompson, W. H.,** Tropisms of La Crosse virus in *Aedes triseriatus* (Diptera:Culicidae) following infective blood meals, *J. Med. Entomol.*, 14, 499, 1978.
85. **Monath, T. P., Nuckolls, J. G., Berall, J., Bauer, H., Chappell, W. A., and Coleman, P. H.,** Studies on California encephalitis in Minnesota, *Am. J. Epidemiol.*, 92, 40, 1970.
86. **Bardos, V., Ryba, J. M., and Hubalek, Z.,** Isolation of Tahyna virus from the blood of sick children, *Acta Virol.*, 19, 446, 1975.
87. **Bardos, V. and Sluka, F.,** Acute human infections caused by Tahyna virus, *Cas. Lek. Cesk.*, 102, 394, 1963.
88. **Simkova, A. and Sluka, F.,** Isolation of Tahyna virus from the blood of a case of influenza-like disease, *Acta Virol.*, 17, 94, 1973.

89. **Sluka, F.,** The clinical picture of the Tahyna virus infection, in *Arboviruses of the California Complex and the Bunyamwera Group,* Bardos, V., et al., Eds., Slovak Academy of Science, Bratislavia, 1969, 311.
90. **Balfour, H. H., Jr., Siem, R. A., Bauer, H., and Quie, P. G.,** California arbovirus (La Crosse) infections. I. Clinical and laboratory findings in 66 children with meningoencephalitis, *Pediatrics,* 52, 680, 1973.
91. **Chun, R. W. M., Thompson, W. H., Grabow, J. D., and Mathews, C. G.,** California arbovirus encephalitis in children, *Neurology,* 18, 369, 1968.
92. **Cramblett, H. G., Stegmiller, H., and Spencer, C.,** California encephalitis virus infections in children: clinical and laboratory studies, *JAMA,* 198, 108, 1966.
93. **Hilty, M. D., Haynes, R. E., Azimi, P. H., and Cramblett, H. G.,** California encephalitis in children, *Am. J. Dis. Child.,* 124, 530, 1972.
94. **Thompson, W. H. and Inhorn, S. L.,** Arthropod-borne California group viral encephalitis in Wisconsin, *Wisc. Med. J.,* 66, 250, 1967.
95. **Young, D. J.,** California encephalitis virus — report of three cases and review of the literature, *Ann. Intern. Med.,* 65, 419, 1966.
96. **Gundersen, C. B.,** personal communication, 1982.
97. **Embil, J. A., Camfield, P. R., and Artsob, H.,** California encephalitis in New Brunswick, *Can. Med. Assoc. J.,* 132, 1166, 1985.
98. **Mahdy, M. S., McLaughlin, B., Paul, N. R., and Surgeoner, G.,** Surveillance of arboviruses in Ontario in 1983 — the increased detection of seropositive cases to the California group viruses, *Ont. Dis. Surveill. Rep.,* 5, 394, 1984.
99. **Quick, D. T., Smith, A. G., Lewis, A. L., Sather, G. E., and Hammon, W. McD.,** California encephalitis virus infection: a case report, *Am. J. Trop. Med. Hyg.,* 14, 456, 1965.
100. **Lewis, A. L., Hammon, W. McD., Sather, G. E., Taylor, D. J., and Bond, J. O.,** Isolation of a California group arbovirus from Florida mosquitoes, *Am. J. Trop. Med. Hyg.,* 14, 451, 1965.
101. **Jennings, W. L., Lewis, A. L., Sather, G. E., Hammon, W. McD., and Bond, J. O.,** California-encephalitis-group viruses in Florida rabbits: report of experimental and sentinel studies, *Am. J. Trop. Med. Hyg.,* 17, 781, 1968.
102. **Deibel, R.,** personal communication, 1985.
103. **Kalfayan, B.,** Pathology of La Crosse virus infection in humans, in *California Serogroup Viruses,* Calisher, C. H. and Thompson, W. H., Eds., Alan R. Liss, New York, 1983, 179.
104. **Deering, W. M.,** Neurologic aspects and treatment of La Crosse encephalitis, in *California Serogroup Viruses,* Calisher, C. H. and Thompson, W. H., Eds., Alan R. Liss, New York, 1983, 187.
105. **Grabow, J. D., Matthews, C. G., Chun, R. W. M., and Thompson, W. H.,** The electroencephalogram and clinical sequelae of California arbovirus encephalitis, *Neurology,* 19, 394, 1969.
106. **Shope, R. E.,** Arboviruses, in *Manual of Clinical Microbiology,* 4th ed., Lennette, E. H., Balows, A., Hausler, W. J., Jr., and Shadomy, H. J., Eds., American Society for Microbiology, Washington, D.C., 1985, 785.
107. **Calisher, C. H. and Bailey, R. E.,** Serodiagnosis of La Crosse virus infections in humans, *J. Clin. Microbiol.,* 13, 344, 1981.
108. **Lindsey, H. S., Calisher, C. H., and Mathews, J. H.,** Serum dilution neutralization test for California group virus identification and serology, *J. Clin. Microbiol.,* 4, 503, 1976.
109. **Dykers, T. I., Brown, K. L., Gundersen, C. B., and Beaty, B. J.,** Rapid diagnosis of La Crosse encephalitis: detection of specific immunoglobulin M in cerebrospinal fluid, *J. Clin. Microbiol.,* 22, 740, 1985.
110. **Beaty, B. J., Jamnback, T. L., Hildreth, S. W., and Brown, K. L.,** Rapid diagnosis of La Crosse virus infections: evaluation of serologic and antigen detection techniques for the clinically relevant diagnosis of La Crosse encephalitis, in *California Serogroup Viruses,* Calisher, C. H. and Thompson, W. H., Eds., Alan R. Liss, New York, 1983, 293.
111. **Jamnback, T. L., Beaty, B. J., Hildreth, S. W., Brown, K. L., and Gundersen, C. B.,** Capture immunoglobulin M system for rapid diagnosis of La Crosse (California encephalitis) virus infections, *J. Clin. Microbiol.,* 16, 577, 1982.
112. **Grimstad, P. R., Ross, Q. E., and Craig, G. B., Jr.,** *Aedes triseriatus* (Diptera:Culicidae) and La Crosse virus. II. Modification of mosquito feeding behavior by virus infection, *J. Med. Entomol.,* 17, 1, 1980.
113. **Grimstad, P. R., Craig, G. B., Jr., Ross, Q. E., and Yuill, T. M.,** *Aedes triseriatus* and La Crosse virus: geographic variation in vector susceptibility and ability to transmit, *Am. J. Trop. Med. Hyg.,* 26, 990, 1977.
114. **Patrican, L. A. and DeFoliart, G. R.,** Lack of adverse effect of transovarially acquired La Crosse virus infection on the reproductive capacity of *Aedes triseriatus, J. Med. Entomol.,* 22, 602, 1985.
115. **Gu, H. X., Spence, L., Artsob, H., Chia, W. K., Th'ng, C., and Lampotang, V.,** Serological evidence of infection with California serogroup viruses (family Bunyaviridae) in residents of Long Hua, suburb of Shanghai, People's Republic of China, *Trans. R. Soc. Trop. Med. Hyg.,* 78, 780, 1984.

116. **Artsob, H., Spence, L. P., Calisher, C. H., Sekla, L. H., and Brust, R. A.,** Isolation of California encephalitis serotype from mosquitoes collected in Manitoba, Canada, *J. Am. Mosq. Control Assoc.*, 1, 257, 1985.
117. **Artsob, H.,** Distribution of California serogroup viruses and virus infections in Canada, in *California Serogroup Viruses*, Calisher, C. H. and Thompson, W. H., Eds., Alan R. Liss, New York, 1983, 277.
118. **Hoff, G. L., Issel, C. J., Trainer, D. O., and Richards, S. H.,** Arbovirus serology in North Dakota mule and white-tailed deer, *J. Wildl. Dis.*, 9, 291, 1973.
119. **Rowley, W. A., Wong, Y. W., Dorsey, D. C., Hausler, W. J., Jr., and Currier, R. W.,** California serogroup viruses in Iowa, in *California Serogroup Viruses*, Calisher, C. H. and Thompson, W. H., Eds., Alan R. Liss, New York, 1983, 237.
120. **Wilson, M. L. and Grimstad, P. R.,** unpublished data, 1985.
121. **Aspock, H., Graefe, G., and Kunz, C.,** Studies on the periodicity of the occurrence of Tahyna and Calovo viruses, *Zentralbl. Bakteriol. I. Orig. Ser. A*, 217, 431, 1971.
122. **Bardos, V., Medek, M., Kania, V., and Hubalek, Z.,** Isolation of Tahyna virus from the blood of sick children, *Acta Virol.*, 19, 447, 1975.
123. **Artsob, H., Spence, L., Surgeoner, G., Th'ng, C., Lampotang, V., Grant, L., and McCreadie, J.,** A focus of California group virus activity in southern Ontario, *Mosq. News*, 43, 449, 1983.
124. **Bellonick, S., Aubin, A., Maire, A., Boisvert, J., Gagnon, J., Th'ng, C., Trudel, C., and Artsob, H.,** Arbovirus studies in the Trois-Rivieres area, Province of Quebec, Canada, *Mosq. News*, 43, 426, 1983.
125. **Mokry, J., Artsob, H., and Butler, R.,** Studies on California serogroup virus activity in Newfoundland, Canada, 1980 to 1983, *Mosq. News*, 44, 310, 1984.
126. **McLean, D. M.,** Yukon isolates of snowshoe hare virus, 1972 to 1982, in *California Serogroup Viruses*, Calisher, C. H. and Thompson, W. H., Eds., Alan R. Liss, New York, 1983, 247.
127. **Clark, G. G., Pretula, H. L., Roher, W. H., Harroff, R. N., and Jakubowski, T.,** Persistence of La Crosse virus (California encephalitis serogroup) in north-central Illinois, *Am. J. Trop. Med. Hyg.*, 32, 175, 1983.
128. **Watts, D. M., Thompson, W. H., Yuill, T. M., DeFoliart, G. R., and Hanson, R. P.,** Overwintering of La Crosse virus in *Aedes triseriatus*, *Am. J. Trop. Med. Hyg.*, 23, 694, 1974.
129. **Lisitza, M. A., DeFoliart, G. R., Yuill, T. M., and Karandinos, M. G.,** Prevalence rates of La Crosse virus (California encephalitis group) in larvae from overwintered eggs of *Aedes triseriatus*, *Mosq. News*, 37, 745, 1977.
130. **Kappus, K. D., Calisher, C. H., Baron, R. C., Davenport, J., Francy, D. B., and Williams, R. M.,** La Crosse virus infection and disease in western North California, *Am. J. Trop. Med. Hyg.*, 31, 556, 1982.
131. **Danielova, V. and Marhoul, Z.,** The incidence of antibodies of some arboviruses in men, domestic and wild animals living in the natural focus of the Tahyna virus in the south of Moravia, *Cesk. Epidemiol. Mikrobiol. Immunol.*, 17, 155, 1968.
132. **Zarnke, R. L., Calisher, C. H., and Kerschner, J.,** Serologic evidence of arbovirus infections in humans and wild animals in Alaska, *J. Wildl. Dis.*, 19, 175, 1983.
133. **Bardos, V.,** Immunological study of antibodies neutralizing Tahyna virus in the sera of inhabitants of Czechoslovakia, *J. Hyg. Epidemiol. Microbiol. Immunol.*, 4, 54, 1960.
134. **Bardos, V., Adamcova, J., Sefcovicova, L., and Cervenka, J.,** Antibodies neutralizing Tahyna virus in different age-groups of inhabitants of an area with mass prevalence of mosquitoes, *Cesk. Epidemiol. Mikrobiol. Immunol.*, 11, 238, 1962.
135. **Heinz, F., Herzig, P., Asmera, J., Gawlas, W., and Sedenka, B.,** A contribution to the problem of the Tahyna virus importance in the North Moravia region, *Cesk. Epidemiol. Mikrobiol. Immunol.*, 21, 149, 1972.
136. **Bardos, V. and Sefcovicova, L.,** The presence of antibodies neutralizing Tahyna virus in the sera of inhabitants of some European, Asian, African, and Australian countries, *J. Hyg. Epidemiol. Microbiol. Immunol.*, 5, 501, 1961.
137. **Gresikova, M., Reeves, W. C., and Scrivani, R. P.,** California encephalitis virus: an evaluation of its continued endemic status in Kern County, California, *Am. J. Hyg.*, 80, 229, 1964.
138. **Reeves, W. C., Emmons, R. W., and Hardy, J. L.,** Historical perspectives on California encephalitis virus in California, in *California Serogroup Viruses*, Calisher, C. H. and Thompson, W. H., Eds., Alan R. Liss, New York, 1983, 19.
139. **Watts, D. M., LeDuc, J. W., Bailey, C. L., Dalrymple, J. M., and Gargan, T. P., II,** Serologic evidence of Jamestown Canyon and Keystone virus infection in vertebrates of the Delmarva peninsula, *Am. J. Trop. Med. Hyg.*, 31, 1245, 1982.
140. **Elbel, R. E., Crane, G. T., Stipe, L. E., Van Nosdol, G. B., and Smart, K. L.,** Arbovirus isolations from mosquitoes collected at Callao, Utah, 1966 and 1967, *Mosq. News*, 31, 61, 1971.
141. **Turell, M. J., Reeves, W. C., and Hardy, J. L.,** Transovarial and transstadial transmission of California encephalitis virus in *Aedes dorsalis* and *Aedes melanimon*, *Am. J. Trop. Med. Hyg.*, 31, 1021, 1982.

142. **Grayson, M. A., Srihongse, S., Deibel, R., and Calisher, C. H.,** California serogroup viruses in New York State: a retrospective analysis of subtype distribution patterns and their epidemiologic significance, 1965 to 1981, in *California Serogroup Viruses,* Calisher, C. H. and Thompson, W. H., Eds., Alan R. Liss, New York, 1983, 257.
143. **Boromisa, R. D.,** Dynamics of Jamestown Canyon Virus in a Northern Indiana Focus, Ph.D. thesis, University of Notre Dame, Notre Dame, Ind., 1985.
144. **Berry, R. L.,** personal communication, 1986.
145. **Nasci, R. S.,** Differences in host choice between the sibling species of treehole mosquitoes, *Aedes triseriatus* and *Aedes hendersoni, Am. J. Trop. Med. Hyg.,* 31, 411, 1982.
146. **Nasci, R. S.,** Local variation in blood feeding by *Aedes triseriatus* and *Aedes hendersoni* (Diptera:Culicidae), *J. Med. Entomol.,* 22, 619, 1985.
147. **Burkot, T. R. and DeFoliart, G. R.,** Blood meal sources of *Aedes triseriatus* and *Aedes vexans* in a southern Wisconsin forest endemic for La Crosse encephalitis virus, *Am. J. Trop. Med. Hyg.,* 31, 376, 1982.
148. **Grimstad, P. R., Williams, D. G., and Schmitt, S. M.,** Infection of white-tailed deer *(Odocoileus virginianus)* in Michigan with Jamestown Canyon virus (California serogroup) and the importance of maternal antibody in the viral maintenance, *J. Wildl. Dis.,* 23, 12, 1987.
149. **Issel, C. J., Trainer, D. O., and Thompson, W. H.,** Serologic evidence of infections of white-tailed deer in Wisconsin with three California group arboviruses (La Crosse, trivittatus, and Jamestown Canyon), *Am. J. Trop. Med. Hyg.,* 21, 985, 1972.
150. **Issel, C. J.,** Isolation of Jamestown Canyon virus (a California group arbovirus) from a white-tailed deer, *Am. J. Trop. Med. Hyg.,* 22, 414, 1973.
151. **Issel, C. J., Hoff, G. L., and Trainer, D. O.,** Serologic evidence of infection of white-tailed deer in Texas with three California group arboviruses (Jamestown Canyon, San Angelo, and Keystone), *J. Wildl. Dis.,* 9, 245, 1973.
152. **Watts, D. M., Tammariello, R. F., Dalrymple, J. M., Eldridge, B. F., Russell, P. K., and Top, F. H., Jr.,** Experimental infection of vertebrates of the Pokomoke Cypress Swamp, Maryland, with Keystone and Jamestown Canyon viruses, *Am. J. Trop. Med. Hyg.,* 28, 344, 1979.
153. **Grimstad, P. R., Schmitt, S. M., and Williams, D. G.,** Prevalence of neutralizing antibody to Jamestown Canyon virus (California group) in elk and moose populations in northern Michigan and Ontario, Canada, *J. Wildl. Dis.,* 22, 453, 1986.
154. **McFarlane, B. L., Embree, J. E., Embil, J. A., and Rozee, K. R.,** Antibodies to the California group of arboviruses in animal populations of New Brunswick, *Can. J. Microbiol.,* 28, 200, 1981.
155. **Trainer, D. O. and Hanson, R. P.,** Serologic evidence of arbovirus infections in wild ruminants, *Am. J. Epidemiol.,* 90, 354, 1960.
156. **Issel, C. J.,** Maternal antibody to Jamestown Canyon virus in white-tailed deer, *Am. J. Trop. Med. Hyg.,* 23, 242, 1974.
157. **Fine, P. E. M. and LeDuc, J. W.,** Towards a quantitative understanding of the epidemiology of Keystone virus in the eastern United States, *Am. J. Trop. Med. Hyg.,* 27, 322, 1978.
158. **Darsie, R. F. and Ward, R. A.,** *Identification and Geographical Distribution of the Mosquitoes of North America, North of Mexico,* Mosq. Syst. Suppl. No. 1, American Mosquito Control Association, Fresno, Calif., 1981.
159. **Pantuwatana, S., Thompson, W. H., Watts, D. M., and Hanson, R. P.,** Experimental infection of chipmunks and squirrels with La Crosse and trivittatus viruses and biological transmission of La Crosse virus by *Aedes triseriatus, Am. J. Trop. Med. Hyg.,* 21, 476, 1972.
160. **Watts, D. M., Grimstad, P. R., DeFoliart, G. R., Yuill, T. M., and Hanson, R. P.,** Laboratory transmission of La Crosse encephalitis virus by several species of mosquitoes, *J. Med. Entomol.,* 10, 583, 173.
161. **Watts, D. M., Morris, C. D., Wright, R. E., DeFoliart, G. R., and Hanson, R. P.,** Transmission of La Crosse virus (California encephalitis group) by the mosquito *Aedes triseriatus, J. Med. Entomol.,* 9, 125, 1972.
162. **Watts, D. M., Pantuwatana, S., DeFoliart, G. R., Yuill, T. M., and Thompson, W. H.,** Transovarial transmission of La Crosse virus (California encephalitis group) in the mosquito, *Aedes triseriatus, Science,* 182, 1140, 1973.
163. **Pantuwatana, S., Thompson, W. H., Watts, D. M., Yuill, T. M., and Hanson, R. P.,** Isolation of La Crosse virus from field collected *Aedes triseriatus* larvae, *Am. J. Trop. Med. Hyg.,* 23, 246, 1974.
164. **Ksiazek, T. G. and Yuill, T. M.,** Viremia and antibody response to La Crosse virus in sentinel gray squirrels *(Sciurus carolinensis)* and chipmunks *(Tamias striatus), Am. J. Trop. Med. Hyg.,* 26, 815, 1977.
165. **Gauld, L. W., Hanson, R. P., Thompson, W. H., and Sinha, S. K.,** Observations on a natural cycle of La Crosse virus (California group) in southwestern Wisconsin, *Am. J. Trop. Med. Hyg.,* 23, 983, 1974.

166. **Walker, E. D. and Edman, J. D.,** Influence of defensive behavior of eastern chipmunks and gray squirrels (Rodentia:Sciuridae) on feeding success of *Aedes triseriatus* (Diptera:Culicidae), *J. Med. Entomol.*, 23, 1, 1986.
167. **Mather, T. N. and DeFoliart, G. R.,** Reduced blood feeding success on squirrels and chipmunks by older *Aedes triseriatus* females, *Mosq. News*, 44, 471, 1984.
168. **Thompson, W. H.,** Vector-virus relationships, in *California Serogroup Viruses*, Calisher, C. H. and Thompson, W. H., Eds., Alan R. Liss, New York, 1983, 57.
169. **Thompson, W. H. and Beaty, B. J.,** Venereal transmission of La Crosse (California encephalitis) arbovirus in *Aedes triseriatus* mosquitoes, *Science*, 196, 530, 1977.
170. **Grimstad, P. R., Paulson, S. L., and Craig, G. B., Jr.,** Vector competence of *Aedes hendersoni* (Diptera:Culicidae) for La Crosse virus and evidence of a salivary-gland escape barrier, *J. Med. Entomol.*, 22, 447, 1985.
171. **Craig, G. B., Jr.,** Biology of *Aedes triseriatus:* some factors affecting control, in *California Serogroup Viruses*, Calisher, C. H. and Thompson, W. H., Eds., Alan R. Liss, New York, 1983, 329.
172. **Grimstad, P. R., Garry, C. E., and DeFoliart, G. R.,** *Aedes hendersoni* and *Aedes triseriatus* (Diptera:Culicidae) in Wisconsin: characterization of larvae, larval hybrids, and comparison of adult and hybrid mesoscutal patterns, *Ann. Entomol. Soc. Am.*, 67, 795, 1974.
173. **McLean, D. M., Grass, P. N., and Judd, B. D.,** Bunyavirus infection rates in Canadian arctic mosquitoes, *Mosq. News*, 39, 364, 1979.
174. **McLintock, J., Curry, P. S., Wagner, R. J., Leung, M. K., and Iversen, J. O.,** Isolation of snowshoe hare virus from *Aedes implicatus* larvae in Saskatchewan, *Mosq. News*, 36, 233, 1976.
175. **McLean, D. M., Grass, P. N., and Judd, B. D.,** California encephalitis virus transmission by arctic and domestic mosquitoes, *Arch. Virol.*, 55, 39, 1977.
176. **Danielova, V.,** Quantitative relationships of Tahyna virus and the mosquito, *Aedes vexans*, *Acta Virol.*, 10, 62, 1966.
177. **Danielova, V. and Ryba, J.,** Laboratory demonstration of transovarial transmission of Tahyna virus in *Aedes vexans* and the role of this mechanism in overwintering of this arbovirus, *Folia Parasitol.*, 26, 361, 1979.
178. **Rodl, P., Bardos, V., and Ryba, J.,** Experimental transmission of Tahyna virus (California group) to wild rabbits *(Oryctolagus cuniculus)* by mosquitoes, *Folia Parasitol.*, 26, 61, 1979.
179. **Simkova, A., Danielova, V., and Bardos, V.,** Experimental transmission of the Tahyna virus by *Aedes vexans* mosquitoes, *Acta Virol.*, 4, 341, 1960.
180. **Malkova, D., Danielova, V., Minar, J., and Ryba, J.,** Virological investigations of mosquitoes in some biotopes of southern Moravia in summer season 1972, *Folia Parasitol.*, 21, 363, 1974.
181. **Bardos, V. and Danielova, V.,** The Tahyna virus — a virus isolated from mosquitoes in Czechoslovakia, *J. Hyg. Epidemiol. Microbiol. Immunol.*, 3, 264, 1959.
182. **Bardos, V.,** The role of mammals in the circulation of Tahyna virus, *Folia Parasitol.*, 22, 257, 1975.
183. **Minar, J.,** Food sources of some mosquito species in the natural focus of Tahyna virus in southern Moravia, *Folia Parasitol.*, 16, 81, 1969.
184. **Simkova, A.,** Tahyna virus-neutralizing antibodies in naturally infected domestic rabbits and hares, *Cesk. Epidemiol. Mikrobiol. Immunol.*, 15, 304, 1966.
185. **Bardos, V., Ryba, J., Hubalek, Z., and Olejnicek, V.,** Virological examination of mosquito larvae from southern Moravia, *Folia Parasitol.*, 25, 75, 1978.
186. **Danielova, V., Minar, J., and Rosicky, B.,** Experimental survival of the virus Tahyna in hibernating mosquitoes *Theobaldia [Culiseta] annulata* (Schrk), *Folia Parasitol.*, 15, 183, 1968.
187. **Simkova, A. and Danielova, V.,** Experimental infection of chimpanzees with Tahyna virus by *Culiseta annulata* mosquitoes, *Folia Parasitol.*, 16, 255, 1969.
188. **Danielova, V.,** The vector efficiency of *Culiseta annulata* mosquito in relation to Tahyna virus, *Folia Parasitol.*, 19, 259, 1972.
189. **Moreau, J. P., Bihan-Faou, P., and Sinegre, G.,** Tahyna virus transovarial transmission; trials in *Aedes caspius*, *Med. Trop. (Marseilles)*, 36, 441, 1976.
190. **Simkova, A.,** Quantitative study of experimental Tahyna virus infection in hibernating hedgehogs, *J. Hyg. Epidemiol. Microbiol. Immunol.*, 10, 499, 1966.
191. **Christensen, B. M., Rowley, W. A., Wong, Y. W., Dorsey, D. C., and Hausler, W. J., Jr.,** Laboratory studies of transovarial transmission of trivittatus virus by *Aedes trivittatus*, *Am. J. Trop. Med. Hyg.*, 27, 184, 1978.
192. **Andrews, W. N., Rowley, W. A., Wong, Y. W., Dorsey, D. C., and Hausler, W. J., Jr.,** Isolation of trivittatus virus from larvae and adults reared from field-collected larvae of *Aedes trivittatus* (Diptera:Culicidae), *J. Med. Entomol.*, 13, 699, 1977.
193. **Kokernot, R. H., Hayes, J., Chan, D. H. M., and Boyd, K. R.,** Arbovirus studies in the Ohio-Mississippi basin, 1964—1967. V. Trivittatus and western equine encephalomyelitis viruses, *Am. J. Trop. Med. Hyg.*, 18, 774, 1969.

194. **Pinger, R. R., Rowley, W. A., Wong, Y. W., and Dorsey, D. C.,** Trivittatus virus infections in wild mammals and sentinal rabbits in central Iowa, *Am. J. Trop. Med. Hyg.*, 24, 1006, 1975.
195. **Pinger, R. R. and Rowley, W. A.,** Host preferences of *Aedes trivittatus* (Diptera:Culicidae) in central Iowa, *Am. J. Trop. Med. Hyg.*, 24, 889, 1975.
196. **Sudia, W. D., Newhouse, V. F., Calisher, C. H., and Chamberlain, R. W.,** California group arboviruses: isolations from mosquitoes in North America, *Mosq. News*, 31, 576, 1971.
197. **Horsfall, W. R.,** *Mosquitoes: Their Bionomics and Relation to Disease*, Hafner Press, New York, 1972.
198. **Fish, D. and Carpenter, S. R.,** Leaf litter and larval mosquito dynamics in tree-hole ecosystems, *Ecology*, 63, 283, 1982.
199. **McCombs, S. D.,** Effect of Differential Nutrition of Larvae on Adult Fitness of *Aedes triseriatus*, M.S. thesis, University of Notre Dame, Notre Dame, Ind., 1980.
200. **Haramis, L. D.,** Larval nutrition, adult body size, and the biology of *Aedes triseriatus*, in *Ecology of Mosquitoes: Proceedings of a Workshop*, Lounibos, L. P., Rey, J. R., and Frank, J. H., Eds., Florida Medical Entomology Laboratory, Vero Beach, 1985, 431.
201. **Grimstad, P. R. and Haramis, L. D.,** *Aedes triseriatus* (Diptera:Culicidae) and La Crosse virus. III. Enhanced oral transmission by nutrition-deprived mosquitoes, *J. Med. Entomol.*, 21, 249, 1984.
202. **Hardy, J. L., Houk, E. J., Kramer, L. D., and Reeves, W. C.,** Intrinsic factors affecting vector competence of mosquitoes for arboviruses, *Ann. Rev. Entomol.*, 28, 229, 1983.
203. **Horsfall, W. R., Fowler, H. W., Jr., Moretti, L. J., and Larsen, J. R.,** *Bionomics and Embryology of the Inland Floodwater Mosquito Aedes vexans*, University of Illinois Press, Urbana, 1973.
204. **Wood, D. M., Dang, P. T., and Ellis, R. A.,** *The Insects and Arachnids of Canada. Part 6. The Mosquitoes of Canada, Diptera:Culicidae*, Biosystematics Research Institute, Ottawa, Canada, 1979.
205. **Sinsko, M. J. and Craig, G. B., Jr.,** Dynamics of an isolated population of *Aedes triseriatus* (Diptera:Culicidae). I. Population size, *J. Med. Entomol.*, 15, 89, 1979.
206. **Scholl, P. J. and DeFoliart, G. R.,** *Aedes triseriatus and Aedes hendersoni:* vertical and temporal distribution as measured by oviposition, *Environ. Entomol.*, 6, 355, 1977.
207. **Sinsko, M. J. and Grimstad, P. R.,** Habitat separation by differential vertical oviposition of two treehole *Aedes* in Indiana, *Environ. Entomol.*, 6, 485, 1977.
208. **Turell, M. J., Hardy, J. L., and Reeves, W. C.,** Stabilized infections of California encephalitis virus in *Aedes dorsalis* and its implications for viral maintenance in nature, *Am. J. Trop. Med. Hyg.*, 31, 1252, 1982.
209. **Mather, T. N. and DeFoliart, G. R.,** Effect of host blood source on the gonotrophic cycle of *Aedes triseriatus*, *Am. J. Trop. Med. Hyg.*, 32, 189, 1983.
210. **Patrican, L. A. and DeFoliart, G. R.,** Lack of adverse effect of transovarially acquired La Crosse virus infection on the reproductive capacity of *Aedes triseriatus* (Diptera:Culicidae), *J. Med. Entomol.*, 22, 604, 1985.
211. **Nasci, R. S.,** Variations in the blood-feeding patterns of *Aedes vexans* and *Aedes trivittatus* (Diptera:Culicidae), *J. Med. Entomol.*, 21, 95, 1984.
212. **Washino, R. K. and Tempelis, C. H.,** Mosquito host blood meal identification: methodology and data analysis, *Ann. Rev. Entomol.*, 28, 179, 1983.
213. **Day, J. F., Ebert, K. M., and Edman, J. D.,** Age of murine hosts determined by binding of estradiol to mosquito blood meals, *J. Med. Entomol.*, 19, 357, 1982.
214. **Craig, G. B., Jr.,** personal communication, 1986.
215. **Thompson, W. H., Trainer, D. O., Allen, V., and Hale, J. B.,** The exposure of wildlife workers in Wisconsin to ten zoonotic diseases, *Trans. 28th N. Am. Wildl. Res. Conf.*, Wildlife Management Institute, Washington, D.C., 1963, 215.
216. **Thompson, W. H.,** Lower rates of oral transmission of La Crosse virus by *Aedes triseriatus* venereally exposed after engorgement on immune chipmunks, *Am. J. Trop. Med. Hyg.*, 32, 1416, 1983.
217. **Mather, T. N. and DeFoliart, G. R.,** Dispersion of gravid *Aedes triseriatus* (Diptera:Culicidae) from woodlands into open terrain, *J. Med. Entomol.*, 21, 384, 1984.
218. **Taylor, W. P., Ed.,** *The Deer of North America*, The Stackpole Co., Harrisburg, Pa., and The Wildlife Management Institute, Washington, D.C., 1956.
219. **Mitchell, C. J.,** Mosquito vector competence and arboviruses, in *Current Topics in Vector Research*, Vol. 1, Harris, K. F., Ed., Praeger, New York, 1983, 63.
220. **Tabachnick, W. J., Aitken, T. H. G., Beaty, B. J., Miller, B. R., Powell, J. R., and Wallis, G. P.,** Genetic approaches to the study of vector competency of *Aedes aegypti*, in *Recent Developments in the Genetics of Insect Disease Vectors*, Steiner, W. W. M., Tabachnick, W. J., Rai, K. S., and Narang, S., Eds., Stipes Publishing, Champaign, Ill., 1982, 413.
221. **Hardy, J. L., Reeves, W. C., and Asman, S. M.,** Arbovirus research program at the University of California, Berkeley, *Proc. Pap. Annu. Conf. Calif. Mosq. Control Assoc.*, 43, 15, 1975.
222. **Trainer, D. O.,** Wildlife as monitors of disease, *Am. J. Public Health*, 63, 201, 1973.

223. **Boromisa, R. D. and Grimstad, P. R.,** Seroconversion rates to Jamestown Canyon virus among six populations of white-tailed deer *(Odocoileus virginianus)* in Indiana, *J. Wildl. Dis.*, 23, 12, 1987.
224. **Service, M. W.,** *Mosquito Ecology Field Sampling Methods*, Applied Science, London, 1976.
225. **Artsob, H., Wright, R., Shipp, L., Spence, L., and Th'ng, C.,** California encephalitis virus activity in mosquitoes and horses in southern Ontario, *Can. J. Microbiol.*, 24, 1544, 1978.
226. **McFarlane, B. L., Embree, J. E., Embil, J. A., Artsob, H., Weste, J. B., and Rozee, K. R.,** Antibodies to snowshoe hare virus of the California group in the snowshoe hare *(Lepus americanus)* and domestic animal populations of Prince Edward Island, Canada, *Can. J. Microbiol.*, 27, 1224, 1981.
227. **Francy, D. B.,** Mosquito control for prevention of California (La Crosse) encephalitis, in *California Serogroup Viruses*, Calisher, C. H. and Thompson, W. H., Eds., Alan R. Liss, New York, 1983, 365.
228. **Berry, R. L.,** *Aedes triseriatus*, man-made containers, and La Crosse encephalitis in Ohio, *Proc. Ohio Mosq. Control Assoc.*, 13, 13, 1983.
229. **Reeves, W. C.,** Summary of recommendations from the international symposium on California serogroup viruses, in *California Serogroup Viruses*, Calisher, C. H. and Thompson, W. H., Eds., Alan R. Liss, New York, 1983, 379.
230. **DeFoliart, G. R., Watts, D. M., and Grimstad, P. R.,** Changing patterns in mosquito-borne arboviruses, *J. Am. Mosq. Control Assoc.*, 2, 437, 1986.
231. **Beaty, B. J., Miller, B. R., Shope, R. E., Rozhon, E. J., and Bishop, D. H. L.,** Molecular basis of bunyavirus per os infection of mosquitoes *(Aedes triseriatus)*: role of the middle-sized RNA segment, *Proc. Natl. Acad. Sci. U.S.A.*, 79, 1295, 1982.

Chapter 20

CHIKUNGUNYA VIRUS DISEASE

P. G. Jupp and B. M. McIntosh

TABLE OF CONTENTS

I. HISTORICAL BACKGROUND

A. Discovery of Agent and Vectors

The word "chikungunya" was first used by the indigenous people of Southern Province, Tanganyika Territory (Tanzania), in reference to a disease which afflicted them in epidemic form in 1952—1953.[1-3] The disease was characterized mainly by a sudden onset, fever, rash, and joint pains. The latter were often severe and sometimes persisted as a recurrent arthralgia for some time. The word is Swahili meaning "that which bends up" and refers to the stooping posture adopted by patients because of the severity of the joint pains. The incidence of the disease was highest on the Makonde Plateau, but it also occurred to a lesser extent in the adjoining lowlands of the Ruvuma River valley. From there it was believed at the time that the disease had been introduced into the highly susceptible plateau population by movement of infected humans and subsequently spread throughout the plateau by the same means. Man-biting *Aedes aegypti*, the suspected vector, was abundant in the villages as were their larvae in water storage jars in the huts. During the outbreak, a previously unknown virus was isolated from humans and mosquitoes, including *Ae. aegypti*,[3,4] and its etiological role confirmed serologically by the demonstration of specific antibodies in recovered patients.[3] In 1956, experimental transmission of the virus by *Ae. aegypti* was reported,[5] and its identification as a group A arbovirus (*Alphavirus*), with a close antigenic relationship to Mayaro and Semliki Forest viruses, was reported in 1957 and 1959.[6,7] The Makonde Plateau is virtually devoid of woodland and, in retrospect, it seems certain that the virus did in fact originate in the lowland woodland and was transmitted on the plateau by a highly efficient domiciliary vector which would explain the high incidence of the disease in the plateau population which was only infrequently exposed to infection, and consequently had only a low immunity.

Studies in Africa have uncovered a sylvan transmission cycle between wild primates and aedine mosquitoes of the *Stegomyia* and *Diceromyia* subgenera in the tree canopy of moist forest and semiarid savannah-woodland. First indications of this were reported from Uganda in 1958 when chikungunya (CHIK) virus was isolated from the primatophilic *Ae. africanus* collected in the forest canopy and from a mosquito catcher who may have been infected in the canopy.[8] Later studies during the 1960s in Uganda further implicated *Ae. africanus* in forest redtail monkey transmission cycles, but also suggested that the virus was not being maintained in the isolated forests around Entebbe with their small monkey populations.[9] In West Africa, the same mosquito, as well as *Ae. luteocephalus* and the *Ae. furcifer*/*Ae. taylori* species pair, were implicated in both feral and human transmission in the 1960s and 1970s.[10-13] In southern Africa, where the *Ae. furcifer-taylori* pair appears to be the only important rural vector, it was implicated in human and wild primate transmission in Zimbabwe in 1963,[14] and in South Africa in studies during 1964 to 1967[15] and again in 1976.[16] In West Africa, *Ae. aegypti* was implicated as an urban vector during an outbreak in 1969 in Ibadan, Nigeria,[17] and in 1970—1971 in Luanda, Angola.[18] Since 1954, the virus has been identified as the cause of epidemics in the Philippines,[19,20] Thailand,[19,21] Kampuchea,[22]

Table 1
LOCATION AND YEAR OF HUMAN OUTBREAKS OF CHIK DIAGNOSED ON CLINICAL AND/OR LABORATORY BASES

Tanzania (Makonde)	1952—1953[12]
Philippines (Manila, Amlan)	1954, 1956,[19,20] 1968[42]
South Africa (Transvaal Province)	1956,[43] 1975, 1976,[16] 1977[44,45]
Thailand (Bangkok)	1958,[19] 1962, 1963, 1964[21]
Zaire	1958[30]
Zambia	1959[46]
Senegal	1960,[47] 1966,[10] 1982[48]
Kampuchea	1961, 1962[22]
Uganda (Entebbe)	1961—1962,[9] 1968[9]
Zimbabwe	1962,[49] 1971[50]
India (various localities)	1963,[23] 1964,[51] 1965,[25] 1973[52]
Burma	1963,[53] 1970—1972[29]
Nigeria (Ibadan)	1964,[17] 1969,[17] 1974[54]
Vietnam	1964[55]
Sri Lanka	1965[27]
Angola (Luanda)	1970—1971[18]
Central African Republic	1978—1979[13]

India,[23-26] Sri Lanka,[27] Vietnam,[28] and Burma.[29] In at least some of these outbreaks, *Ae. aegypti* was implicated as the main vector.

Evidence of infection in wild primates in Africa concerned both the presence of antibodies as well as the isolation of virus from several primate species. Reports on the presence of antibodies emanated from Zaire (1960),[30] Zimbabwe (1964),[14] Nigeria (1968),[31] South Africa (1970, 1977),[16,32] and Uganda (1971).[9] Virus isolations were reported from Senegal in 1969, 1979, and 1984.[12,33,34]

Previous recent reviews on CHIK were published in 1975,[35] 1981,[36-38] and 1984.[39]

B. History of Epidemics

Carey,[40] after study of the historical literature, advanced the thesis that there has been confusion between dengue (DEN) and CHIK, but since the isolation of the causal viruses, it is now possible to differentiate these illnesses on clinical grounds alone. DEN is typically characterized by a diphasic fever lasting about 1 week, headache, retroorbital pain, backache with generalized body pains, and rash, the acute illness being followed by residual asthenia. CHIK differs in that the pains are predominantly located in the joints rather than the muscles; the febrile period is shorter and usually not diphasic; there is an absence of mortality; and some patients experience persistent arthralgia following the acute episode, but no asthenia. Using these clinical differences as a basis, Carey suggested that epidemics of CHIK can be said to have occurred in 1779 (Batavia-Jakarta; Cairo), 1823 (Zanzibar), 1824—1825 (India), 1827—1828 (West Indies; New Orleans; Charleston, S.C.), 1870 (Zanzibar), 1871—1872 (India), 1901—1902 (Hong Kong; Burma; Madras), and 1923 (Calcutta). It was in Cuba in 1828 that the term "dengue" was first used, which is possibly derived from the word "dinga" used in Zanzibar in 1923. Since Mayaro virus is now known to cause an illness very similar to CHIK virus, including residual arthralgia,[41] it seems that the outbreaks in the Western Hemisphere might have been caused by that virus.

Table 1 lists the epidemics due to CHIK which have occurred since the isolation of the prototype CHIK virus in 1952—1953 until the present. CHIK appears to be enzootic throughout much of tropical Africa from where it has apparently spread to other parts of the world; there is historic evidence for this. Such evidence includes the chronology of the outbreaks, their infrequency in Asia and the very high morbidity prevailing there when they did occur,[40] and the failure so far to find evidence of a feral transmission cycle outside of Africa.

II. THE VIRUS

A. Antigenic Relationships

The initial antigenic classification of CHIK virus was based on reactions obtained with hemagglutination-inhibition (HI) and complement-fixation (CF) tests as well as neutralization (N) tests in mice, which demonstrated its relationship to certain group A viruses available at the time, particularly to Mayaro and Semliki Forest viruses.[6,7] These studies were confirmed and extended by others using variations of the N test in cell cultures,[56-60] which gave better definition of antigenic complexes within the genus *Alphavirus*. At present, 6 complexes are recognized among the 25 members of the genus.[60] CHIK virus is one of the four species of the Semliki Forest complex, the others being Semliki Forest, Getah, and Mayaro. CHIK and o'nyong-nyong (ONN) viruses are regarded as subtypes of the CHIK virus species. In the N test, there is a partial one-way relationship between CHIK and ONN with CHIK antisera neutralizing both viruses equally and ONN antisera neutralizing the homologous virus to a significantly greater degree.[41,57]

B. Vertebrate Host Range

In the laboratory, 1- to 4-day-old mice die 2 to 3 days after intracerebral (i.c.) or intraperitoneal (i.p.) inoculation, while 3- to 4-week-old mice usually survive and develop antibodies,[61] although a variant strain of the virus has been developed which is lethal to adult mice by the i.c. route.[61] Guinea pigs, hamsters, and rabbits do not develop viremia after inoculation, but survive with development of antibodies.[61] African rodents of the genera *Mastomys, Arvicanthis,* and *Aethomys* develop a low viremia, while *Mystromys* does so with high-level viremia, and they all develop antibodies.[62] Viremia followed by antibodies develops in rhesus monkeys, vervet monkeys, and baboons (*Macaca radiata, Cercopithecus aethiops,* and *Papio ursinus*) and day-old chickens.[61,63,64] Adult fowls,[65] domestic sparrows, and pigeons[66] are refractory to infection by the virus. One of nine Indian fruit-eating bats, *Rousettus leschenaulti*, inoculated with virus, showed a trace of viremia and eight developed low titer hemagglutinating and/or neutralizing antibodies.[66] Furthermore, two African bats of the genera *Tadarida* and *Pipistrellus* circulated virus after inoculation.[67] Adult cats are refractory to infection, although inoculation of newborn kittens may lead to viremia.[65]

C. Strain Variation

African and Asian strains of CHIK virus differ biologically, viz. differences in plaque size and heat stability, although these differences may possibly be due to the relatively greater number of mouse passages of the African strains.[68] Strains from Calcutta have been shown by the kinetic HI test to be more closely related to a strain from Thailand than to an African strain, but differences are slight and probably of little importance in vaccines.[23,68] The KLA-16 strain of CHIK virus which was recovered from a child with hemorrhagic fever in Thailand[69] differs from other strains by its effects on infant mice, rats, and hamsters. These animals develop a hemorrhagic syndrome, mainly associated with the intestine, and also characterized by thrombocytopenia, prolonged bleeding and clotting times, and a decrease in the time for formation of prothrombin.[69,70] Strains isolated from Calcutta also occasionally produced intestinal hemorrhage in the course of their isolation in mice.[71]

Small and large plaque (SP and LP) variants of CHIK virus have been produced experimentally.[72] These have been compared serologically with ONN virus to elucidate the one-way antigenic relationship between ONN and CHIK viruses.[59] The results suggest that similar antigens are present on both ONN and CHIK variants, but in the latter, the antigens are differently distributed which prevents them from participating in the neutralization of the virus by ONN antibodies. ONN virus could not be distinguished from CHIK SP on the basis of plaque morphology, although it is clearly distinct because of its inability to multiply in

Ae. aegypti cells. CHIK SP virus differs in its greater ability to replicate in *Ae. aegypti* mosquitoes than either CHIK prototype virus of CHIK LP virus. All three viruses had the ability to multiply in *An. gambiae* cells, although ONN was the only virus to infect *An. gambiae* mosquitoes. The similarities between ONN and CHIK viruses might be due to convergent evolution.[59]

D. Methods for Assay

For primary isolation and cultivation of laboratory strains of CHIK virus, both suckling mice and a wide range of vertebrate and mosquito cell lines are available. Cell lines which have been used successfully include VERO,[73] BHK-21,[74] and the C6-36 *Ae. albopictus*[75] and *Toxorhynchites amboiensis*[76] cell lines. Mosquito cells do have a disadvantage because of the absence or irregular production of cytopathology,[77] although this is irrelevant where immunofluorescence is used for the detection of viral antigens.[78] The enzyme-linked immunosorbent assay (ELISA) has not been used yet for detection of CHIK antigen, but it should be applicable for the assay of antigen in sera or in individual mosquitoes as has been done in the case of Rift Valley fever virus.[79] It does seem that cell lines from *Aedes* mosquitoes are more sensitive to CHIK than those from *Culex*.[77] For primary isolation, suckling mice still have certain advantages over cell cultures because of their high susceptibility, the ease by which CF and hemagglutinating antigens can be obtained from their brains, the lower level of technique required in their usage, and the higher tolerance they possess to the toxic effects of inoculating pools consisting of large numbers of mosquitoes. Also, because of their high susceptibility to a wide spectrum of arboviruses, mice are certainly better in infectivity surveys in which as many arboviruses as possible are being investigated. The sensitivity of mice can be increased in primary isolation if the inocula are first inoculated into male *Ae. aegypti* or species of *Toxorhynchites* mosquitoes.[80]

III. DISEASE ASSOCIATIONS

A. Humans

CHIK is a feverish illness characterized by sudden onset, chills, flushed face, nausea, vomiting, backache, headache, photophobia, lymphadenopathy, arthralgia, and rash.[44,55,81-83] The acute illness lasts 3 to 5 days, with recovery in 5 to 7 days. The incubation period is usually about 2 to 4 days and the single most significant symptom is the arthralgia, presenting in 70% of cases. It may be severe, affecting one joint or several. Reddening and swelling of the joint may occur. The arthritis may persist in a small proportion of cases for months or years and mimic rheumatoid arthritis.

Patients with persistent joint pain and stiffness have shown high viral antibody titers. The rash, appearing most commonly on the trunk, is macular or maculopapular, and rarely petechiae may be present. It may be pruritic and occur in short-lived episodes.[43,44,49,55,82] Biopsy specimens from the skin lesions of CHIK patients show a perivascular lymphocytic infiltrate in the upper half of the dermis and red blood cell extravasation is seen around the superficial capillaries.[45] In Asia, comparison of the symptomatology of DEN and CHIK infections indicated that hemorrhagic manifestations rarely occur in cases of CHIK and that CHIK should not be listed as a hemorrhagic fever.[36] Severe hemorrhagic symptoms have not been reported in CHIK cases in Africa. Viruses associated with severe hemorrhagic fever in south and Southeast Asia from 1956 to 1963 were either DEN alone or DEN and CHIK.[84] In Africa, ONN fever has similar symptoms to CHIK, but ONN may be distinguished by the presence of lymphadenitis which is absent in CHIK patients.[85]

B. Domestic Animals and Wildlife

There are no records of clinical disease in domestic animals or wildlife due to CHIK virus.

C. Applicable Diagnostic Procedures

CHIK fever can be confused clinically with ONN, DEN, Sindbis, and West Nile infections, so diagnosis should be confirmed by virus isolation as well as serologically. Virus is most readily isolated from the blood within 48 hr of the onset of illness, but has been isolated as late as 6 days afterwards using infant mice.[71] In Africa, it may be necessary to differentiate CHIK from ONN virus which is difficult to do serologically (see Section II.A). However, the two can be separated because CHIK has a greater pathogenicity for infant mice.[86] Apart from ONN, the HI test will readily differentiate CHIK from other *Alphavirus* infections. In India, the virus was readily isolated and identified in 72 hr using mice and CF or HI tests with antigens prepared from the mice.[24] Diagnosis is thus based on either virus isolation or a fourfold or greater rise in antibody titer following the illness. Usually a serum sample collected within 5 days of onset of fever will be free of HI, CF, and N antibodies.[21,26] A sample collected 2 weeks or more after onset should have HI and low-level N antibodies; CF antibodies develop more slowly.[21,26] N antibodies can be measured by the virus dilution method in infant mice, by the serum dilution method in cell cultures by CPE, or by plaque assay methods. Radial hemolysis has been shown to correlate well with the HI test.[87] Methods for virus isolation are by inoculating acute-phase serum or other tissues i.c. into infant mice[3] or into cell cultures. CHIK virus produces cytopathogy in a variety of vertebrate cell lines, but VERO or BHK-21 cells have been commonly used. For the rapid identification of virus in sera, these may be inoculated into cell cultures and a diagnosis made after 9 hr by the indirect immunofluorescence test. Alternatively, after 24 hr the infected cells are negatively stained and examined with the electron microscope.[78] Indirect immunofluorescence has also been used to distinguish CHIK from the closely related Mayaro and Semliki Forest viruses.[88]

IV. EPIDEMIOLOGY

A. Geographic Distribution

Table 1 lists the countries that have experienced CHIK outbreaks from 1952 onwards. The virus seems to be enzootic throughout tropical Africa. Outbreaks, sometimes major epidemics, have been documented for East Africa (Tanzania and Uganda), southern Africa (northeastern Transvaal and Zimbabwe), West Africa (Senegal, Nigeria, and Angola), and Central Africa (Central African Republic, Zambia, and Zaire). According to antibody surveys, human infection has also occurred in Mozambique, northern Botswana, and northeastern Namibia.[89,90]

CHIK appears to have spread to other parts of the world from Africa to cause pandemics in both the American and Asian tropics.[40] India has had a history of epidemics from 1824 until 1965 when the virus spread to Sri Lanka. CHIK became established endemically in Southeast Asia during the late 1950s to early 1960s and was continuously transmitted in the towns and cities in Thailand, Kampuchea, and Vietnam, probably largely by *Ae. aegypti*. Outbreaks have been recorded in Burma in 1963, 1970, and 1973 and in Manila in 1954, 1956, and 1968. Serological surveys alone have recorded the presence of the virus in South Vietnam (Laos),[91] Vietnam (Saigon),[28] the Malaysian Peninsular,[92] Pakistan,[93] and the Pacific Islands.[94]

B. Incidence and Serologic Epidemiology

In the various outbreaks studied, human infection rates have varied depending upon factors relating to the immune status of the population, the size and density of the population, whether infection was urban or rural, the extent to which human infection was dependent upon the feral transmission cycles, and the efficiency of the particular vector system. In Asia, where most recognized outbreaks have been in large urban populations with transmission effected by *Ae. aegypti*, outbreaks have been on a large scale. In contrast, in Africa they have been smaller, have tended mainly to involve rural populations, transmission has often been by feral vectors and evidently less efficient, and outbreaks have often been conditional upon viral activity in the wild primate transmission cycles.

In India, where infection rates were obviously very high, it was estimated that during the outbreak in Madras in 1964, nearly 400,000 cases occurred.[25] In the town of Barsi in Central India in 1973, the overall morbidity was 37.5%.[52] In Thailand during 1962 to 1964, CHIK was common in Bangkok. During the 1962 outbreak alone it was estimated that between 44,000 and 70,000 cases were seen in outpatient children, and the infection rate, based on 1887 persons bled before and after the 1962 rainy season, was 31%.[95] Immune rates by age in 1962 were 10 to 20% in 1 to 2 year olds, 70% in 15 to 19 year olds, and 70 to 85% in 20 to 70 year olds.[95] Antibody surveys among adults from different regions outside of Bangkok indicated that immune rates were the same as in Bangkok, i.e., 80 to 100%, while conversion rates of 30 to 50% were obtained in 337 school children; it was evident from both studies that infection was widespread in rural Thailand.[96] After the 1964 outbreak in Vietnam, immune rates among 472 children were 33.3% in those less than 2 years old and 81.6% in the 10- to 15-year age group.[28] However, recent evidence indicates that the virus had virtually disappeared from Bangkok by the early 1980s despite the continued presence of DEN and abundance of *Ae. aegypti*. The reason for this decline was not evident.[97]

The 1952 outbreak in Tanzania involved a population of 150,000 living in small villages scattered over the Makonde Plateau, with many villages more than 8 km from water, thus giving rise to a need to store domestic water which led to large numbers of *Ae. aegypti*.[2] Once virus had been introduced, the outbreak spread rapidly to affect eventually most of the villages, The overall morbidity on the plateau was 47.6% among adults and 50.8% among children. *Ae. aegypti* was also the main vector in Ibadan, Nigeria, in 1969[17] and in Luanda, Angola, in 1970.[98] In Luanda, an antibody survey after the outbreak showed that 65% of sera reacted only with CHIK antigen; 33.7% with both CHIK and ONN, but usually to a greater degree with CHIK; and 1.3% with ONN alone. In the Central African Republic in 1978—1979, CHIK virus was active in several areas.[13] An antibody survey at this time revealed an immune rate of 17% suggestive of a poor vector system. Prior to this, virus had not been active since 1975 despite continuous surveillance. Studies in Uganda over several years revealed a very low level of human infection, also apparently the result of an inefficient vector system.[9] A sylvan vector *Ae. africanus* was involved and human infection resulted from incidental leakage of virus from transmission cycles involving monkeys in small isolated forests. It seemed that there was a 5- to 7-year — perhaps even a 10-year — periodicity in monkey epizooticity because of the need to replace immune monkeys by susceptibles. In southern Africa, where human infection has been caused by the sylvan vectors *Ae. furcifer-taylori*, a somewhat similar periodicity in human infection has been evident, although human infection has sometimes been quite intense among small localized groups.[16,49] During an outbreak in South Africa in 1976, postepidemic immune rates in a locality with a large baboon population were 77.7% among 18 children and 55.7% among 106 adults, while all of the 92 baboons tested were immune.[16] Antibody surveys in Mozambique in 1957 showed immune rates in adult humans of 37.5% and in children of 4%,[89] in Angola in 1960 of 32.5% in adults and 3.3% in children;[99] in Botswana in 1959 of 25.3% in adults and 13.4% in children;[90] and in northern Namibia in 1966 of 1%.[100]

C. Seasonal Distribution

Outbreaks of CHIK depend upon sufficient rainfall filling the tree-holes or artificial containers preferred for oviposition by the aedine mosquito vectors, resulting in high densities of the vector species. Sufficient nonimmune humans must be present. Rural outbreaks in Africa are also related to the presence of adequate nonimmune populations of wild primates which in turn will also depend to a degree on the rainfall.

In Africa, the 1962 epidemic in Zimbabwe and the 1976 epidemic in South Africa, both rural outbreaks, occurred following unusually heavy rains in the wooded savannah which caused high densities of *Ae. furcifer-taylori* mosquitoes. Furthermore, the original epidemic in 1952—1953 in Tanzania was associated with the wettest year (1952) for a decade.[2] This heavy rainfall probably increased the numbers of *Ae. furcifer-taylori* in the woodland habitat in the lowlands of the Ruvuma River from where infection was brought into the villages on the Makonde Plateau. In 1975, a rural epizootic of CHIK occurred in western Senegal which peaked in November at the end of the rainy season with the highest number of virus isolations obtained from *Ae. furcifer-taylori* and *Ae. luteocephalus* during that month[12] when mosquito populations were actually declining. The explanation for this unexpected peaking of virus transmission so long after the heavy rainfall in July is unclear.

In urban outbreaks where *Ae. aegypti* is the vector, possibly supplemented by *Ae. albopictus* in Asia, the relationship with rainfall pattern has been recorded for several countries. In the 1969 Ibadan, Nigeria, outbreak, the frequency of infections increased and decreased parallel with the rainfall pattern, but the virus did not subsequently reappear during the late rains in October, probably because of high immunity in the human population by that time.[17] A similar pattern occurred in the 1964 epidemic in Vellore, India.[26] The highest densities of *Ae. aegypti* have occurred in Thailand just after the first rains in May and the epidemic peak has followed 2 months later. CHIK virus is only active in the rainy season in Kampuchea,[22] while in Sri Lanka, maximum infection occurred in June and July 1 to 2 months after the heavy rains in May.[27] In other areas of south and Southeast Asia where rainfall is not markedly seasonal, cases of CHIK may occur throughout the year.[84]

D. Risk Factors

Risk factors are related mainly to the degree of exposure of humans to the mosquito vectors. Age, sex, and occupation are irrelevant except insofar as they may increase this exposure. On the other hand, place of residence or temporary visits to high-risk areas have been of primary importance. Thus, risk factors are largely dictated by ecological factors determining the geographical distribution and abundance of the vectors. Where the vector has been the domestic man-biting *Ae. aegypti*, risk to man has been highest among urban populations, especially those of the lower socioeconomic class, where the container habitat for mosquito larvae is usually most abundant and where, in some rural villages, the need to store water is great. Furthermore, the houses occupied by this class of the population are less likely to be mosquito-proof. This probably accounts for the threefold greater infection rate in the lower compared to the higher socioeconomic class in Mandalay, Burma, shown by the 1973—1974 serological survey.[101] Similarly, in the 1965 outbreak in Colombo, Sri Lanka, the lowest income group had the highest attack rate.[27] In Africa, where proximity to the feral transmission cycles has been of importance, high-risk areas are either forest or riverine and savannah-woodland where the habitats of the sylvan vectors and the wild primate hosts occur. In southern Africa, this effect of the wild transmission cycles has been clearly evident as recognized human infection there has been restricted exclusively to the putative distribution of *Ae. furcifer-taylori*,[14,16] and, consequently, human outbreaks have been confined to the tropical region of southern Africa. This region embraces the extreme north of Namibia and Botswana (Ovamboland and the Okavango), the lowlands of Zimbabwe, the northeastern Transvaal and northern Natal in South Africa, and the Mozambique lowlands.[32] Within this tropical region, human infection has been most intense in habitats chosen by wild primates for their dormitories, i.e., riverine woodland and high granite outcrops.

Table 2
CHIK VIRUS ISOLATIONS FROM *AEDES* IN AFRICA

Country	Year	Species (no. isolations >1)
Tanzania	1953	*aegypti* (2)[2,61]
Uganda	1956	*africanus*[8]
	1961	*africanus* (5)[103]
	1968	*africanus* (9)[9]
Nigeria	1963	*aegypti* + *taylori*[a31]
	1969	*aegypti* (13)[17]
Senegal	1966—1967	*luteocephalus* (2), *aegypti*, *irritans*[33]
	1975	*furcifer-taylori* (28), *luteocephalus* (8), *dalzieli*[12]
	1983	*furcifer* (29), *luteocephalus* (2), *africanus*, *neoafricanus*, *vittatus*[104]
South Africa	1976	*furcifer-taylori* (16)[16]
Angola	1971	*aegypti*[18]
Central African Republic	1978—1979	*africanus* (33)[13]

[a] This pool also contained 3 *Eretmapodites grahami*.

V. TRANSMISSION CYCLES

A. Evidence from Field Studies

1. Vectors

Field evidence implicating vectors largely involves that obtained by isolating the virus from wild mosquito populations complemented by studies on the biology of the suspected vector species. This section deals with virus isolation studies; those concerning vector biology are described in Section VI.

Virus isolation studies have indicated that *Aedes* species are the main CHIK vectors in both wild primate and human transmission cycles. More specifically, the evidence is strongest for species of the *Diceromyia* and *Stegomyia* subgenera, although odd isolations have come from *Aedimorphus* species. Based on isolation frequency in Africa, it is clear that the main vectors in the Afrotropical Region are in probable order of importance the *Ae. (Dic.) furcifer-taylori* pair, *Ae. (Stg.) africanus*, *Ae. (Stg.) luteocephalus*, and *Ae. (Stg.) aegypti*. The evidence in Africa concerning the last species relates solely to domestic populations; in fact, the indications are that the feral form is not involved. Table 2 shows the numbers of virus isolations from African aedine species according to country since the isolation of the prototype strain in 1953. Of the 156 total isolations, 74% came from *Ae. furcifer-taylori*. Since it became possible to separate the females of these species in Senegal, isolations of CHIK virus have been made only from *Ae. furcifer*. Evidence from studies in southern Africa suggests that this species is also the main vector there. There have been a few isolations from *Mansonia* species, (Nigeria, Uganda) and, since *Mansonia africana* is a proven experimental vector,[102] it is possible that this species, at least, could at times be a covector.

In Asia, virus isolations have been obtained only from *Ae. aegypti*, and it is reliably certain that this species has been responsible for the Asian urban epidemics. There were seven isolations in Thailand in 1962 and nine in 1964.[95] In India, there were 2 isolations in Calcutta in 1964[71] and 111 isolations in South India in 1964.[24] There were no isolations from *Cx. quinquefasciatus* or from the *Cx. vishnui* complex. The highest minimum field infection rate in mosquitoes was obtained over 1 week in Vellore, when it was 18/1000.

Occasional isolations have also been made from *Culex* species in Asia and Africa, but as experimental transmission studies indicate that *Culex* species are not biological vectors, any role these species might have in transmission would evidently be limited to mechanical transmission.

2. Vertebrate Hosts

There have been single isolations of CHIK virus from various vertebrates including a golden sparrow (*Auripasser luteus*) in Nigeria,[17] a ground squirrel (*Xerus erythropus*),[33] and a bat (*Scotophilus* sp.) in Senegal,[34] but the only group of vertebrates for which there is a strong body of field evidence implicating them as primary hosts in CHIK transmission cycles are the wild primates. The evidence comes mainly from studies undertaken in several countries in Africa. The first recorded indirect evidence of the involvement of wild primates in sylvatic infection of CHIK was the isolation of virus from primatophilic *Ae. africanus* in 1956 in Uganda.[8] The first recorded direct evidence was published by Osterrieth and co-workers in 1960[30] who demonstrated immunity in chimpanzees in Zaire. Studies in southern Africa have demonstrated antibodies in vervet monkeys (*Cercopithecus aethiops*) and Chacma baboons (*Papio ursinus*). A year after the 1962 epidemic in Zimbabwe, all the baboons and vervets in small samples tested were seropositive.[14] Similarly, after the 1976 epidemic in the northeastern Transvaal, 70% of the sera from 92 baboons contained antibodies.[16] Immune rates of 25% were found in vervets in northern Natal, South Africa, and of 16% in baboons in the northeastern Transvaal and Zimbabwe sampled in regions which had not experienced epidemics.[32] In Senegal, isolations have been made from vervet monkeys (two), the baboon *P. papio* (one), and the bushbaby *Galago senegalensis* (one).[12,33] Serological studies in Nigeria have shown immune rates in three monkey species: *Erythrocebus patus* (10%), *C. aethiops tantalus* (7%), and *C. mona* (20%).[31] Immunity has been shown in monkeys sampled at various localities in Uganda,[29] and in the Zika forest near Entebbe observations indicated the occurrence of an intense sylvan epizootic of CHIK in a cycle between red-tailed monkeys. (*C. ascanius schmidti*) and *Ae. africanus*. These long-term studies at Zika have provided evidence that suggests a 5- to 7-year periodicity in CHIK activity in the forest related to the reproductive replacement of immune monkeys by nonimmune monkeys.[9]

Serological tests done on a variety of wild primates imported into the U.S. have shown significant immune rates in gorillas, chimpanzees, baboons, vervets from Africa, and orangutans and macaque monkeys (*Macaca mulatta*) from Asia.[105] Other evidence for infection in Asian monkeys was obtained in a serological survey from 1962 to 1970 in Malaysia which showed a low immunity rate.[106]

In Uganda, no antibodies were found in 224 *Tadarida* bats and 78 other bats of 7 other species, nor in samples of birds, rodents, and cattle.[9,107] In Bankok, however, antibodies were reported in cattle, water buffalo, horses, pigs, dogs, rabbits, and bats, but not in cats or rodents.[91] Titers of virus recorded in patients during the Calcutta epidemic of 1963—1964 ranged from 1 to 7.0 log/mℓ.[23,71] The higher titers would allow significant numbers of *Ae. aegypti* and other vectors to become infected and subsequently transmit virus.[108]

B. Evidence from Experimental Infection Studies

1. Vectors

The first vector competence test with CHIK virus was done by Ross[5] with *Ae. aegypti* from Uganda: 87% of the mosquitoes became infected after feeding on a virus suspension through a bat wing membrane, and 25% of the infected mosquitoes transmitted the virus into blood suspensions through other membranes.[5] Since this experiment, reported in 1956, there have been several tests with *Ae. aegypti*, frequently attempting to transmit the virus to infant mice. Whether the population of *Ae. aegypti* is the feral form (subspecies *formosus*) or the domestic form has not been stated in most reports. The 50% infection thresholds determined range from 6.7 to 9.2 log/mℓ, and transmission rates were 13, 78, 27, and 90 to 100% in mosquitoes previously fed on infective meals with titers 7.3, 8.2, 8.9, and 9.2 log/mℓ, respectively.[64,108-112] Since this virus probably does not circulate much above 7.0 log in humans, these figures suggest that some populations of *Ae. aegypti* would be rather inefficient vectors during man-to-man transmission. Mechanical transmission between mice

Table 3
RESULTS OF VECTOR COMPETENCE TESTS WITH VARIOUS MOSQUITO SPECIES AND CHIK VIRUS

Species	Country/continent	50% infection threshold (log_{10}/mℓ)	Transmission (host animal)	Ref.
Ae. furcifer	South Africa	<6.2	+, 25%, 32% (monkeys, hamsters)	102
Ae. vittatus	South Africa	<6.7	+ (hamsters)	112
Ae. fulgens	South Africa	ca. 7.0	+ (hamsters)	102
Ma. africana	South Africa	ca. 7.0	+, 29% (monkeys)	102
Ae. metallicus	South Africa	>7.2	± (hamsters)	112
Ae. ledgeri	South Africa	>7.2	± (hamsters)	112
Ae. simpsoni	South Africa	<5.2	− (monkeys)	110
Ae. circumluteolus	South Africa	>7.3	− (monkey)	110
Ae. apicoargenteus	Uganda	+	+ (monkey)	116
Ae. togoi	Asia	+	+	111
Er. chrysogaster	Tropical Africa	+	+	109, 111

Note: + = infection or transmission; − = no transmission; % = transmission rate (proportion of infected mosquitoes transmitting virus).

by *Ae. aegypti* has been shown to occur up to 8 hr after the infective feed and the most efficient transmission occurred within the first hour.[113] All *Culex* species so far tested — *Cx. quinquefasciatus*,[14,102] *Cx. molestus*,[109] *Cx. horridus*,[102] *Cx. poicilipes*, and *Cx. univittatus*[110] — have been found refractory to infection with CHIK virus. Tests have shown *Anopheles* species to be barely susceptible to infection (*An. gambiae*[63,115] and *An. stephensi*[114]) or susceptible but unable to transmit (*An. albimanus*[109]).

Tests on other aedine mosquitoes that have been incriminated as vectors in field studies have been done in the case of *Ae. furcifer* in South Africa[102,115] and *Ae. africanus* in Uganda.[116] *Ae. furcifer* has a 50% infection threshold of <6.2 log/mℓ and a transmission rate of 25 to 32% which would enable it to participate in transmission cycles involving vervet monkeys and baboons which circulate virus at levels up to about 7.0 and 8.0 log/mℓ, respectively. Furthermore, two laboratory personnel were accidentally infected by bite by *Ae. furcifer* which had previously fed on viremic monkeys.[115] *Ae. africanus* has also experimentally transmitted the virus between rhesus monkeys.[116]

The results of the experiments done on *Ae. furcifer* as well as those done on other species are shown in Table 3. The 50% infection threshold for *Ae. vittatus*, *Ae. fulgens*, and *Mansonia africana* would permit them to become infected in nature by feeding on viremic vervets or baboons so that the measure of their susceptibility and ability to transmit the virus makes them potential vectors. *Ae. metallicus*, *Ae. ledgeri*, and *Ae. circumluteolus* were poorly susceptible and largely failed to transmit; *Ae. simpsoni*, although susceptible, did not transmit the virus. Some populations of *Ae. albopictus*, the Asian species, have been shown as better vectors than populations of *Ae. aegypti*,[64,108,111] while other populations appear to be less efficient vectors.[114] Hence, *Ae. albopictus* can be regarded as a potential feral vector in India and Thailand.

2. Vertebrate Hosts

In one study, cattle, sheep, goats, horses, and various species of birds showed no viremia after inoculation of CHIK virus. There also was no antibody response except for low-titer hemagglutinating antibodies in some of the goats and sheep.[63] In another study, a significant number of horses and cattle developed hemagglutinating antibodies after inoculation, but the positive reactions in horses may have resulted from cross reactions with antibodies to Getah virus.[117]

In Uganda, inoculation of the fruit-eating bat *Eidolon heluum* failed to elicit either viremia or an antibody response.[118] However, inoculation of the insectivorous bats *Pipestrellus nanus* and *Tadarida aegyptiaca* in South Africa caused viremias lasting at least 3 days ranging from 4.9 to 6.7 and 3.0 to 5.3 log/mℓ, respectively, and the bushbaby or night ape (*Galago senegalensis*) circulated virus for at least 3 days at concentrations of from 4.1 to 7.2 log/mℓ.[67] Inoculation of virus into South African rodents showed that only one species, *Mystromys albicaudatus*, the white-tailed rat, circulated virus at high levels of from 4.2 to 8.1 log/mℓ.[62] The distribution of this rodent does not, however, coincide with the distribution of CHIK virus.

Studies with Indian rhesus monkeys (*Macaca radiata*) and South African vervet monkeys (*Cercopithecus aethiops*) and baboons (*Papio ursinus*) demonstrated that all circulated the virus at high concentrations and developed hemagglutinating and N antibodies. Viremia in rhesus monkeys lasted up to 6 days and titers were 3.3 to 7.5 log/mℓ by mouse titration.[64] Similar tests done with African and Asian strains of the virus using the less sensitive monkey kidney cells for titration of virus showed viremia of up to 5 days duration and titers of 3.5 to 5.0 and 4.5 to 5.5 log/mℓ, respectively.[119] Titers in vervet monkeys and baboons ranged from 3.5 to 7.0 and 4.6 to 8.2 log/mℓ, respectively, and the viremia lasted up to 4 days.[63]

C. Maintenance/Overwintering Mechanisms

From observations in a small, isolated, moist equatorial forest at Zika near Entebbe, Uganda, it was postulated in 1971 that CHIK virus, together with Zika virus, were not being maintained in a monkey *Ae. africanus* cycle within this forest.[9] Instead, it seemed that viral activity in the monkeys followed the introduction of virus, and the subsequent monkey epizootics were in time terminated because of rising monkey immunity and could not recur until adequate replacement of immunes had taken place. This resulted in a 5- to 7-year periodicity in their activity and, during the interepizootic years, virus was absent as became apparent by the failure to isolate the viruses from *Ae. africanus*. It was concluded that CHIK virus could only be maintained in a very large forested area by a series of epizootic foci moving continually because of the changing pattern of wild primate immunity. Movement of virus within forests could be accounted for by long-range dispersal of *Ae. africanus* and extra-forest *Ae. africanus* could transfer virus between forests.

Observations on yellow fever virus in West Africa by workers in Senegal, the Ivory Coast, Upper Volta, and the Central African Republic led them to postulate that this virus is maintained at low levels in an enzootic cycle between *Ae. africanus* and monkeys in moist equatorial forests.[13,39,120-122] Furthermore, the evidence indicated that after heavy rains, yellow fever virus spills over from this enzootic forest cycle into the surrounding savannah-woodland so that high rates of transmission occur involving monkeys, man, and other sylvan *Aedes* species. This region surrounding the forests was thus named the zone of emergence.[122] The general belief now is that CHIK virus has a similar ecology.[39] After heavy rains, CHIK virus also appears to spill over from the enzootic forest cycle to the savannah-woodland where an epizootic cycle is initiated between *Ae. furcifer taylori* or *Ae. luteocephalus* and monkeys.[10-12] This extension of infection to the savannah-woodland region is thought to be affected by either viremic monkeys or infected *Ae. furcifer-taylori*.

In southern Africa where there are no extensive evergreen forests with *Ae. africanus* or an equivalent species as vector, it does not seem likely that the virus could be maintained by a similar enzootic forest cycle. The occurrence of adult females of *Ae. furcifer-taylori* in the wooded savannah areas is relatively short lived and dependent upon adequate summer rains with sometimes long intervals of up to several years during drought periods when adults are either absent or present in only small numbers. This suggests that the only mechanism that would permit the virus to persist is survival in the dormant egg stage of *Ae. furcifer-taylori* mosquitoes followed by vertical (transovarial and transstadial) transmission to the next generation. Studies in southern Africa have so far failed to provide evidence of vertical transmission in *Ae. furcifer-taylori* either experimentally in the laboratory,[102] or in the field by testing newly emerged adults which were the progeny of naturally infected females of the previous summer, or by exposing in the field eggs laid by experimentally infected females and testing the adults reared out the following spring.[112] There is, in fact, only the smallest of evidence for the natural occurrence of transovarial transmission in the case of any *Alphavirus*, and the only evidence in support of aedine transovarial transmission is the laboratory transovarial transmission of Ross River virus in *Aedes vigilax*.[139] There is, however, evidence that *Ae. furcifer-taylori* may be involved in the vertical transmission of yellow fever as isolations of this virus have been made from males of this mosquito in nature.[123]

In both Africa and Asia, the virus can doubtless survive for considerable periods in the epidemic man-*Ae. aegypti* cycle, moving from locality to locality according to the availability of sufficient numbers of susceptible human hosts.[22] Whether this is the only way by which it has survived in Asia or whether there is a feral maintenance cycle, similar to the African ones, which occurs there remains to be ascertained. The apparent disappearance of CHIK virus from Bangkok in the presence of *Ae. aegypti*[97] does not support transovarial transmission in this species.

VI. ECOLOGICAL DYNAMICS

A. Climate and Environment

CHIK virus is strictly tropical in distribution which is clear from its geographical distribution pattern in southern Africa where the virus is absent from the temperate areas. The location of past human outbreaks and antibody surveys in humans and wild primates in South Africa suggest that low temperature, and not low rainfall, is the main factor excluding virus from the temperate areas. This is particularly evident in Natal where the southern limits of the virus apparently correspond with the 18°C midwinter isotherm.[32] It seems that temperature exerts its effect through the ecological requirements of the vector which, in southern Africa, is *Ae. furcifer-taylori* rather than the wild primate hosts. For the same reasons, it is essentially a virus of lowlands and woodland, the latter being necessary because the sylvan vectors breed exclusively in tree-holes. Within the tropics of Africa, the virus occurs in a variety of woodland. In Central and West Africa, it is apparently enzootic in large rainforests from where it probably emerges periodically during the rainy season into the adjoining semihumid savannah-woodland and subsequently even beyond that into semiarid savannah-woodland; the geographical limits of emergence are dependent upon the degree of precipitation and availability of susceptible wild primates and humans. In regions where forest is limited, e.g., southern Africa, the identity of the enzootic environment remains obscure, although wild primate epizooticity and human outbreaks have been linked unquestionably to the occurrence of abnormally wet summers in mixed savannah-woodland. In tropical Asia, where outbreaks have been temporally associated with heavy rainfall, the original source of the virus has not been determined.

B. Virus Transmission

A computer-simulated model of a CHIK transmission cycle indicated that provided there was a reasonably large number of vertebrate hosts present, the most critical factor in the transmission cycle is the survival probability of the vector.[124] It seems that vector longevity could decisively influence transmission in both human and wild primate cycles transmitted by *Ae. furcifer-taylori* in the savannah-woodlands of Africa, since these species are only active for a relatively short period coinciding with the warm rainy season. During the outbreak in Zimbabwe in 1962, which started in January, the outbreak ended dramatically with the onset of cold weather in early May.[14,49] In retrospect, it also seems likely that short-lived *Ae. furcifer-taylori* terminated a monkey epizootic in gallery forest in South Africa.[15,32] However, it would seem that CHIK transmission cycles are influenced more frequently by rising immune rates in the vertebrate hosts. This could well be the limiting factor in *Ae. aegypti*-transmitted urban outbreaks where very high postepidemic immune rates have been recorded in Asia and where outbreaks have ended without any vector control measures being applied. Rising immune rates are also believed to be the limiting factor in localized *Ae. africanus*-transmitted monkey epizootics in rain forest,[9] and probably also frequently end *Ae. furcifer-taylori*-transmitted wild primate epizootics in savannah-woodland.[16]

C. Vector Species

1. Aedes africanus

This vector is the most consistently numerically dominant arboreal mosquito species in the rainforests of West and Central Africa.[120,125] Studies in Uganda showed that this species had a maximum density at 12 to 18 m in the forest canopy where its single sharp peak of biting activity occurred from 1800 to 2200 hr at the time when the monkeys had settled for the night in the canopy of their dormitory trees.[125,126] In Uganda, this species can also survive as an aestivating adult during periods of marked drought. In both West Africa and Uganda it was observed that during the day the mosquitoes moved from the canopy to the ground and that at this time feeding on humans was greater on the ground than at night.[125,127] Mosquitoes oviposited at about the same frequency in tree-holes as in the fruits of *Saba senegalensis* hollowed out by monkeys and dropped on the forest floor. In the semihumid savannah-woodland of the Central African Republic, biting females with a remarkably constant parous rate were collected throughout the year, although their numbers dropped to very low levels by the end of the dry season.[128] Observations showed that the gonotrophic cycle lasts 4 to 8 days and daily adult female survival rates vary from 0.925 in June in the first half of the rainy season to 0.95 in October at the end of the rainy season. Hence, longevity increases in the second half of the rainy season.[129]

2. Aedes luteocephalus

In the savannah of eastern Senegal, *Ae. luteocephalus* is the predominant species in mangrove gallery forest, where humidity is high and temperature is fairly constant, and it is still abundant in the adjoining drier forest zone but is rare in more distant arid zones. In the mangroves, *Ae. luteocephalus* is active for 5 months of the year compared to only 3 months in drier forest. This mangrove forest was thus regarded as the ''zone of intensification'' of virus transmission.[11] It seems that this species has a major role in amplification of CHIK virus in the early part of the rainy season, but that *Ae. furcifer* assumes this role later in the season.[130]

3. Aedes furcifer-taylori

The ecology of *Ae. furcifer-taylori* has been studied in the savannah-woodland of eastern Senegal[47] where it is the numerically dominant mosquito. These mosquitoes prefer to ovipost in tree holes with small openings and have a low preference for bamboo pots.[131] The eggs hatch in installments after successive immersion by successive rains. There is possibly only one generation a season with long intervals between the hatching of different groups of eggs. More recent work in eastern Senegal, done since it became possible to separate the females of *Ae. furcifer* and *Ae. taylori*, has shown that *Ae. taylori* is more active in the canopy, biting little at ground level in villages,[104,130] and suggests that this species is more exclusively simiophilic than *Ae. furcifer*. It seems to persist longer than *Ae. furcifer* after the end of the rains probably because of a greater longevity. *Ae. furcifer*, in spite of a marked canopy activity, is also quite common at ground level, particularly in the nearby villages. Biting is mainly nocturnal with a peak between 1800 and 2100 hr — the first 3 hr after sunset. Observations suggest that *Ae. furcifer-taylori* takes two blood meals in one gonotrophic cycle, which would enhance transmission of virus. Mosquitoes probably use resting sites in the canopy rather than on the ground. The proportions of *Ae. taylori* to *Ae. furcifer* in Senegal was 1:3.[104]

The observations described above agree largely with those made in the wooded savannah of southern Africa where it has not as yet been possible to separate females of the two species.[16] Studies there indicated that *Ae. furcifer* and *Ae. taylori* usually coexist and that *Ae. furcifer* is probably more abundant than *Ae. taylori*. For example, in several series of collections based on the collection of males coming to human and wild primate bait, the *furcifer* to *taylori* ratio was 32:1. *Ae. furcifer-taylori* was by far the most prevalent species feeding on wild primates in the canopy and was 10 times more prevalent in the canopy than on the ground for 2 hr after sunset, although at times large numbers were collected on the ground on human bait.[16] In western Zimbabwe, the only occasion when *Ae. taylori* was evidently the numerically dominant species, it was not collected on the ground at a time when large numbers were collected in the canopy, suggesting that *Ae. taylori* may play a relatively smaller role in human infection.

4. Aedes aegypti

The man-biting habits of domestic *Ae. aegypti* appear to vary in different countries where CHIK epidemics have occurred. On the Makonde Plateau in Tanzania in 1952—1953 it fed on man mainly inside the huts but also outside on the verandahs,[2] while in Bangkok in 1962 biting was largely indoors. In Bangkok, the epidemic dissemination of CHIK virus accompanied the annual monsoon and varied directly with the density of *Ae. aegypti*.[132] Since about 1963, populations of *Ae. aegypti* in Calcutta appear to a large extent to have competitively displaced those of *Ae. albopictus* in the urban areas.[133]

If *Ae. aegypti* were to be eradicated from these areas, *Ae. albopictus* might well come in from the adjoining rural areas to occupy the vacated urban habitat. A feature of the 1964 epidemic in Madras and Vellore was the rapid build-up of human cases so that within 12 weeks nearly 40% of the population had become infected. It was suggested that this rapid spread of virus might have been due to the daytime biting habit of the local *Ae. aegypti*. During the day the human hosts are more likely to disturb the mosquitoes, thus interrupting their feeds, so that a mosquito moves to another host to finish feeding thus transmitting the virus mechanically.[113]

The present population growth in Africa with its associated urbanization could lead to large-scale future epidemics on the scale already experienced in Asia, with the possible maintenance of the virus in an *Ae. aegypti*-man cycle.

D. Movements and Migrations of Vectors and Hosts

Monkeys in the savannah-woodland of eastern Senegal sometimes move from woodland to the nearby villages in search of food.[11] In this way, they probably take infection from the feral to the domestic transmission cycles. An alternative way for virus to reach the villages is probably by *Ae. furcifer* flying from woodland, as it has been shown to fly distances of up to 3 km.[104] This strong flying ability, together with its habit of feeding both in the canopy and on the ground in woodland and on the ground in the villages[130] would enable it to fulfill this role. In evergreen forests in West Africa and Uganda,[127,134] the propensity of *Ae. africanus* to move from the canopy to the ground in the daytime and to migrate above the forest canopy would enable this species to disseminate virus between forests and from sylvan transmission cycles to man.

VII. PREVENTION AND CONTROL

Although several experimental inactivated vaccines have been produced,[135-138] vaccines have not yet been used to control outbreaks of CHIK. However, with the possible increase in human infection in the future, there may well be a need for a live virus vaccine. Control measures would therefore center around the avoidance of mosquito bites and the reduction in the density of vectors. In rural areas of Africa, the feral vectors are most active for only a few hours after sunset which facilitates the avoidance of these mosquitoes. In urban areas infested by *Ae. aegypti* but free of disease, quarantine measures could be applied to prevent the introduction of virus. Mosquito control may be needed in urban epidemics. The breeding sites of *Ae. aegypti* should be eliminated by reducing the number of water containers in- and outside homes and supplying these with piped water. Insecticidal control of adult mosquitoes and perhaps larvae may be necessary. Surveillance of *Ae. aegypti* densities by regular collection of larvae should form the background of any such control program.

VIII. FUTURE RESEARCH

Aspects of CHIK epidemiology requiring investigation include the possible role of bats in viral dissemination and maintenance; the identity of the wild primate species in the forests of Africa primarily concerned with viral maintenance or amplification; the possible existence of a feral transmission cycle in Asia; the mechanism of viral survival in nature, whether it is in fact through continuous transmission in moving foci of wild primate infection in large rain forests or whether it is primarily dependent upon vertical transmission in the mosquito host; the biology and taxonomy of the four important vectors; and the development of a live virus vaccine.

REFERENCES

1. **Robinson, M. C.,** An epidemic of virus disease in Southern Province, Tanganyika Territory, in 1952—53. I. Clinical features, *Trans. R. Soc. Trop. Med. Hyg.*, 49, 28, 1955.
2. **Lumsden, W. H. R.,** An epidemic of virus disease in Southern Province, Tanganyika Territory, in 1952—53. II. General description and epidemiology, *Trans. R. Soc. Trop. Med. Hyg.*, 49, 33, 1955.
3. **Ross, R. W.,** The Newala epidemic. III. The virus: isolation, pathogenic properties and relationship to the epidemic, *J. Hyg.*, 54, 177, 1956.
4. **Mason, P. J. and Haddow, A. J.,** An epidemic of virus disease in Southern Province, Tanganyika Territory, *Trans. R. Soc. Trop. Med. Hyg.*, 51, 238, 1957.
5. **Ross, R. W.,** A laboratory technique for studying the insect transmission of animal viruses, employing a bat-wing membrane, demonstrated with two African viruses, *J. Hyg.*, 54, 192, 1956.
6. **Casals, J.,** Viruses: the versatile parasites. I. The arthropod-borne group of animal viruses, *Trans. N.Y. Acad. Sci. Ser. II*, 19, 219, 1957.
7. **Spence, L. P. and Thomas, L.,** Application of hemagglutination and complement-fixation technique to the identification and serological classification of arthropod-borne viruses. Studies on chikungunya and Makonde viruses, *Trans. R. Soc. Trop. Med. Hyg.*, 53, 248, 1959.
8. **Weinbren, M. P., Haddow, A. J., and Williams, M. C.,** The occurrence of chikungunya virus in Uganda. I. Isolations from mosquitoes, *Trans. R. Soc. Trop. Med. Hyg.*, 52, 253, 1958.
9. **Macrae, A. W. R., Henderson, B. E., Kirya, B. G., and Sempala, S. D. K.,** Chikungunya virus in the Entebbe area of Uganda: isolations and epidemiology, *Trans. R. Soc. Trop. Med. Hyg.*, 65, 152, 1971.
10. **Taufflieb, R., Cornet, M., and Camicas, J. L.,** Les vecteurs d'arbovirus au Senegal, *Cah. ORSTOM Ser. Entomol. Med. Parasitol.*, 6, 221, 1968.
11. **Cornet, M. and Chateau, R.,** Quelques donnees biologiques sur *Aedes (Stegomyia) luteocephalus* (Newstead), 1907 en zone de savane soudanienne dans l'ouest du Senegal, *Cah. ORSTOM Ser. Entomol. Med. Parasitol.*, 12, 97, 1974.

12. **Cornet, M., Robin, Y., Chateau, R., Heme, G., Adam, C., Valade, M., Le Gonidec, G., Jan, C., Renaudet, J., Dieng, P. L., Bangoura, J., and Loraud, A.,** Isolements d'arbovirus au Senegal oriental a partir de moustiques (1972—1977) et notes sur l'epidemiologic des virus par les *Aedes*, en particulier du virus amaril, *Cah ORSTOM Ser. Entomol. Med. Parasitol.*, 17, 149, 1979.
13. **Saluzzo, J. F., Gonzalez, J. P., Herve, J. P., and Georges, A. J.,** Contribution a l'etude epidemiologique des arbovirus en centrafrique: manifestation du virus chikungunya au cours des annees 1978 et 1979, *Bull. Soc. Pathol. Exot.*, 73, 390, 1980.
14. **McIntosh, B. M., Paterson, H. E., McGillivray, G., and de Sousa, J.,** Further studies on the chikungunya outbreak in southern Rhodesia in 1962. I. Mosquitoes, wild primates and birds in relation to the epidemic, *Ann. Trop. Med. Parasitol.*, 58, 45, 1964.
15. **McIntosh, B. M., Jupp, P. G., and de Sousa, J.,** Mosquitoes feeding at two horizontal levels in gallery forest in Natal, South Africa, with reference to possible vectors of chikungunya virus, *J. Entomol. Soc. S. Afr.*, 35, 81, 1972.
16. **McIntosh, B. M., Jupp, P. G., and dos Santos, I.,** Rural epidemic of chikungunya in South Africa with involvement of *Aedes (Diceromyia) furcifer* (Edwards) and baboons, *S. Afr. J. Sci.*, 73, 267, 1977.
17. **Moore, D. L., Reddy, S., Akinkugbe, F. M., Lee, V. H., David-West, T. H., Causey, O. R., and Carey, D. E.,** An epidemic of chikungunya fever at Ibadan, Nigeria, 1969, *Ann. Trop. Med. Parasitol.*, 68, 59, 1974.
18. **Filipe, A. R. and Pinto, M. R.,** Arbovirus studies in Luanda, Angola. II. Virological and serological studies during an outbreak of dengue-like disease caused by the chikungunya virus, *Bull. WHO*, 49, 37, 1973.
19. **Hammon, W. McD., Rudnick, A., and Sather, G. E.,** Viruses associated with epidemic haemorrhagic fevers of the Philippines and Thailand, *Science*, 131, 1102, 1960.
20. **Basaca-Sevilla, V. and Halstead, S. B.,** Recent virological studies of haemorrhagic fever and other arthropod-borne virus infections in the Philippines, *J. Trop. Med. Hyg.*, 69, 203, 1966.
21. **Nimmannitya, S., Halstead, S. B., Cohen, S. N., and Margiotta, M. R.,** Dengue and chikungunya virus infection in man in Thailand, 1962—64. I. Observations on hospitalized patients with hemorrhagic fever, *Am. J. Trop. Med. Hyg.*, 18, 954, 1969.
22. **Chastel, C.,** Human infections in Cambodia with chikungunya or a closely allied virus. III. Epidemiology, *Bull. Soc. Pathol. Exot.*, 57, 65, 1964.
23. **Shah, K. V., Gibbs, C. J., and Banerjee, G.,** Virological investigation of the epidemic of haemorrhagic fever in Calcutta: isolation of three strains of chikungunya virus, *Indian J. Med. Res.*, 52, 676, 1964.
24. **Myers, R. M., Carey, D. E., Reuben, R., Jesudass, E. S., de Ranitz, C. D., and Jadhav, M.,** The 1964 epidemic of dengue-like fever in South India: isolation of chikungunya virus from human sera and from mosquitoes, *Indian J. Med. Res.*, 53, 694, 1965.
25. **Rao, T. R.,** Recent epidemics caused by chikungunya virus in India, 1963—65, *Sci. Cult.*, 32, 215, 1966.
26. **Carey, D. E., Myers, R. M., de Ranitz, C. M., Jadhav, M., and Reuben, R.,** The 1964 chikungunya epidemic at Vellore, South India, including observations on concurrent dengue, *Trans. R. Soc. Trop. Med. Hyg.*, 63, 434, 1969.
27. **Mendis, N. M. P.,** Epidemiology of dengue-like fever in Ceylon, *Ceylon Med. J.*, 12, 67, 1967.
28. **Vu-Qui, D., Nguyen-Thi, K. T., and Ly, Q. B.,** Enquete sur les anticorps anti-chikungunya chez des enfants Vietnamiens de Saigon, *Bull. Soc. Pathol. Exot.*, 60, 353, 1967.
29. **Ming, C. K., Thein, S., Thaung, U. T,. Myint, R. S., Swe, T., Halstead, S. B., and Diwan, A. R.,** Clinical laboratory studies on haemorrhagic fever in Burma, 1970—1972, *Bull. WHO*, 51, 227, 1974.
30. **Osterrieth, P., Deleplanque-Liegeois, P., and Renoirte, R.,** Recherche sur le virus chikungunya au Congo Belge. II. Enquette serologique, *Ann. Soc. Belge Med. Trop.*, 40, 205, 1960.
31. **Boorman, J. P. T. and Drafer, C. C.,** Isolations of arboviruses in the Lagos area of Nigeria and a survey of antibodies to them in men and animals, *Trans. R. Soc. Trop. Med. Hyg.*, 62, 269, 1968.
32. **McIntosh, B. M.,** Antibody against chikungunya virus in wild primates in southern Africa, *S. Afr. J. Med. Sci.*, 35, 65, 1970.
33. **Brès, P., Camicas, J. L., Cornet, M., Robin, Y., and Taufflieb, R.,** Considerations sur l'epidemiologie des arboviruses au Senegal, *Bull. Soc. Pathol. Exot.*, 62, 253, 1969.
34. Centre Collaborateur OMS de reference et de recherche pour les arbovirus, Annual Report, Institut Pasteur, Dakar, Senegal, 1984.
35. **McIntosh, B. M. and Gear, J. H. S.,** Mosquito-borne arboviruses, primarily in the eastern hemisphere, in *Diseases Transmitted from Animals to Man*, 6th ed., Hubbert, W. T., McCulloch, W. F., and Schnurrenberger, P. R., Eds., Charles C Thomas, Springfield, Ill., 1975, 939.
36. **Halstead, S. B.,** Chikungunya virus, in *Handbook Series in Zoonoses*, Section B, *Viral Zoonoses*, Vol. 1, Steele, J. H. and Beran, G. W., Eds., CRC Press, Boca Raton, Fla., 1981, 437.
37. **McIntosh, B. M. and Gear, J. H. S.,** Arboviral zoonoses in southern Africa, in *Handbook Series in Zoonoses*, Section B, *Viral Zoonoses*, Vol. 1, Steele, J. H. and Beran, G. W., Eds., CRC Press, Boca Raton, Fla., 1981, 217.

38. **Robin, Y.**, Arboviral zoonoses in East, Central and West Africa: Chikungunya fever, in *Handbook Series in Zoonoses*, Section B, *Viral Zoonoses*, Vol. 1, Steele, J. H. and Beran, G. W., Eds., CRC Press, Boca Raton, Fla., 1981, 235.
39. **Monath, T. P.**, Impact of arthropod-borne virus diseases on Africa, in *Applied Virology*, Kurstak, E., Ed., Academic Press, Orlando, 1984, 392.
40. **Carey, D. E.**, Chikungunya and dengue: a case of mistaken identity?, *J. Hist. Med. Allied Sci.*, 26, 243, 1971.
41. **Pinheiro, F. P., Freitas, R. B., Travassos da Rosa, J. F., Gabbau, Y. B., Mellow, W. A., and Le Duc, J. W.**, An outbreak of Mayaro virus disease in Belterra, Brazil. I. Clinical and virological findings, *Am. J. Trop. Med. Hyg.*, 30, 674, 1981.
42. **Macasaet, F. F., Villamil, P. T., Wexler, S., and Beran, G. W.**, Epidemiology of arbovirus infections in Negro Oriental. II. Serologic findings of the epidemic in Amlam, *J. Philipp. Med. Assoc.*, 45, 311, 1969.
43. **Gear, J. and Reid, F. P.**, The occurrence of a dengue-like fever in the north-eastern Transvaal. I. Clinical features and isolation of virus, *S. Afr. Med. J.*, 31, 253, 1957.
44. **Fourie, E. D. and Morrison, J. G. L.**, Rheumatoid arthritic syndrome after chikungunya fever, *S. Afr. Med. J.*, 56, 130, 1979.
45. **Morrison, J. G. L.**, Chikungunya fever, *Int. J. Dermatol.*, 18, 628, 1979.
46. **Rodger, L. M.**, An outbreak of suspected chikungunya fever in northern Rhodesia, *S. Afr. Med. J.*, 35, 126, 1961.
47. **Cornet, M., Chateau, R., Valade, M., Dieng, P. L., Raymond, H., and Lorand, A.**, Donnees bio-ecologiques sur les vecteurs potentiels du virus amaril an Senegal oriental. Role des differentes especes dans la transmission du virus, *Cah. ORSTOM Ser. Entomol. Med. Parasitol.*, 16, 315, 1978.
48. **Saluzzo, J. F., Cornet, M., and Digoutte, J. P.**, Une poussee epidemique du virus chikungunya dans l'ouest du Senegal en 1982, *Dakar Med.*, 28, 497, 1983.
49. **McIntosh, B. M., Harwin, R. M., Paterson, H. E., and Westwater, M. L.**, An epidemic of chikungunya in South-eastern southern Rhodesia, *Cent. Afr. J. Med.*, 9, 351, 1963.
50. **Swanepoel, R. and Cruickshank, J. G.**, Arthropod-borne viruses of medical importance in Rhodesia, *Cent. Afr. J. Med.*, 20, 71, 1974.
51. **Jadhav, M., Nambooripad, M., Carman, R. H., Carey, D. E., and Myers, R. M.**, Chikungunya disease in infants and children in Vellore: a report of clinical and haematological features of virologically proved cases, *Indian J. Med. Res.*, 53, 764, 1965.
52. **Padbidri, V. S. and Gnaneswar, T. T.**, Epidemiological investigations of chikungunya epidemic at Barsi, Maharashtra State, India, *J. Hyg. Epidemiol. Microbiol. Immunol.*, 23, 445, 1979.
53. **Ming, C. K., Thein, S., Thaung, U. T., Myint, R. S., Swe, T., Halstead, S. B., and Diwan, A. R.**, Clinical laboratory studies on haemorrhagic fever in Burma, 1970—1972, *Bull. WHO*, 51, 227, 1974.
54. **Tomori, O., Fagbami, A., and Fabiyi, A.**, The 1974 epidemic of chikungunya fever in children in Ibadan, *Trop. Geogr. Med.*, 27, 413, 1975.
55. **Deller, J. J. and Russell, P. K.**, Chikungunya disease, *Am. J. Trop. Med. Hyg.*, 17, 107, 1968.
56. **Porterfield, J. S.**, Cross-neutralization studies with group A arthropod-borne viruses, *Bull. WHO*, 24, 735, 1961.
57. **Karabatsos, N.**, Antigenic relationships of group A arboviruses by plaque reduction neutralization testing, *Am. J. Trop. Med. Hyg.*, 24, 527, 1975.
58. **Chanas, A. C., Johnson, B. K., and Simpson, D. I. H.**, Antigenic relationships of *Alphavirus* by a simple micro-culture cross-neutralization method, *J. Gen. Virol.*, 32, 295, 1976.
59. **Chanas, A. C., Hubalek, Z., Johnson, B. K., and Simpson, D. I. H.**, A comparative study of O'nyong-nyong virus with chikungunya virus and plaque variants, *Arch. Virol.*, 59, 231, 1979.
60. **Calisher, C. H., Shope, R. E., Brandt, W., Casals, J., Karabatsos, N., Murphy, F. A., Tesh, R. B., and Wiebe, M.**, Proposed antigenic classification of registered arboviruses. I. Togaviridae, *Alphavirus, Intervirology*, 14, 229, 1980.
61. **Karabatsos, N., Ed.**, *International Catalogue of Arboviruses*, 3rd ed., American Society of Tropical Medicine and Hygiene, San Antonio, Tex., 1985, 327.
62. **McIntosh, B. M.**, Susceptibility of some African wild rodents to infection with various arthropod-borne viruses, *Trans. R. Soc. Trop. Med. Hyg.*, 55, 63, 1961.
63. **McIntosh, B. M., Paterson, H. E., Donaldson, J. M., and de Sousa, J.**, Chikungunya virus: viral susceptibility and transmission studies with some vertebrates, *S. Afr. J. Med. Sci.*, 28, 45, 1963.
64. **Paul, S. D. and Singh, K. R. P.**, Experimental infection of *Macaca radiata* with chikungunya virus and transmission of virus by mosquitoes, *Indian J. Med. Res.*, 56, 802, 1968.
65. **Chakravarty, S. K. and Sarkar, J. K.**, Susceptibility of newborn and adult laboratory animals to chikungunya virus, *Indian J. Med. Res.*, 57, 1157, 1969.
66. **Bedekar, S. D. and Pavri, K. M.**, Studies with chikungunya virus. I. Susceptibility of birds and small mammals, *Indian J. Med. Res.*, 57, 1181, 1969.

67. **McIntosh, B. M.,** unpublished data, 1976.
68. **Nakao, E.,** Biological and immunological studies on chikungunya virus: a comparative observation of two strains of African and Asian origins, *Kobe J. Med. Sci.*, 18, 133, 1972.
69. **Halstead, S. B. and Buescher, E. L.,** Hemorrhagic disease in rodents infected with virus associated with Thai hemorrhagic fever, *Science*, 134, 475, 1961.
70. **Weiss, H. J., Halstead, S. B., and Russ, S. B.,** Hemorrhagic disease in rodents caused by chikungunya virus. I. Studies of hemostasis, *Proc. Soc. Exp. Biol. Med.*, 119, 427, 1965.
71. **Sarkar, J. K.,** Virological studies of haemorrhagic fever in Calcutta, *Bull. WHO*, 35, 59, 1966.
72. **Chanas, A. C., Johnson, B. K., and Simpson, D. I. H.,** Chikungunya *in vivo* interference associated with a small plaque variant, *Acta Virol.*, 22, 485, 1978.
73. **Umrigar, M. D. and Kadam, S. S.,** Comparative sensitivity of suckling mice and Vero cells for primary isolation of chikungunya virus, *Indian J. Med. Res.*, 62, 1893, 1974.
74. **Davis, J. L., Hodge, H. M., and Campbell, W. E.,** Growth of chikungunya virus in baby hamster kidney cell (BHK-21-clone 13) suspension cultures, *Appl. Microbiol.*, 21, 338, 1971.
75. **Igarashi, A.,** Isolation of a Singh's *Aedes albopictus* cell clone sensitive to dengue and chikungunya viruses, *J. Gen. Virol.*, 40, 531, 1978.
76. **Sanchez Legrand, F. and Hotta, S.,** Susceptibility of cloned *Toxorhynchites amboinensis* cells to dengue and chikungunya viruses, *Microbiol. Immunol.*, 27, 101, 1983.
77. **Buckley, S. M., Singh, K. R., and Bhat, U. K.,** Small and large plaque variants of chikungunya virus in two vertebrate and seven invertebrate cell lines, *Acta Virol.*, 19, 10, 1975.
78. **El Mekki, A. A. and van der Groen, G.,** A comparison of indirect immunofluorescence and electron microscopy for the diagnosis of some haemorrhagic viruses in cell cultures, *J. Virol. Methods*, 3, 61, 1981.
79. **Niklasson, B., Grandien, M., Peters, C. J., and Gargan, T. P.,** Detection of Rift Valley fever virus antigen by enzyme-linked immunosorbent assay, *J. Clin. Microbiol.*, 17, 1026, 1983.
80. **Coz, J., Valade, M., Cornet, M., Lemoine, M., and Lorand, A.,** Utilization du moustique pour la multiplication des arbovirus, *Cah. ORSTOM Ser. Entomol. Med. Parasitol.*, 15, 209, 1977.
81. **Ranitz, C. M., Myers, R. M., Varkey, M. J., Isaac, Z. H., and Carey, D. E.,** Clinical impressions of chikungunya in Vellore gained from study of adult patients, *Indian J. Med. Res.*, 53, 756, 1965.
82. **Brighton, S. W., Prozesky, O. W., and de la Harpe, A. L.,** Chikungunya virus infection. A retrospective study of 107 cases, *S. Afr. Med. J.*, 63, 313, 1983.
83. **Brighton, S. W.,** Chloroquine phosphate treatment of chronic chikungunya arthritis, *S. Afr. Med. J.*, 66, 217, 1984.
84. **Halstead, S. B.,** Mosquito-borne haemorrhagic fevers of South and South-East Asia, *Bull. WHO*, 35, 3, 1966.
85. **Shore, H.,** O'nyong-nyong fever: an epidemic virus disease in East Africa. III. Some clinical and epidemiological observations in the Northern Province of Uganda, *Trans. R. Soc. Trop. Med. Hyg.*, 55, 361, 1961.
86. **Williams, M. C. and Woodall, J. P.,** O'nyong-nyong fever: an epidemic virus disease in East Africa. II. Isolation and some properties of the virus, *Trans. R. Soc. Trop. Med. Hyg.*, 55, 135, 1961.
87. **Odelola, H. A.,** Application of single-radial-haemolysis test for the detection of antibodies to *Togavirus*, *Arch. Virol.*, 60, 325, 1979.
88. **Buckley, S. M. and Clarke, D. H.,** Differentiation of group A arboviruses, chikungunya, Mayaro, and Semliki Forest by the fluorescent antibody technique, *Proc. Soc. Exp. Biol. Med.*, 135, 533, 1970.
89. **Kokernot, R. H., Smithburn, K. C., Gandara, A. F., McIntosh, B. M., and Heymann, C. S.,** Provas de neutralizacao com soros de individuos residentes em Mocambique contra determinados virus isolados em Africa transmitidos por artropodes, *An. Inst. Med. Trop.*, 17, 201, 1960.
90. **Kokernot, R. H., Szlamp, E. L., Levitt, J., and McIntosh, B. M.,** Surveys for antibodies against arthropod-borne viruses in the sera of indigenous residents of the Caprivi Strip and Bechuanaland Protectorate, *Trans. R. Soc. Trop. Med. Hyg.*, 59, 553, 1965.
91. **Halstead, S. B. and Udomsakdi, S.,** Vertebrate hosts of chikungunya virus, *Bull. WHO*, 35, 89, 1966.
92. **Marchette, N. J., Rudnick, A., and Garcia, R.,** Alphaviruses in Peninsular Malaysia. II. Serological evidence of human infection, *Southeast Asian J. Trop. Med. Public Health*, 11, 14, 1980.
93. **Darwish, M. A., Hoogstraal, H., Roberts, T. J., Ahmed, I. P., and Omar, F.,** A sero-epidemiological survey for certain arboviruses (Togaviridae) in Pakistan, *Trans. R. Soc. Trop. Med. Hyg.*, 77, 442, 1983.
94. **Tesh, R. G., Gadjusek, D. C., Carrato, R. M., Cross, J. H., and Rosen, L.,** The distribution and prevalence of group A arbovirus neutralizing antibodies among human populations in Southeast Asia and Pacific Islands, *Am. J. Trop. Med. Hyg.*, 24, 664, 1975.
95. **Halstead, S. B., Scanlon, J. E., Umpairit, P., and Umdomsakdi, S.,** Dengue and chikungunya virus infection in man in Thailand, 1962—64. IV. Epidemiologic studies in the Bangkok metropolitan area, *Am. J. Trop. Med. Hyg.*, 18, 997, 1969.

96. **Halstead, S. B., Udomsakdi, S., Scanlon, J. E., and Rohitayodhin, S.,** Dengue and chikungunya virus infections in man in Thailand, 1962—64. V. Epidemiologic observations outside Bangkok, *Am. J. Trop. Med. Hyg.*, 18, 1022, 1969.
97. **Burke, D. S., Nisalak, A., and Nimmannitya, S.,** Disappearance of chikungunya virus from Bangkok, *Trans. R. Soc. Trop. Med. Hyg.*, 79, 419, 1985.
98. **Filipe, A. R., Pinto, M. R., Serra, M. L., and Champalimaud, J. L.,** Yellow fever infection after recent convalescence from chikungunya fever: report of a fatal case, *Trans. R. Soc. Trop. Med. Hyg.*, 67, 400, 1973.
99. **Kokernot, R. H., Casaca, V. M. R., Weinbren, M. P., and McIntosh, B. M.,** Survey for antibodies against arthropod-borne viruses in the sera of indigenous residents of Angola, *Trans. R. Soc. Trop. Med. Hyg.*, 59, 563, 1965.
100. **McIntosh, B. M.,** unpublished data, 1966.
101. **Thaung, U., Ming, C. K., Swe, T., and Thein, S.,** Epidemiological features of dengue and chikungunya infections in Burma, *Southeast Asian J. Trop. Med. Public Health*, 6, 276, 1975.
102. **Jupp, P. G., McIntosh, B. M., Dos Santos, I., and de Moor, P.,** Laboratory vector studies on six mosquito and one tick species with chikungunya virus, *Trans. R. Soc. Trop. Med. Hyg.*, 75, 15, 1981.
103. **Haddow, A. J., Williams, M. C., and Woodall, J. P.,** Virus isolations from biting flies, East African Virus Research Institute Annual Rep. No. 11, East African High Commission, 1961, 16.
104. Annual Report Institut Pasteur, Dakar, Senegal, 1984, 77.
105. **Harrison, V. R., Marshall, J. M., and Guilloud, N. B.,** The presence of antibody to chikungunya and other serologically related arboviruses in the sera of subhuman primate imports to the United States, *J. Immunol.*, 98, 979, 1967.
106. **Marchette, N. J., Rudnick, A., Garcia, R., and MacVean, D. W.,** Alphaviruses in Peninsular Malaysia. I. Virus isolations and animal serology, *Southeast Asian J. Trop. Med. Public Health*, 9, 317, 1978.
107. **Simpson, D. I. H., Williams, M. C., O'Sullivan, J. P., Cunningham, J. C., and Mutere, F. A.,** Studies on arboviruses and bats (Chiroptera) in East Africa. II. Isolation and haemagglutination-inhibition studies on bats collected in Kenya and throughout Uganda, *Ann. Trop. Med. Parasitol.*, 62, 432, 1968.
108. **Singh, K. R. P. and Pavri, K. M.,** Experimental studies with chikungunya virus in *Aedes aegypti* and *Aedes albopictus*, *Acta Virol.*, 11, 517, 1967.
109. **Gilotra, S. K. and Shah, K. V.,** Laboratory studies on transmission of chikungunya virus by mosquitoes, *Am. J. Epidemiol.*, 86, 379, 1967.
110. **McIntosh, B. M. and Jupp, P. G.,** Attempts to transmit chikungunya virus with six species of mosquito, *J. Med. Entomol.*, 7, 615, 1970.
111. **Mangiafico, J. A.,** Chikungunya virus infection and transmission in five species of mosquito, *Am. J. Trop. Med. Hyg.*, 20, 642, 1971.
112. **Jupp, P. G. and McIntosh, B. M.,** unpublished data, 1980 to 1982.
113. **Rao, T. R., Devi, P. S., and Singh, K. R. P.,** Experimental studies on the mechanical transmission of chikungunya virus by *Aedes aegypti*, *Mosq. News*, 28, 406, 1968.
114. **Rao, T. A., Singh, K. R. P., and Pavri, K. M.,** Laboratory transmission of an Indian strain of chikungunya virus, *Curr. Sci.*, 33, 235, 1964.
115. **Paterson, H. E. and McIntosh, B. M.,** Further studies on the chikungunya outbreak in southern Rhodesia in 1962. II. Transmission experiments with the *Ae. furcifer-taylori* group of mosquitoes and with a member of the *Anopheles gambiae* complex, *Ann. Trop. Med. Parasitol.*, 58, 52, 1964.
116. **Sempala, S. D. K. and Kirya, B. G.,** Laboratory transmission of chikungunya virus by *Aedes (Stegomyia) apicoargenteus* Theobald, *Am. J. Trop. Med. Hyg.*, 22, 263, 1973.
117. **Bedekar, S. D. and Pavri, K. M.,** Studies with chikungunya virus. II. Serological survey of humans and animals in India, *Indian J. Med. Res.*, 57, 1193, 1969.
118. **Simpson, D. I. H. and O'Sullivan, J. P.,** Studies on arboviruses and bats (Chiroptera) in East Africa. I. Experimental infection of bats and virus transmission attempts in *Aedes (Stegomyia) aegypti* (Linnaeus), *Ann. Trop. Med. Parasitol.*, 62, 422, 1968.
119. **Binn, L. N., Harrison, V. R., and Randall, R.,** Patterns of viraemia and antibody observed in rhesus monkeys inoculated with chikungunya and other serologically related group A arboviruses, *Am. J. Trop. Med. Hyg.*, 16, 782, 1967.
120. **Germain, M., Sureau, P., Hervé, J. P., Fabre, J., Mouchet, J., Robin, Y., and Geoffroy, B.,** Isolements du virus de la fievre jaune a partir d' *Aedes* du groupe *Ae. africanus* (Theobald) en Republique Centrafricaine. Importance des savanes humides et semi-humides en tant que zone d'emergence du virus amaril, *Cah. ORSTOM Ser. Entomol. Med. Parasitol.*, 14, 125, 1976.
121. **Saluzzo, J. F., Hervé, J. P., Germain, M., Geoffroy, B., Huard, M., Fabre, J., Salaun, J. J., Heme, G,. and Robin, Y.,** Second serie d'isolements du virus de la fievre jaune, a partir d'*Aedes africanus* (Theobald), dans une galerie forestiere des savanes semi-humides du Sud de l'Empire Centrafricain, *Cah. ORSTOM Ser. Entomol. Med. Parasitol.*, 17, 19, 1979.

122. **Germain, M., Cornet, M., Mouchet, J., Hervé, J. P., Robert, V., Camicas, J. L., Cordellier, R., Hervy, J. P., Digoutte, J. P., Monath, T. P., Salaun, J. J., Deubel, V., Robin, Y., Coz, J., Taufflieb, R., Saluzzo, J. F., and Gonzalez, J. P.,** La fievre jaune selvatique en Afrique: donnes recentes et conceptions actuelles, *Med. Trop., (Marseilles)*, 41, 31, 1981.
123. **Salaun, J. J., Germain, M., Robert, V., Robin, Y., Monath, T. P., Camicas, J. L., and Digoutte, J. P.,** Yellow fever in Senegal from 1976—1980, *Med. Trop. (Marseilles)*, 41, 45, 1981.
124. **de Moor, P. P. and Steffens, F. E.,** A computer-simulated model of an arthropod-borne virus transmission cycle, with special reference to chikungunya virus, *Trans. R. Soc. Trop. Med. Hyg.*, 64, 927, 1970.
125. **Haddow, A. J., Gillet, J. D., and Highton, R. B.,** The mosquitoes of Bwamba County, Uganda. V. The vertical distribution and testing cycle of mosquitoes in rain-forest, with further observations on microclimate, *Bull. Entomol. Res.*, 37, 301, 1947.
126. **Haddow, A. J. and Mahaffy, A. F.,** The mosquitoes of Bwamba County, Uganda. VII. Intensive catching on tree-platforms, with further observations on *Aedes (Stegomyia) africanus*, Theobald, *Bull. Entomol. Res.*, 40, 169, 1949.
127. **Haddow, A. J., Williams, M. C., Woodall, J. P., Simpson, D. I. H., and Goma, L. K. H.,** Twleve isolations of Zika virus from *Aedes (Stegomyia) africanus* (Theobald) taken in and above a Uganda Forest, *Bull. WHO*, 31, 57, 1964.
128. **Hervé, J. P., Germain, M., and Geoffroy, B.,** Bioecologie comparee d'*Aedes (Stegomyia) opok* Corbet et van Someren et *A. (s). africanus* (Theobald) dans une galerie forestiere du sud de l'Empire Centrafricain. II. Cycles sainsonniers d'abondance, *Cah. ORSTOM Ser. Entomol. Med. Parasitol.*, 15, 271, 1977.
129. **Germain, M., Hervé, J. P., and Geoffroy, B.,** Variation du taux de survie des femelles d'*Aedes africanus* (Theobald) dans une galerie forestiere du sud de l'Empire Centrafricain, *Cah. ORSTOM Ser. Entomol. Med. Parasitol.*, 15, 291, 1977.
130. **Cornet, M., Saluzzo, J. F., Hervy, J. P., Digoutte, J. P., Germain, M., Chauvancy, M. F., Eyraud, M., Ferrara, L., Heme, G., and Legros, F.,** Dengue 2 au Senegal oriental: une poussee epizootioque en milieu selvatique; isolements du virus a partir de moustiques et d'un singe et considerations epidemiologiques, *Cah. ORSTOM Ser. Entomol. Med. Parasitol.*, 22, 313, 1984.
131. **Raymond, H. L., Cornet, M., and Dieng, P. Y.,** Etudes sur les vecteurs sylvatiques du virus amaril. Inventaire provisaire des habitats larvaires d'une foret-galerie dans le foyer endemique du Senegal oriental, *Cah. ORSTOM Ser. Entomol. Med. Parasitol.*, 14, 301, 1976.
132. **Halstead, S. B.,** Epidemiological studies of Thai haemorrhagic fever, *Bull. WHO*, 35, 80, 1966.
133. **Gilotra, S. K., Rozeboom, L. E., and Bhattacharya, N. C.,** Observations on possible competitive displacement between populations of *Aedes aegypti* Linnaeus and *Aedes albopictus* Skuse in Calcutta, *Bull. WHO*, 37, 437, 1967.
134. **Mattingly, P. F.,** Studies on West African forest mosquitoes. I. The seasonal distribution and biting cycle and vertical distribution of four of the principal species, *Bull. Entomol. Res.*, 40, 149, 1949.
135. **Harrison, V. R., Binn, L. N., and Randall, R.,** Comparative immunogenicities of chikungunya vaccines prepared in avian and mammalian tissues, *Am. J. Trop. Med. Hyg.*, 16, 786, 1967.
136. **Harrison, V. R., Eckels, K. H., Bartelloni, P. J., and Hampton, C.,** Production and evaluation of a formalin-killed chikungunya vaccine, *J. Immunol.*, 107, 643, 1971.
137. **White, A., Berman, S., and Lowenthal, J. P.,** Comparative immunogenicities of chikungunya vaccines propagated in monkey kidney monolayers and chick embryo suspension cultures, *Appl. Microbiol.*, 23, 951, 1972.
138. **Nakao, E. and Hotta, S.,** Immunogenicity of purified, inactivated chikungunya virus in monkeys, *Bull. WHO*, 48, 559, 1973.
139. **Kay, B. H.,** Three modes of transmission of Ross River virus by *Aedes vigilax* (Skuse), *Aust. J. Exp. Biol. Med. Sci.*, 60, 339, 1982.

Chapter 21

COLORADO TICK FEVER

G. Stephen Bowen

TABLE OF CONTENTS

I. HISTORICAL BACKGROUND AND SOCIAL IMPACT

Settlers and travelers to the Rocky Mountain Region in the early to mid-19th century recognized "mountain fever" as a cause of morbidity and occasionally a cause of mortality. Mountain fever was a general term to describe several illnesses which probably included typhoid fever, Rocky Mountain spotted fever, Colorado tick fever, and a variety of other febrile illnesses. Hygiene reports of medical officers of the U.S. Army and trail diaries from a variety of sources contain convincing accounts of a milder form of mountain fever which are compatible with better-documented descriptions of Colorado tick fever (CTF). References 1 to 7 contain fascinating early trail descriptions and accounts of CTF-like illnesses.

Eventually, the occurrence of fever, chills, headache, severe myalgia (especially in the back and legs), and arthralgia was associated with a disease that occurred primarily in the spring and summer in the Rocky Mountain Region. This disease was ultimately called Colorado tick fever by Becker.[2] Several early authors, including Becker[2] and Topping et al.,[4] suspected CTF to be transmitted by ticks. The viral etiology of CTF and proof of transmission by *Dermacentor andersoni* (Stiles) were confirmed by Florio and co-workers.[8-11] The maintenance cycle of CTF virus in natural foci was first described by Eklund and colleagues[12-14] and was further elucidated by Burgdorfer and Eklund[15-18] and by other studies in different geographical areas.

CTF does not occur in epidemic form, but is endemic in many western states and provinces of southwestern Canada which are within the range of *Dermacentor andersoni*. Since no epidemics occur, no major disruption of tourist or other industries has occurred. Nevertheless, substantial direct and indirect costs of nonfatal cases and occasional fatalities do occur. Mesch[19] estimated the average direct and indirect cost per nonfatal case of CTF in Colorado in 1981 to be $705.17. Current costs are probably at least $1000. The mean number of reported cases per year nationwide (a substantial underestimate of the true number) is more than 219 cases (Table 1). The estimated yearly minimum cost of known CTF cases is $219,000. The true cost is probably several times this amount. A single case requiring prolonged hospitalization significantly increases the cost.

II. THE VIRUS

The virus of CTF is currently classified among the orbiviruses of the family Reoviridae.[20,21] Orbiviruses are divided serologically into 12 groups. CTF is unique among orbiviruses containing 12 segments of double-strand RNA;[22] all other orbiviruses contain 10 segments. Because of this difference it has been proposed, but not generally accepted, that CTF be classified as a separate genus (*Coltivirus*) within the *Reoviridae* family. CTF virus and other orbiviruses replicate in cells of arthropods as well as in vertebrates, and viremia is produced in vertebrate hosts. The virus is unstable under mild acid conditions and is relatively resistant to many lipid solvents. Virus morphology has been studied by Murphy et al.[23] and Oshiro and Emmons.[24] The virus is 75 to 80 μm in diameter, contains a 50-μm core, and has regularly spaced surface projections.

A. Antigenic Relationships

CTF virus is the prototype of the CTF serogroup within the *Orbivirus* genus.[20,21] It is serologically related by plaque reduction neutralization (PRNT) and complement-fixation (CF) tests to one other registered arbovirus, Eyach virus,[25] which was isolated from *Ixodes ricinus* in the Federal Republic of Germany and France and from *I. ventalloi* in France.[26] It is also related to an unregistered virus isolated from the blood of a jackrabbit (*Lepus californicus*) and blood and spleen of a western gray squirrel, *Sciurus griseus*, in California.[27]

Table 1
MEAN REPORTED CTF CASES FROM STATES WHERE THE DISEASE IS REGULARLY REPORTED: 1970 TO 1984

State	No.	Range
Colorado	153.3	72—261
Utah	23.3	3—62
Montana	13.3	5—24
California	7.4	2—23
Wyoming	6.7	0—24
Idaho	6.3	0—17
Oregon	3.4	0—9
South Dakota	3.4	0—9
Washington	0.5	0—2
New Mexico	0.5	0—2
Nevada[a]	0.5	0—1
Total	218.6	

Note: CTF has also been reported irregularly from Kansas, Nebraska, Texas, Missouri, Minnesota, Indiana, Illinois, Massachusetts, Pennsylvania, Florida, New York, and Maine. These represent imported cases that can occur in any state.

[a] Includes California cases imported from Nevada.

B. Host Range

CTF virus is present in all stages of its principal vector *D. andersoni*. It has also been isolated from other ticks including *D. albopictus, D. parumapterus, Otobius lagophilus, I. sculptus, I. spinipalpus,* and *Haemaphysalis leporispalustris*.[28-32,87] Where *D. andersoni* occurs, other ticks play a minor role in the transmission of CTF in its natural cycle or to man. However, other species may transmit CTF in the few locales where CTF has been detected outside the known range for *D. andersoni*.[27,32]

D. andersoni ticks are opportunistic and the different stages of this species feed on mammals of many genera ranging in size from mice and chipmunks to elk. The following genera have been found to be infected by CTF: *Peromyscus, Eutamius, Spermophilus (Citellus), Sciurus, Tamiasciurus, Microtus, Perognathus, Dipodomys, Neotoma, Clethrionomys, Ochontona, Erithizon,* and *Marmota* among the rodents; *Lepus* and *Sylvilagus* among the rabbits and hares; *Canis* among the carnivores; and *Cervus* and *Odocoileus* among the ungulates.[12,15,31,33-37,88,89]

C. Strain Variation

More antigenic variation (demonstrated by PRNT) was found among strains of CTF isolated from humans than among those strains isolated from ticks or mammals.[90] The reason for this observation is unknown. Analysis of the migration of double-stranded RNA genome segments on polyacrylamide gels of numerous CTF isolates obtained over many years from many vertebrate and tick species and from several geographic locations[91] indicates considerable variation in the relative migration of segments in the L and M classes. However, the variations do not correlate with the source of the virus or the geographic location. Strains obtained from a specific area at the same time are more similar than those obtained in different areas or in years separated widely in time. Analysis by nucleic acid hybridization of more than 50 CTF isolates indicated that compared to other orbiviruses, CTF isolates show much less sequence variation.[92]

None of the techniques that have been used to analyze the degree of variation of CTF isolates has allowed them to be clearly separated by host, geography, or time. Neither specific hosts nor geographic isolation appears to produce major changes in genome sequences or in antigenic determinants of CTF viruses to an extent that allows identification of serotypes or genetic variants.

D. Assay Methods

Infectivity assays of CTF virus can be performed by intracerebral (i.c.) inoculation of suckling mice[38,39] or hamsters[11] or inoculation of tissue culture [BHK-21, VERO, KB, L929, FL, chick embryo primary tissue culture, *D. andersoni* primary tissue culture, and even in mosquito (*Aedes albopictus*) cells].[40-44] A plaque assay for the titration of virus[45] and for the performance of N tests[46] can be carried out in VERO or BHK-21 cells. Virus can be demonstrated by electron microscopy[23,24] in preparations of mouse brain, other organs, or from infected tissue culture preparations. The virus can also be demonstrated by direct or indirect immunofluorescence[47-49] or by immunoperoxidase staining.[50]

Virus can be isolated from ticks by grinding them with mortar and pestle in media containing antibiotics and a source of protein for stabilization.[27,35,51] CTF virus can also be isolated from human or vertebrate blood clots or heparinized blood specimens.[49,50,52]

III. DISEASE ASSOCIATIONS

A. Clinical Manifestations in Humans

The following discussion is based primarily on recent laboratory-confirmed case series,[52-54] on a chapter by Poland et al.[55] in *Current Diagnosis,* and from annual reports (1975 to 1984) of the Division of Vector-Borne Viral Diseases, Center of Infectious Diseases, Centers for Disease Control, Ft. Collins, Colo.

Symptoms usually appear suddenly after an incubation period of 1 to 14 days (median 4 days). Initially, the patient may have fever, headache, myalgia or joint pain, and lethargy. More than 20% of patients experience retroorbital pain, chills, photophobia, vomiting, neck pain, or sore throat. Less frequent complaints may include diarrhea or truncal rash. Physical findings may include fever, lymphadenopathy, conjunctival injection, mild hepatosplenomegaly, maculopapular rash, and stiff neck.

From 30 to 50% of patients have a biphasic illness with defervescence of fever and remission of symptoms followed by a second (sometimes more severe) period of illness. Prolonged convalescence of more than 3 weeks duration may occur especially in adults more than 30 years of age.[93] A small number of patients, especially children, experience meningitis or encephalitis,[53,55-58] gastrointestinal (GI) hemorrhage, diffuse intravascular coagulopathy (DIC), or a clinical picture resembling that of acute rheumatic fever. One child has been reported to have had leg weakness following encephalitis.[58] Although an early case series[14] reported complications to occur in 15% of children less than 10 years of age, a recent case series[54] reported only 1 case (3.3%) of confusion and ataxia among 30 CTF patients less than 10 years of age. Adults have occasionally developed pericarditis,[59] orchitis,[54] or a syndrome resembling myocardial infarction.[60] At least four children have died from severe GI hemorrhage and DIC; one child also had meningitis. Based on known cases, the case-fatality ratio, even in the pediatric age group, appears to be less than 0.2%. One instance of repeat infection after loss of N antibody has been documented.[54]

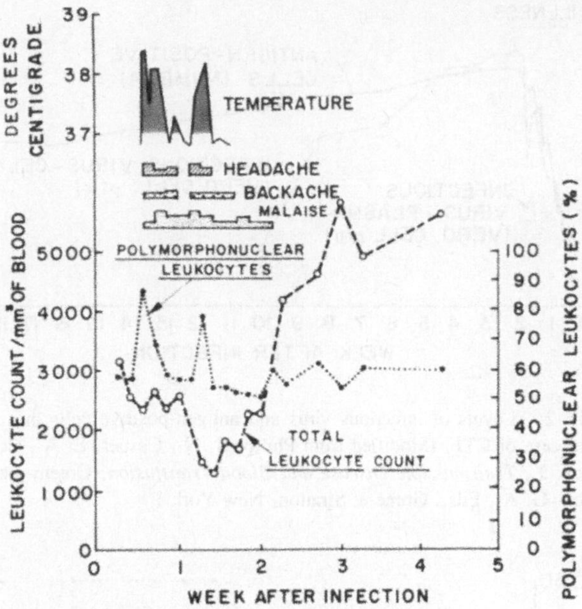

FIGURE 1. Clinical features of CTF by time after infection. (Modified from Philip, R. N., Casper, E. A., Cory, J., and Whitlock, J., *Transmissible Disease and Blood Transfusion*, Greenwalt, T. J. and Jamieson, G. A., Eds., Grune & Stratton, New York.)

A few cases of CTF have been documented in pregnant women.[94] From 1973 to 1986, the Division of Vector-Borne Viral Diseases collected information on 11 such patients. One child, whose mother developed CTF during the 14th week of gestation, was born with multiple congenital anomalies which were not believed to be caused by CTF. A child whose mother developed CTF 6 days before delivery, developed a febrile illness and neutropenia 3 days after delivery. The infant recovered. Another investigator[55] reported a mother who experienced a spontaneous abortion 2 weeks after having CTF.

B. Anatomical and Clinical Pathology

In fatal cases, severe GI hemorrhage, renal failure, DIC, alveolar hyaline membrane and focal necrotic lesions of the brain, liver, spleen, myocardium, and intestine have been reported.[13,55] The virus has been shown to infect bone marrow cells and red blood cell precursors.[61,62] The virus remains visible in mature erythrocytes for as long as 2 to 3 months after illness,[61-63] where it is partially protected from the immune system. CTF has been transmitted by blood transfusion.[64]

Clinical laboratory findings including leukopenia (less than 4500 white blood cells (WBC) per cubic millimeter) in 67% of patients.[54] The mean leukocyte count among persons with leukopenia was 2400/mm^3 in one study.[54] Other studies indicate that 100% of patients with CTF develop leukopenia.[65-67] Thrombocytopenia[68] may also occur; platelet counts may range from 20,000 to 60,000/mm^3.[67,68] No anemia or coagulopathies usually develop. Transient elevations in serum glutamate oxaloacetic transaminase (SGOT) or alkaline phosphatase are noted occasionally. Elevated levels of α-interferon during the first 10 days of illness have been observed in 78% of CTF patients.[69] Interferon levels are positively associated with increasing levels of fever, but not with number or severity of accompanying symptoms.[69] Figure 1 shows the relationship over time between symptoms and leukocyte counts in a case of laboratory-acquired CTF.[66]

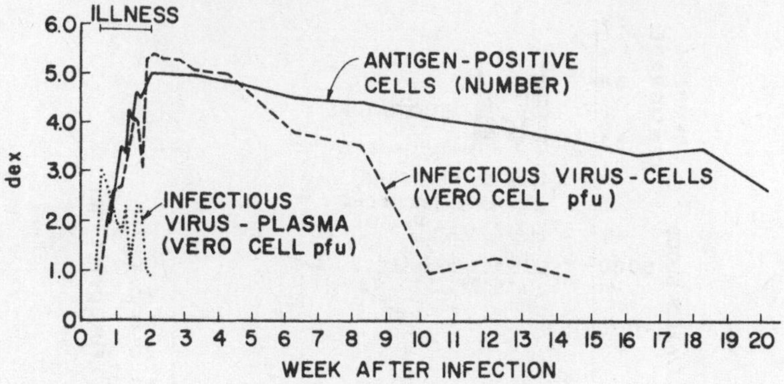

FIGURE 2. Levels of infectious virus and antigen-positive cells in a laboratory-acquired case of CTF. (Modified from Philip, R. N., Casper, E. A., Cory, J., and Whitlock, J., *Transmissible Disease and Blood Transfusion*, Greenwalt, T. J. and Jamieson, G. A., Eds., Grune & Stratton, New York.)

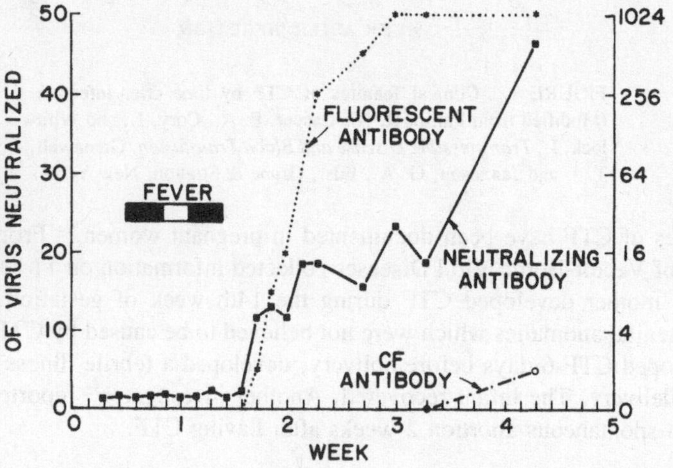

FIGURE 3. Antibody responses to CTF infection. (After Philip, R. N., Casper, E. A., Cory, J., and Whitlock, J., *Transmissible Disease and Blood Transfusion*, Greenwalt, T. J. and Jamieson, G. A., Eds., Grune & Stratton, New York.)

C. Diagnosis

A diagnosis of CTF on clinical grounds alone is not reliable. Laboratory confirmation is required. Viremia is present from the onset of symptoms and persists for up to 12 weeks. Reliable diagnosis is made by isolation of the virus by inoculation of heparinized blood or preparations of blood clot i.c. into 1- to 4-day-old mice or by inoculation of susceptible cell culture systems. Direct immunofluorescence testing of a slide preparation of a patient's blood clot is almost as reliable as the above test systems,[48] and has the advantage of being cheaper and faster. Virus-specific antibody may be demonstrated 1 to 2 weeks after onset of illness by indirect immunofluorescence,[47] 3 weeks after onset by N test,[49] or by enzyme immunoassay,[70] and, less reliably, 4 to 6 weeks after onset by CF.[49,52,70] The time of appearance of infectious virus, viral antigen, and antibodies in a laboratory-acquired case[66] is shown in Figures 2 and 3.

D. Disease in Nonhuman Vertebrates and Effect of Infection on Vectors

No adverse affect of CTF infection has been demonstrated in domestic or wild animals,

Table 2
AGE AND SEX DISTRIBUTION[a] OF CTF CASES

Age	Male	Female	Total
0—9	7.3	4.3	11.6
10—19	11.9	4.5	16.4
20—29	18.2	4.8	23.0
30—39	10.2	2.5	12.7
40—49	8.7	4.1	12.8
50—59	6.1	3.5	9.6
60—69	7.1	3.8	10.9
70+	2.0	1.0	3.0
Total	71.5	28.5	100.0

[a] N = 606.

but this has not been well studied in the former. Significant pathology primarily of the myocardium and brain, but also of the liver, spleen, and lymphoid tissue, occurs in experimentally inoculated mice, hamsters, and guinea pigs.[71-73] The myocardium and brain pathology consists of necrotic patches, lymphocytic and mononuclear cell inflammatory infiltrates, and vascular congestion with endothelial cell swelling.[71,72] Splenic lesions in hamsters include temporary loss of normal follicular structure and disappearance of follicular margins.[73] Some experimentally infected rhesus monkeys developed irregular febrile responses and one animal developed bloody diarrhea.[74] Viremia developed in all monkeys and lasted for 15 to 50 days. Leukopenia developed in two thirds of the monkeys. Teratogenic effects of CTF virus on mouse embryos have been reported.[75] Vectors infected with CTF virus do not appear to be adversely affected. Survival of infected *D. andersoni* for 2 to 3 years has been demonstrated.[76]

IV. EPIDEMIOLOGY

A. Demographic Variables

The age and sex distributions of CTF cases are shown in Table 2. The male:female sex ratio is 2.5:1. People aged 20 to 29 were more likely than people of other ages to get CTF (23% of cases), and 64.9% of cases were aged 10 to 49. However, people of all ages were confirmed to be cases.

B. Geographic Distribution

The geographic distribution of sites of tick exposure of human cases follows that of *D. andersoni*, although in California a few cases may have occurred outside the range of this tick. *D. andersoni* occurs in mountainous parts of Colorado, Montana, Utah, Idaho, Nevada, Wyoming, eastern California, eastern Washington, eastern Oregon, northern New Mexico, western South Dakota, and southern Alberta and British Columbia from elevations of 4000 to 10,500 ft (Figure 1). Isolations of CTF or related viruses and serological evidence of infections of wild vertebrates have also been reported from northwestern California[27] and Ontario[32] outside the recognized range of *D. andersoni*. Laboratory-confirmed cases have been reported in numerous states outside the endemic area (Table 1), including states in the Midwest, South, and Northeast. Physicians nationwide should suspect the diagnosis in any febrile patient with an appropriate travel history in spring and early summer.

C. Incidence

Because of lack of recognition, underreporting, and the nonpathognomonic nature of the clinical manifestations of many patients, the actual incidence of CTF is unknown. CTF is not an officially reportable disease; based on surveys in Rocky Mountain National Park in 1975 to 1976, it appears that only 10% of CTF cases are reported.[95] Laboratory evaluation is required to confirm the diagnosis as 30 to 40% of clinically compatible cases are not CTF. The mean number of reported CTF cases in states regularly reporting cases is shown in Table 1. Colorado has the largest number of cases each year, followed by Utah and Montana. The mean number of reported cases may reflect the diligence of health departments and laboratories in soliciting case reports and obtaining specimens as well as the actual incidence. Even in states where the disease is well recognized by physicians, considerable variation in annual incidence may occur depending on the weather conditions of that year and the conditions favoring or suppressing virus transmission of previous years. For example, in Colorado, many cases were reported from 1973 to 1977, but the number decreased in 1978 and did not rise to former levels until 1982. Recently, the Division of Vector-Borne Viral Diseases has not solicited blood specimens from Colorado physicians and the number of reported cases has decreased. Increased numbers of reported cases in other states such as Utah from 1981 to 1982, California from 1973 to 1975, and Idaho from 1974 to 1977 may reflect locally favorable conditions for increased transmission of virus during these periods since passive surveillance in these states has not changed.

The incidence of CTF-like illness within 1 week of leaving campsites in highly endemic areas in northern Colorado has been studied.[96] In Moraine Park Campground (MPCG) in Rocky Mountain National Park in 1975, the incidence of CTF-like illness was 0.42 cases per 100 camper days (or 238 camper days per CTF-like illness). In Cherokee Park Campground (CPCG), 0.37 cases per 100 camper days (or 343 camper days per illness). In 1976, the incidence of compatible illness at MPCG and CPCG was 0.25 and 0.23 illnesses per 100 camper days, respectively. In 1978, the incidence at MPCG and CPCG was 0.11 and 0.23 cases, respectively, per 100 camper days.

D. Seasonal Distribution

Most cases of CTF occur from April to early July, but cases have been reported from March to November. This seasonal pattern reflects adult tick host-seeking activity patterns, local weather conditions, and seasonal variation in human use of mountain areas for recreational activities. In Colorado, early warm spring weather and light winter snowfall in some years result in CTF cases in March and early April. In most years, the months of highest incidence are May and June. In some years adult ticks are active until late June and early July, whereas in other years adults are inactive after mid-June and a sharp decrease in human cases occurs.

E. Risk Factors

Risk factors include those associated with risk of acquisition of infection, severe clinical illness, and long duration of illness. Factors that increase risk of infection are primarily those that bring people into contact with infected adult *D. andersoni*. Prior infection and immunity, and avoidance of ticks by wearing long trousers with cuffs tied tight or clothes impregnated with acaricides, are possible protective factors. Risk factors for acquisition of infection include male sex, age between 10 and 50, outdoor occupation, frequent and lengthy outdoor recreational activities, exposure in endemic mountain areas between April and June, exposure within a favorable ecological setting and altitude, and exposure during years when environmental conditions favor virus transmission. Receipt of blood from donors within an endemic area[64] and occupational exposure for researchers and laboratory technicians are minor risk factors.

The higher risk of males of acquiring CTF is due to increased outdoor work or recreational activities of males rather than any inherent difference in susceptibility. People with outdoor occupations (rancher, park ranger, logger, cowboy, sheepman, surveyor, geologist, forester, road crew worker, or telephone linesman) in endemic areas are particularly at risk. Hikers, campers, and fishermen in spring and early summer in mountain zones in endemic areas are also at risk. Years of high recreational use of mountain areas, early influx of tourists, late adult tick host-seeking as a result of cooler temperatures or higher humidity, or opening of new camping or hiking areas to the public may lead to increased risk of exposure.

Altitude, geographic location, and ecological setting of residence, campsite, or recreational activity all are important risk factors for CTF illness. For example, a campsite in Colorado below 6000 or above 10,000 ft elevation or in an unfavorable ecological setting such as open grassland or tundra (see detailed discussion later) results in relatively little risk to its users. A similar campsite located at 8300 ft elevation on a south-facing slope with numerous rocky outcroppings and ponderosa pine overstory results in higher risk of CTF for its users. Risk reduction by thoughtful location of residences and campgrounds may be possible.

Nothing is known about increased susceptibility to infection by sex, race, age, nutritional status, or immunologic or genetic background. Virtually all people are susceptible; the antibody prevalence in most populations is low. Young adults have been reported to be more likely to have illness of more than 3 weeks duration than are children.[54] Complications such as pericarditis, orchitis, and pneumonitis have been reported only in adults. However, severe hemorrhagic manifestations — especially GI bleeding and death — have been reported only in children. The case-fatality ratio for reported cases in children is about 0.2% compared to 0 in adults.

F. Seroepidemiology

No careful population-based seroepidemiological studies to determine age-, sex-, and race-specific antibody prevalence rates have been published. Based on data from other endemic diseases, one would expect the antibody prevalence rate to increase with increasing age and longer duration of residence in an endemic area. Although exhaustive studies have not been carried out, the proportion of inapparent CTF infections may be small. Among 12 to 15 people with repeated tick exposure during 5 years of field work at Rocky Mountain National Park, 3 of 3 persons who seroconverted were clinically ill.

V. CYCLE OF TRANSMISSION

A generalized cycle of transmission for CTF virus is shown in Figure 4.

CTF virus was first isolated from *D. andersoni* and from human patients by Florio et al.[8,9] Laboratory studies by Florio et al.[8,11] showed that all stages of *D. andersoni* could be infected by feeding on infected hamsters. The basic transmission cycle was elucidated by Eklund et al.[12-14] and Burgdorfer and Eklund.[15-18] Other field and laboratory studies documenting further details of the cycle,[27,29,30,32,33,35-37,51,76-79] including possible preservation of virus through the winter by hibernating rodents,[80] were carried out by many research groups.

CTF virus is maintained in nature in a cycle in which infected nymphal *D. andersoni* ticks feed upon and transmit CTF to rodent hosts of several genera which in turn are fed upon by larval ticks. Many rodents have a prolonged period of viremia as a result of infection with CTF virus. The virus invades red blood cells and their precursors[61,62] and is ingested by the larvae with the viremic blood meal. The virus is then passed transstadially to the nymphal stage as the larvae molt. Infected nymphs usually overwinter without feeding, then feed again on rodents the following spring, thus perpetuating the cycle. After feeding, engorged nymphs drop off the host, molt to become adults (which are infected by transstadial passage of virus), and subsequently seek hosts for blood meals prior to laying eggs. The

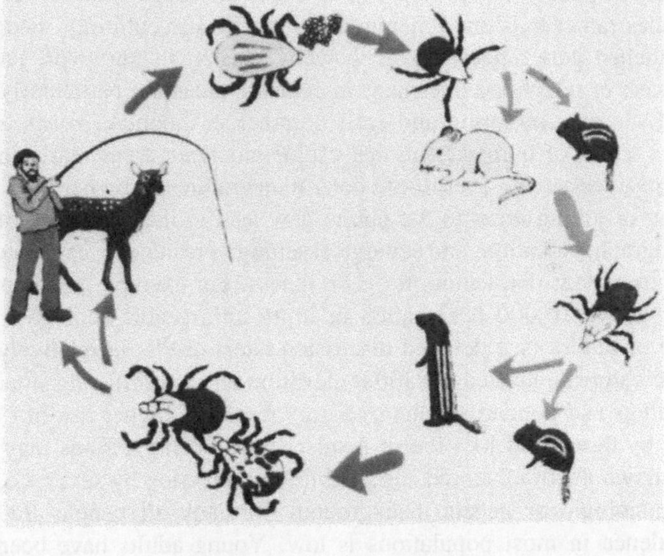

FIGURE 4. Natural history of CTF virus.

virus is not passed transovarially from adult female to larvae,[15,76,81,82] but is passed by bite to certain natural hosts such as the porcupine and to humans. Adult *D. andersoni* are not an essential part of the virus transmission cycle, but do produce eggs which become the next generation of larvae. The entire life cycle of *D. andersoni* lasts 2 to 3 years, but the essential nymph-to-rodent-to-larva virus transmission occurs each year. Although laboratory studies have shown that rodents may preserve the virus through a period of hibernation, this mechanism does not appear essential to virus maintenance.

The cycle of transmission of Eyach virus, which is serologically related to CTF, is poorly understood. Thet virus has been isolated from *I. ricinus* in Germany[25] and France[26] and from *Ix. ventalloi* in France.[26] In France, the ticks which yielded Eyach virus were collected as ectoparasites of a wild rabbit, *Oryctolagus cuniculus*. The cycle of the unregistered CTF serogroup virus from California[27] is also poorly known. This virus has been isolated only from the blood of a jackrabbit, *Lepus californicus*, in Mendocino County, and the blood and spleen of a western gray squirrel, *Sciurus griseus*, from San Luis Obispo County, California.

A. Vertebrate Hosts

1. Evidence from Field Studies

Field studies have shown that although the basic cycle outlined above is true in all areas studied, the contribution of various small- to medium-sized mammals varies from one study area to another. In addition, tick species other than *D. andersoni* may play a role in CTF transmission in some areas. Mammal species may vary in susceptibility and the duration and magnitude of viremia.[51] Indeed, variation may occur between geographic races or subpopulations of the same species. Different mammal species also may serve as hosts for larvae, nymphs, and adults in different areas. Mammal hosts may be important locally as a source of blood meals to one stage of ticks without being susceptible to virus infection. The most important mammal hosts are those which are locally common or abundant, which are fed upon by both larvae and nymphs, and which are susceptible to virus infection resulting in prolonged, high-level viremia. Although immature *D. andersoni* are opportunistic feeders, field studies indicate that one, or at most a small number of, mammal species fit the above criteria in each study area.

In all areas studied, the golden mantled ground squirrel, *Spermophilus (Citellus) lateralis*, has been an important mammal host for immature *D. andersoni* and CTF virus. In Montana, Burgdorfer and Eklund[15,17] found this species to be the principal host and that minimal transmission occurred in habitats where this species was not present. Clark et al.[35] and Sonenshine et al.[79] found the deer mouse (*Peromyscus maniculatus*) and the bushy-tailed woodrat (*Neotoma cinerea*) to be additional important hosts at another Montana study site. In 1957[12] in Pitkin County, Colorado, CTF virus was isolated from *S. lateralis* and porcupines (*Erithizon dorsatum*). In Rocky Mountain National Park from 1974 to 1976, the least chipmunk, *Eutamias minimus*, was the most important small mammal in the CTF cycle,[37] but *S. lateralis* made a significant contribution as well. Virus isolation rates were highest from blood specimens collected from *E. minimus* (18.3%), and this species was heavily infested with tick larvae and nymphs at the same time. Virus isolation rates from blood specimens collected from *S. lateralis* were lower (8.1%), and these rodents had few larvae.[97] However, *S. lateralis* was host for the largest number of infected *D. andersoni* nymphs, and the infection rate was higher for nymphs collected from this species than for nymphs removed from any other species.

The most important natural host for adult *D. andersoni* in Colorado appears to be the porcupine (*Erithizon dorsatum*).[98] Other hosts include elk, marmots, and occasionally deer and coyote. Dispersal of adult *D. andersoni* by migration and local movement of ungulates may occur.

2. *Evidence from Experimental Studies*

Experimental infection studies of Colorado mammals[51] indicated that adult chipmunks of two species (*E. minimus* and *E. umbrinus*) were the best experimental hosts for CTF virus of the species tested. Juvenile *S. lateralis, P. maniculatus,* and *Spermophilus richardsoni* were more susceptible than were adults, and resulting viremia levels were higher and longer lasting than in adults of the same species. Viremia lasted up to 57 days in an individual *S. lateralis* and up to 51 days for an individual *E. minimus*. The mean duration, however, was 12 to 16 days for adult chipmunks, juvenile *S. lateralis,* and *P. maniculatus,* but only 4.5 to 4.9 days for adult *S. lateralis* and *P. maniculatus*. Mean peak viremia titers were 7000 to 8000 plaque-forming units per milliliter (pfu/mℓ) for chipmunks and 600 to 1050 pfu/mℓ for *S. lateralis* and *P. maniculatus*. *Sylvilagus nuttalli* rabbits collected in Colorado and inoculated with Colorado isolates of CTF were resistant. Limited studies (two animals inoculated per species) by Burgdorfer[16,18] of mammals collected in Montana indicated that five species (*S. lateralis, E. amoenus, S. columbianus, P. maniculatus,* and *E. dorsatum*) were susceptible; viremia lasted 16 to 39 days and peak titers ranged from 20,000 to 100,000 mouse intraperitoneal (i.p.) LD_{50}/mℓ. These titers were higher than those found in the studies of Colorado mammals cited above. Differences in susceptibility of geographically distinct populations of mammals, failure to detect individual differences in susceptibility because of small numbers inoculated, and differences in virus strains or virus test systems may account for the discrepant results. The cited laboratory and field studies by many investigators showing a lack of host specificity of immature stages of *D. andersoni* confirm that many rodent species play a role in CTF cycles in different locales.

B. Tick Vectors

1. *Evidence from Field Studies*

Field studies of adult *D. andersoni* have been carried out by dragging, by asking hikers or campers to collect ticks from their persons, or by the use of CO_2-baited traps.[76,82] Immature stages can be collected by trapping wild mammals or by the use of CO_2-baited traps. Adults seek hosts from the time of appearance of free water when the snow melts in February or March, and continue until June or early July. They begin diapause with completion of the spring snow-melt and evaporation of surface water from spring rains. Nymphs can be collected from CO_2-baited traps and from mammals from April to September, but the largest numbers are present in May and June.[76,87] Larvae are difficult to collect from CO_2-baited traps; they are more easily collected from trapped animals from June to September. Field infection rates for adult ticks may range from 0 to 45%[11,15,17,33-35,76,82,99] and may vary considerably among areas quite close to each other. In Colorado in 1975—1976, adult ticks collected in areas of high human recreational use usually had infection rates of 10 to 20%. Unfed nymphs collected by CO_2-baited traps had infection rates of 2.6 to 7.1%.[76] No CTF isolations have been made from unfed larvae. Data are currently insufficient to characterize field infection rates of other tick species or to define their role in CTF cycles.

Eads and Smith[76] have estimated the density of adult *D. andersoni* at Cherokee Park (Larimer County, Colorado) in 1979 to be 1139/ha. Marked ticks survived at least 2 winters and 11% of ticks studied on a grid survived until the following year. Sonenshine et al.[79] estimated the larval *D. andersoni* population density to range from 119,000 to 959,000/ha at their study site in Ravalli County, Montana. Ticks in a particular area are generally produced locally. Small mammals fed on by immature stages of *D. andersoni* have small home ranges; although dispersal of young rodents does occur, distances traveled are usually small. Adult ticks which feed upon porcupines and elk may be dispersed over distances of several miles.

2. Evidence from Experimental Studies

Laboratory transmission studies document that adult and immature stages of *D. andersoni* can become infected as a result of feeding on laboratory-infected animals and that they can transmit to susceptible rodents under laboratory conditions. The threshold of infection for immature stages of *D. andersoni* is estimated to be 100 suckling mouse intraperitoneal (SMIP) 50% lethal doses (SMIP LD_{50})[16,18] or 10 SMIP LD_{50} for larvae and 30 SMIP LD_{50} for nymphs.[78] No exhaustive studies of vector competency or of variation of field populations in vector competency have been carried out.

VI. ECOLOGICAL DYNAMICS

CTF is endemic in mountainous areas over a large geographic range. Foci of especially high virus activity exist within this large range, and are associated with ecological factors that sustain abundant populations of vertebrate hosts which are susceptible to infection with CTF virus. These foci provide conditions for shelter, development, and survival of all stages of ticks, including eggs.

The important features of foci of CTF transmission are soil temperature, moisture, and depth; slope of hill; presence of exposed rock outcroppings; ground cover (including litter); and abundance and diversity of vegetation.[83,84] Latitude and elevation are also critical. In Rocky Mountain National Park, the foci of highest CTF activity (Figure 5) are between 2400 to 2745 m elevation (8000 to 9000 ft) on south-facing drier slopes of lower montane forest, and are covered with scattered shrubs, herbs, and grasses under an open ponderosa pine overstory.[83,84] In the ecological setting above, CTF virus and tick abundance were associated with shallow, relatively dry soil, exposed rock outcroppings with abundant interstices providing shelter to chipmunks and ground squirrels, moderate shrub cover, moderate to steep slopes, abundant pine trees and log litter, and presence of chipmunks and *S. lateralis*, but not other small mammals.[84]

VII. SURVEILLANCE

Surveillance for CTF is carried out primarily by voluntary reporting of suspect cases with optional submission of blood for serological diagnosis. State laws do not require reporting

FIGURE 5. Econiche favoring CTF transmission in the Rocky Mountains: south-facing slopes of lower montane forest covered with scattered shrubs and open ponderosa pine overstory.

Table 3
YEARLY VARIATION IN REPORTED CTF CASES FROM REPRESENTATIVE STATES

	No. of cases					
Year	**Colorado**	**Utah**	**California**	**Wyoming**	**Idaho**	**Oregon**
1970	72	3	2	?	2	4
1971	89	6	6	2	0	6
1972	115	10	3	2	8	6
1973	242	24	23	6	8	9
1974	192	13	13	2	13	1
1975	261	44	12	6	17	3
1976	220	22	9	11	15	1
1977	200	21	9	0	16	0
1978	104	21	6	1	7	1
1979	148	21	3	17	4	2
1980	127	43	6	7	1	5
1981	159	62	6	24	6	5
1982	176	62	1	7	2	1
1983	116	33	3	10	3	3
1984	79	23	0	9	3	?

of CTF. Passive surveillance systems currently in place in most states do indicate fluctuations in numbers of human cases from year to year (Table 3). Since the disease does not occur in epidemics, little incentive exists for early notification leading to vector control efforts. Active surveillance by health departments of primary care practitioners in endemic areas would be required to increase the proportion of reported cases. Laboratory evaluation is required to eliminate the approximately 30% of cases with similar clinical manifestations that are not CTF.

Surveillance for illness in cohorts of campers can be carried out, but incomplete reporting of illness and the nonspecific nature of the clinical syndrome of CTF make this technique somewhat inaccurate unless adequate laboratory analysis of blood specimens is done for all suspect cases.

Research laboratories can do surveillance by carrying out regular tick collections in standard locations with CO_2-baited traps. The years with increased risk of transmission to humans can roughly by correlated with tick density or numbers of adult ticks collected by tick traps, but adult tick infection rates do not appear to be highly correlated with human risk.

VIII. EPIDEMIC INVESTIGATION

CTF does not occur in epidemics. Local health authorities may wish to identify areas of high levels of endemic virus transmission in public parks and campgrounds to carry out local prevention and control efforts.

IX. TREATMENT, PREVENTION, AND CONTROL

Current treatment is entirely supportive, but in the future antiviral agents such as ribavirin and others[85] may provide effective therapy. A formalin-inactivated vaccine prepared in the brains of suckling mice was tested and found to be safe and antigenic in 1963,[86] but was never licensed. No new vaccine is under development. CTF is transmitted over a large geographic area in very rugged terrain. Control or eradication of the virus cycle is not possible. Risk reduction is possible through public education to reduce exposure to tick bites. Risk may possibly be decreased by moving campgrounds to ecologically low-risk locations, permitting camping in ecologically high-risk settings only during months of low adult tick host-seeking activity, and through mammal and tick population control.

Public education efforts should be directed toward hikers, campers, people with outdoor occupations, and residents of endemic areas. In public recreation areas, educational materials can be distributed to people using these areas. People should be encouraged to wear long pants with cuffs tied or clipped tight at the ankles with elastic or rubber bands. Adults should inspect themselves and their children at frequent intervals to remove ticks before they attach. Clothing soaked in or sprayed with permethrin, diethyltoluamide, pyrethrin, or other tick repellents or acaricides may be effective in reducing exposure to ticks.

In public parks or campgrounds where previous field studies have shown large numbers of infected ticks to be present and where CTF cases occur regularly, local mammal or vector control efforts may be considered. Mammal populations can be reduced by trapping or shooting. Both methods are time consuming and expensive. Poisoning is not recommended since poisoning of nontarget species including pets or children may occur. Local applications of acaricides by back-pack spraying have been successful in reducing tick populations in other areas (carried out for local reduction of *Ix. dammini* in New Jersey),[100] but have not been tried for *D. andersoni*. The risk of poisoning other species would have to be evaluated. An efficient method of killing immature stages of ticks is the application of carbaryl to rodents at bait stations. This method of tick reduction has a 1-year lag time and is also costly and labor intensive.

Manipulation of times of campground use and movement of campgrounds to sites which are ecologically unfavorable for CTF transmission may be attractive options for risk reduction in parks. High-risk campgrounds can be closed until mid-June or July when adult ticks are no longer seeking hosts. Moving campgrounds to lower-risk locations may permit camping earlier in the year.

X. FUTURE RESEARCH

Areas for fruitful future research might include

1. Development of better methods for molecular and immunological characterization of strains of CTF virus
2. Comparison of large numbers of CTF strains from several sources, different geographic locations, and over many years to determine the degree of distinctness or relatedness of virus populations in different foci and to look at trends in evolution
3. Development of a suitable primate or other animal model for studies of illness, pathogenesis, treatment, and immunity
4. Characterization of the phenomenon of viral persistence in vertebrates and the role, if any, of this phenomenon in cycles of transmission and in virus evolution
5. Characterization of the vector competence of the different stages of *D. andersoni* including the degree of natural variation of different field populations
6. Definition of the role of tick species other than *D. andersoni* in CTF cycles by field and laboratory studies
7. Characterization of geographic differences in susceptibility of the principal mammal hosts by experimental inoculation studies
8. Determination of differences in the ability of field isolates of CTF from different states to produce infection in vectors and vertebrate hosts

REFERENCES

1. **Patzki, M.** (Assistant Surgeon), Sanitary Report, Fort Steele (U.S. Army), Wyoming Territory, 1875.
2. **Becker, F. E.,** Tick-borne infections in Colorado. II. A survey of occurrence of infections transmitted by the wood tick, *Color. Med.*, 27, 87, 1930.
3. **Toomey, N.,** American mountain tick fever — semiography and nosology with remarks on pathology and treatment, *Ann. Intern. Med.*, 5, 912, 1932.
4. **Topping, N. H., Cullyford, J. S., and Davis, G. E.,** Colorado tick fever, *Public Health Rep.*, SS, 2224, 1940.
5. **Lloyd, L. W.,** Colorado tick fever, *Med. Clin. North Am.*, 2, 587, 1951.
6. **Drevets, C. C.,** Colorado tick fever, observations on eighteen cases and review of the literature, *J. Kans. Med. Soc.*, 58, 448, 1957.
7. **Olch, P. D.,** Treading the elephant's tail: medical problems on the overland trails, *Bull. Hist. Med.*, 59, 196, 1985.
8. **Florio, L., Stewart, M. D., and Mugrage, E. R.,** The experimental transmission of Colorado tick fever, *J. Exp. Med.*, 80, 165, 1944.
9. **Florio, L., Stewart, M. D., and Mugrage, E. R.,** The etiology of Colorado tick fever, *J. Exp. Med.*, 83, 1, 1946.
10. **Florio, L. and Miller, M. S.,** Epidemiology of Colorado Tick fever, *Am. J. Public Health*, 38, 211, 1948.
11. **Florio, L., Miller, M. S., and Mugrage, E. R.,** Colorado tick fever. Isolation of the virus from *Dermacentor andersoni* in nature and a laboratory study of the transmission of the virus in the tick, *J. Immunol.*, 64, 257, 1950.
12. **Eklund, C. M., Kohls, G. M., and Jellison, W. L.,** Isolation of Colorado tick fever virus from rodents in Colorado, *Science*, 128, 413, 1958.
13. **Eklund, C. M.,** Natural history of Colorado tick fever virus, *Lancet*, 82, 172, 1962.
14. **Eklund, C. M., Kohls, G. M., Jellison, W. L., Burgdorfer, W., Kennedy, R. C., and Thomas, L.,** The clinical and ecological aspects of Colorado tick fever, *Proc. 6th Int. Congr. Trop. Med. Malariol.*, 5, 197, 1959.
15. **Burgdorfer, W. and Eklund, C. M.,** Studies on the ecology of Colorado tick fever virus in western Montana, *Am. J. Hyg.*, 69, 127, 1959.
16. **Burgdorfer, W.,** Colorado tick fever. The behavior of CTF virus in the porcupine, *J. Infect. Dis.*, 104, 101, 1959.

17. **Burgdorfer, W. and Eklund, C. M.,** Colorado tick fever. I. Further ecological studies in western Montana, *J. Infect. Dis.*, 107, 379, 1960.
18. **Burgdorfer, W.,** Colorado tick fever. II. The behavior of Colorado tick fever virus in rodents, *J. Infect. Dis.*, 107, 384, 1960.
19. **Mesch, K. E.,** Decision Analysis for the Control of Colorado Tick Fever, unpublished M. P.A. thesis, University of Colorado, Denver, 1983.
20. **Berge, T. O., Ed.,** *International Catalogue of Arboviruses Including Certain Other Viruses of Vertebrates*, 2nd ed., DHEW Publ. No. (CDC) 75-8301, Public Health Service, Washington, D.C., 1975.
21. **Fenner, F.,** Classification and nomenclature of viruses. Second report of the International Committee on Taxonomy of Viruses, *Intervirology*, 7, 1, 1976.
22. **Knudson, D. L.,** Genome of Colorado tick fever virus, *Virology*, 112, 361, 1981.
23. **Murphy, F. A., Coleman, P. H., Harrison, A. K., and Gary, G. W.,** Colorado tick fever virus: an electron microscopic study, *Virology*, 35, 28, 1968.
24. **Oshiro, L. S. and Emmons, R. W.,** Electron microscopic observations of Colorado tick fever virus in BHK 21 and KB cells, *J. Gen. Virol.*, 3, 279, 1968.
25. **Rehse-Kupper, B., Casals, J., Rehse, E., and Ackerman, R.,** Eyach — an arthropod-borne virus related to Colorado tick fever virus in the Federal Republic of Germany, *Acta Virol.*, 20, 339, 1976.
26. **Chastel, C., Main, A. J., Couatarmanac, H. A., Le Lay, G., Knudsen, D. L., Quillen, M. C., and Beaucournu, J. C.,** Isolation of Eyach virus (reoviridae, Colorado tick fever group) from *Ixodes ricinus* and *Ixodes ventalloi* ticks in France, *Arch. Virol.*, 82, 161, 1984.
27. **Lane, R. S., Emmons, R. W., Devlin, V., Dondero, D. V., and Nelson, B. C.,** Survey for evidence of Colorado tick fever virus outside of the known endemic area in California, *Am. J. Trop. Med. Hyg.*, 31, 837, 1982.
28. **Eklund, C. M., Kohls, G. M., and Brennan, J. M.,** Distribution of Colorado tick fever and virus-carrying ticks, *JAMA*, 157, 335, 1955.
29. **Philip, C. B., Bell, J. F., and Larson, C. L.,** Evidence of infectious diseases and parasites in a peak population of black-tailed jack rabbits in Nevada, *J. Wildl. Manage.*, 19, 225, 1955.
30. **Kohls, G. M.,** Colorado tick fever discovered in California, *Calif. Vector Views*, 2, 17, 1955.
31. **Eklund, C. M.,** Colorado tick fever, in *Diseases Transmitted from Animals to Man*, Hull, T. G., Ed., Charles C Thomas, Springfield, Ill., 1963, chap. 36.
32. **Newhouse, V. F., McKiel, J. A., and Burgdorfer, W.,** California encephalitis, Colorado tick fever and Rocky Mountain spotted fever in eastern Canada, *Can. J. Public Health*, 55, 257, 1964.
33. **Johnson, H. N.,** The ecological approach to the study of small mammals in relation to arboviruses, in *Anais de Microbiologia*, Vol. 11 (Part A), Proc. 7th Int. Congr. Trop. Med. Malaria, Bruno-Lobo, M. and Shope, R., Eds., Rio de Janeiro, 1963.
34. **Eklund, C. M.,** Role of mammals in maintenance of arboviruses, in *Anais de Microbiologia*, Vol. 11 (Part A), Proc. 7th Int. Congr. Trop. Med. Malaria, Bruno-Lolbo, M. and Shope, R., Eds., Rio de Janeiro, 1963, 99.
35. **Clark, G. M., Clifford, C. M., Fadness, L. V., and Jones, E. K.,** Contributions to the ecology of Colorado tick fever virus, *J. Med. Entomol.*, 7, 189, 1970.
36. **Johnson, H. N.,** Keynote address: the ecological approach to the study of zoonotic disease, *J. Wildl. Dis.*, 6, 194, 1970.
37. **Bowen, G. S., McLean, R. G., Shriner, R. B., Francy, D. B., Pokorny, K. S., Trimble, J. M., Bolin, R. A., Barnes, A. M., Calisher, C. H., and Muth, D. J.,** The ecology of Colorado tick fever in Rocky Mountain National Park in 1974. II. Infections in small animals, *Am. J. Trop. Med. Hyg.*, 30, 490, 1981.
38. **Koprowski, H. and Cox, H. R.,** Adaptation of Colorado tick fever virus to mouse and developing chick embryo, *Proc. Soc. Exp. Biol. Med.*, 62, 320, 1946.
39. **Oliphant, J. W. and Tibbs, R. D.,** Colorado tick fever. Isolation of virus strains by inoculation of suckling mice, *Public Health Rep.*, 65, 521, 1950.
40. **Pickens, E. G. and Luoto, L.,** Tissue culture studies with Colorado tick fever virus. I. Isolation and propagation of virus in KB cultures, *J. Infect. Dis.*, 103, 102, 1958.
41. **Trent, D. W. and Scott, L. V.,** Colorado tick fever in cell culture. I. Cell-type susceptibility and interaction with L cells, *J. Bacteriol.*, 88, 702, 1964.
42. **Trent, D. W. and Scott, L. V.,** Colorado tick fever virus in cell culture. II. Physical and chemical properties, *J. Bacteriol.*, 91, 1282, 1966.
43. **Yunker, C. E. and Cory, J.,** Growth of Colorado tick fever (CTF) virus in primary tissue cultures of its vector, *Dermacentor andersoni* Stiles (Acarina: Ixodidae), with notes on tick tissue culture, *Exp. Parasitol.*, 20, 267, 1967.
44. **Yunker, C. E. and Cory, J.,** Colorado tick fever virus: growth in a mosquito cell line, *J. Virol.*, 3, 631, 1969.
45. **Deig, E. F. and Watkins, H. M. S.,** Plaque assay procedure for Colorado tick fever virus, *J. Bacteriol.*, 88, 42, 1964.

46. **Gerloff, R. K. and Eklund, C. M.,** A tissue culture neutralization test for Colorado tick fever antibody and use of the test for serologic surveys, *J. Infect. Dis.*, 104, 174, 1959.
47. **Burgdorfer, W. and Lackman, D.,** Identification of the virus of Colorado tick fever in mouse tissues by means of fluorescent antibodies, *J. Bacteriol.*, 80, 131, 1960.
48. **Emmons, R. W. and Lennette, E. H.,** Immunofluorescent staining in the laboratory diagnosis of Colorado tick fever, *J. Lab. Clin. Med.*, 68, 923, 1966.
49. **Emmons, R. W., Dondero, D. V., Devlin, V., and Lennette, E. W.,** Serologic diagnosis of Colorado tick fever, a comparison of complement-fixation, immunofluorescence, and plaque-reduction methods, *Am. J. Trop. Med. Hyg.*, 18, 796, 1969.
50. **Desmond, E. P., Schmidt, N. J., and Lennette, E. H.,** Immunoperoxidase staining for detection of Colorado tick fever virus, and a study of congenital infection in the mouse, *Am. J. Trop. Med. Hyg.*, 28, 729, 1979.
51. **Bowen, G. S., Shriner, R. B., Pokorny, K. S., Kirk, L. J., and McLean, R. G.,** Experimental Colorado tick fever virus infection in Colorado mammals, *Am. J. Trop. Med. Hyg.*, 30, 224, 1981.
52. **Earnest, M. P., Breckinridge, J. C., Barr, R. J., Francy, D. B., and Mollohan, C. S.,** Tick fever. Clinical and epidemiological features and evaluation of diagnostic methods, *Rocky Mount. Med. J.*, 68, 60, 1971.
53. **Spruance, S. L. and Bailey, A.,** Colorado tick fever. A review of 115 laboratory confirmed cases, *Arch. Intern. Med.*, 131, 288, 1973.
54. **Goodpasture, H. C., Poland, J. D., Francy, D. B., Bowen, G. S., and Horn, K. A.,** Colorado tick fever: clinical, epidemiologic, and laboratory aspects of 228 cases in Colorado from 1973 to 1974, *Ann. Intern. Med.*, 88, 303, 1978.
55. **Poland, J. D.,** Colorado tick fever, in *Current Diagnosis*, Conn, R. B., Ed., W.B. Saunders, Philadelphia, 1985, 195.
56. **Fraser, C. H. and Schiff, D. W.,** Colorado tick fever encephalitis. Report of a case, *Pediatrics*, 29, 187, 1962.
57. **Draughn, D. E., Sieber, D. E., Jr., and Umlauf, H. J., Jr.,** Colorado tick fever encephalitis, *Clin. Pediatr.*, 4, 626, 1965.
58. **Silver, H. J., Meiklejohn, G., and Kempe, C. H.,** Colorado tick fever, *Am. J. Dis. Child.*, p. 30, 1961.
59. **Hierholzer, W. J. and Barry, D. W.,** Colorado tick fever pericarditis, *JAMA*, 217, 825, 1971.
60. **Emmons, R. W.,** Colorado tick fever simulating acute myocardial infarction, *JAMA*, 222, 87, 1972.
61. **Emmons, R. W., Oshiro, L. S., Johnson, H. N., and Lennette, E. H.,** Intra-erythrocytic location of Colorado tick fever virus, *J. Gen. Virol.*, 17, 185, 1972.
62. **Oshiro, L. S., Dondero, D. V., Emmons, R. W., and Lennette, E. H.,** The development of Colorado tick fever virus within cells of the haemopoietic system, *J. Gen. Virol.*, 39, 73, 1978.
63. **Hughes, L. E., Casper, E. A., and Clifford, C. M.,** Persistence of Colorado tick fever virus in red blood cells, *Am. J. Trop. Med. Hyg.*, 23, 350, 1974.
64. **Randall, W. H., Simmons, J., Casper, E. A., and Philip, R. N.,** Transmission of Colorado tick fever virus by blood transfusion, *Mont. Morb. Mort. Weekly Rep.*, 24, 422, 1975.
65. **Anderson, R. A., Entringer, M. A., and Robinson, W. A.,** Virus-induced leukopenia: Colorado tick fever as a human model, *J. Infect. Dis.*, 151, 449, 1985.
66. **Philip, R. N., Casper, E. A., Cory, J., and Whitlock, J.,** The potential for transmission of arboviruses by blood transfusion with particular reference to Colorado tick fever, in *Transmissible Disease and Blood Transfusion*, Greenwalt, T. J. and Jamieson, G. A., Eds., Grune & Stratton, New York, 1975.
67. **Stewart, M. D.,** The white blood cell picture in virus diseases with special reference to Colorado tick fever, oral presentation, Annu. Meet. Am. Soc. Med. Technol., Chicago, 1944.
68. **Markovitz, A.,** Thrombocytopenia in Colorado tick fever, *Arch. Intern. Med.*, 111, 307, 1963.
69. **Ater, J. L., Overall, J. C., Jr., Tze-Jou Yeh, O'Brien, R. T., and Bailey, A.,** Circulating interferon and clinical symptoms in Colorado tick fever, *J. Infect. Dis.*, 151, 966, 1985.
70. **Calisher, C. H., Poland, J. D., Calisher, S. B., and Wormoth, L. A.,** Diagnosis of Colorado tick fever infection by enzyme immunoassays for immunoglobulin M and G antibodies, *J. Clin. Microbiol.*, 22, 84, 1986.
71. **Black, W. C., Florio, L., and Stewart, M. O.,** A histologic study of the reaction in the hamster spleen produced by the virus of Colorado tick fever, *Am. J. Pathol.*, 23, 217, 1947.
72. **Hadlow, W. J.,** Histopathologic changes in suckling mice infected with the virus of Colorado tick fever, *J. Infect. Dis.*, 101, 158, 1957.
73. **Miller, J. K., Tompkins, V. N., and Sieracki, J. C.,** Pathology of Colorado tick fever in experimental animals, *Arch. Pathol.*, 72, 149, 1961.
74. **Gerloff, R. K. and Larson, C. L.,** Experimental infection of rhesus monkeys with Colorado tick fever virus, *Am. J. Pathol.*, 35, 1043, 1959.
75. **Harris, R. E., Morahan, P., and Coleman, P.,** Teratogenic effects of Colorado tick fever virus in mice, *J. Infect. Dis.*, 131, 397, 1975.

76. **Eads, R. B. and Smith, G. C.,** Seasonal activity and Colorado tick fever virus infection rates in Rocky Mountain wood ticks, *Dermacentor andersoni* (Acari:Ixodidae), in north central Colorado, USA, *J. Med. Entomol.*, 20, 49, 1983.
77. **Hall, R. R., McKiel, J. A., and Gregson, J. D.,** Occurrence of Colorado tick fever virus in *Dermacentor andersoni* ticks in British Columbia, *Can. J. Public Health*, 59, 273, 1968.
78. **Rozeboom, L. E. and Burgdorfer, W.,** Development of Colorado tick fever virus in the Rocky Mountain wood tick, *Dermacentor andersoni, Am. J. Hyg.*, 69, 138, 1959.
79. **Sonenshine, D. E., Yunker, C. E., Clifford, C. E., Clark, G. M., and Rudbach, J. A.,** Contributions to the ecology of Colorado tick fever virus. II. Population dynamics and host utilization of immature stages of the Rocky Mountain wood tick, *Dermacentor andersoni, J. Med. Entomol.*, 12, 651, 1976.
80. **Emmons, R. W.,** Colorado tick fever: prolonged viremia in hibernating *Citellus lateralis, Am. J. Trop. Med. Hyg.*, 15, 428, 1966.
81. **Eklund, C. M., Kohls, G. M., and Kennedy, R. C.,** Lack of evidence of transovarial transmission of Colorado tick fever virus in *Dermacentor andersoni*, in *Biology of Viruses of the Tick-Borne Encephalitis Complex*, Proc. Symp., Libikova, H., Ed., Academic Press, New York, 1962, 401.
82. **Garcia, R.,** Collection of *Dermacentor andersoni* (Stiles) with carbon dioxide and its application in studies of Colorado tick fever virus, *Am. J. Trop. Med. Hyg.*, 14, 1090, 1965.
83. **McLean, R. G., Francy, D. B., Bowen, G. S., Bailey, R. B., Calisher, C. H., and Barnes, A. M.,** The ecology of Colorado tick fever in Rocky Mountain National Park in 1974. I. Objectives, study design, and summary of principal findings, *Am. J. Trop. Med. Hyg.*, 30, 483, 1981.
84. **Carey, A. B., McLean, R. G., and Maupin, G. D.,** The structure of a Colorado tick fever ecosystem, *Ecol. Monogr.*, 50, 131, 1980.
85. **Smee, D. F., Sidwell, R. W., Clark, S. M., Barnett, B. B., and Spendlove, R. S.,** Inhibition of bluetongue and Colorado tick fever orbiviruses by selected antiviral substances, *Antimicrob. Agents Chemother.*, 20, 533, 1981.
86. **Thomas, L. A., Eklund, C. M., Philip, R. N., and Casey, M.,** Development of a vaccine against Colorado tick fever for use in man, *Am. J. Trop. Med. Hyg.*, 12, 678, 1963.
87. **Francy, D. B. and McLean, R. G.,** unpublished data.
88. **Emmons, R. W.,** unpublished data.
89. **McLean, R. G.,** unpublished data.
90. **Karabatsos, N.,** unpublished data, 1979.
91 **Trent, D. and Roehrig, J.,** unpublished data, 1979.
92. **Knudson, D.,** unpublished data, 1986.
93. **Poland, J. D.,** personal communication, 1985.
94. **Poland, J.,** unpublished data, 1985.
95. **Poland, J.,** unpublished data, 1986.
96. **Poland, J.,** unpublished data, 1978.
97. **McLean, R. G. and Francy, D. B.,** unpublished data, 1974 to 1976.
98. **McLean, R. G.,** unpublished data, 1974 to 1977.
99. **Francy, D. B., et al.,** unpublished data, 1974 to 1976.
100. **Schulze, T. L.,** personal communication, 1986.

Chapter 22

CRIMEAN-CONGO HEMORRHAGIC FEVER

Douglas M. Watts, Thomas G. Ksiazek, Kenneth J. Linthicum, and Harry Hoogstraal*

TABLE OF CONTENTS

*** The views of the authors do not purport to reflect the positions of the Department of the Army or the Department of Defense.**

I. HISTORICAL BACKGROUND

A. Discovery of the Agent

Crimean hemorrhagic fever (CHF) was described as a clinical entity in 1944 and 1945 during an epidemic in the western steppe region of Crimea, U.S.S.R.[1-4] A viral etiology was suggested by reproducing a similar disease syndrome in psychiatric patients undergoing pyrogenic therapy after inoculation with a filterable agent from the blood of CHF patients. Evidence of a viral etiology and of a suspected tickborne route of transmission for the agent was also demonstrated by inducing a mild, but characteristic, clinical course of CHF in healthy human volunteers after their inoculation with suspensions of nymphal ticks, *Hyalomma marginatum marginatum*, in the presence of antibiotics.[4-6] Subsequent attempts to determine the etiology of clinically diagnosed CHF during epidemics in Eurasia culminated in 1967 with the discovery that the agent replicated in newborn white mice.[4,7-10] Intracerebral (i.c.) inoculation of mice with blood from clinically diagnosed CHF patients and corpses led to the isolation of a virus subsequently designated CHF virus.[3,4]

CHF virus was shown to be antigenically indistinguishable from Congo virus,[11,12] originally isolated in 1956 from a febrile patient in Belgian Congo (Zaire).[13] In addition, an antigenic conspecificity was demonstrated between Eurasian strains of CHF virus and several strains of Congo virus[11] isolated from the initial Zaire patient's physician; additional febrile patients, including laboratory workers in Uganda;[13-15] wild and domestic animals; ticks and biting gnats in Nigeria;[16,17] and *Hyalomma* ticks in Pakistan.[18] Observations that CHF and Congo virus were antigenically indistinguishable gave rise to a new proposed name, CHF-Congo,[19] or as employed[20] previously and herein, Crimean-Congo hemorrhagic fever (CCHF) virus.

B. History of Epidemics and Outbreaks

An excellent description of the history of CCHF epidemics/outbreaks prior to 1980 was presented in a comprehensive review of the epidemiology of CCHF virus.[20] This review, and more recent observations, indicated that CCHF epidemics/outbreaks have occurred in Asia, Africa, Europe, and the Middle East (Table 1).

The first documented epidemic occurred during 1944 and 1945 in World War II-devastated areas of the western steppes in the Crimean Oblast.[2,4,20] Approximately 200 Soviet military troops and an unstated number of peasants were infected while harvesting crops. Morbidity and mortality estimates during 1944 portrayed a severe and frequently fatal hemorrhagic disease as evidenced by 9 fatalities among 92 hospitalized military troops.

Almost a decade after the Crimean epidemic, annual outbreaks, characterized by sporadic and widely scattered cases, began in the floodplains and delta of the Volga River in the Astrakhan Oblast.[4,20,21] Among an estimated 104 cases that affected primarily recently recruited agricultural and dairy workers from 1953 to 1969, approximately 17% were fatal. The Astrakhan Oblast outbreaks were associated with extensive conversion of the marshy Volga River floodplains and delta to model dairy farms and agricultural fields.

The CCHF outbreaks in Astrakhan Oblast were paralleled by a series of outbreaks that began during 1953 in Bulgaria.[4,20,22,23] The first cases were recognized shortly after the beginning of a massive, nationwide plan to collectivize agriculture in the foothills and plains of the Balkian terrain. A total of 717 cases were reported between 1953 and 1965; the mortality rate among these cases was 17%. An additional 42 nosocomial cases (including 17 fatalities) occurred during this period, and 129 cases (including 20 fatalities) were recognized during a series of outbreaks between 1968 and 1973. Affected persons were primarily forest, agricultural, and animal husbandry workers.

The last CCHF epidemic recorded in the southwestern Palearctic faunal region of the U.S.S.R. was a series of outbreaks between 1963 and 1971 in forest and steppe areas along the eastern slopes of Donets Ridge and in the Don River floodplains of Rostov Oblast.[4,20,24,25] About 325 cases with an estimated morbidity rate of 15% were reported between 1963 and 1971. CCHF cases were more commonly recognized among dairy and farm workers, housewives, school children, and unclassified employees. The Rostov Oblast outbreaks were attributed to large-scale changes in agricultural and animal husbandry practices.

Epidemics/outbreaks of CCHF in the southwestern Palearctic faunal region were paralleled by scattered small-scale localized outbreaks in the southeastern Palearctic faunal region of the U.S.S.R. Numerous episodes occurred mainly in Uzbekistan,[4,20,26,27] Kazakhstan,[4,20,28,29] and Tadzhikistan.[4,20,30] Index CCHF cases associated with tick bites, or acquired by contact with infected domestic animals, such as butchering cattle or shearing sheep, frequently served as sources of nosocomial and homestead-associated episodes.

An outbreak of CCHF in Bachu southern Xinjiang, China in 1965, with a case fatality rate of 80%, was alluded to in a recent report, but the details were not presented.[31] Between 1954 and 1967, 8 cases of CCHF were recognized in Yugoslavia,[32] and in 1970, 13 members of a single family suffered from CCHF with 2 deaths.[33] A nosocomial episode of CCHF occurred during January and February 1976, in the Central Government Hospital at Ra-

Table 1
SUMMARY OF THE EPIDEMIOLOGICAL CHARACTERISTICS OF MOST CCHF EPIDEMICS, OUTBREAKS, AND SPORADIC CASES, 1944—1986

Location	Date	No. of human cases	Mortality rates	Source of infection	Occupation	Evidence for CCHF viral infection	Ref.
Eurasia[a]							
U.S.S.R.							
Crimean Oblast (Ukrainian, SSR)	Spring—summer, 1944—1945	200	10 (1944)	Tick bite	Military	Clinical and suspected viral agent	2, 4, 20
Astrakhan Oblast (RSFSR)	Spring—summer, 1953—1963	104	17	Tick bite, crushing ticks, and patient contact	Agricultural and related workers	Clinical	4, 20, 21
Rostov Oblast (RSFSR)	Spring—summer, 1963—1969	323	15	Tick bite and crushing ticks	Agricultural and related workers	Clinical, virus isolations, and antibody	4, 20, 24, 25
Uzbekistan SSR	Spring—summer, 1944—?	Unknown	Unknown	Tick bite, domestic animal, and patient contact	Agricultural and related workers	Clinical, virus isolations, and antibody	4, 20, 26, 27
Kazakhstan SSR	Spring—summer, 1948—1968	75	50	Tick bite and patient contact in home and hospital	Agricultural and related workers	Clinical, virus isolations, and antibody	4, 20, 28, 29
Tadzhikistan SSR	Spring—summer, 1943—1970	97	23	Tick bite and patient contact in home and hospital	Agricultural and laboratory workers	Clinical, virus isolations, and antibody	4, 20, 30
Africa							
Zaire	March—April 1956	2	0	Unknown (1) Exposure to virus in laboratory (1)	Unknown Physician	Clinical and virus isolations	13—15
	November 1985	1	100	Livestock/ticks?	Business executive	Clinical and virus isolations	55
Uganda	1958—1977	12	8	Unknown (6) Exposure to virus in laboratory (6)	Unknown Laboratory workers	Clinical and virus isolations	13—15
Mauritania	May 1983	1	0	Unknown, tick bite likely	Camel herd owner	Clinical and antibody	43

Burkina Faso	October—November 1983	1	0	Unknown, tick bite likely	Unknown	Clinical and virus isolations	44
Republic of South Africa	February 1981—May 1986	32	31	Tick bite, domestic, animals, and patient contact in hospital	Farmers, medical workers, others	Clinical, virus isolations, and antibody	45—55
Tanzania	January 1986	1	0	Tick bite	Student	Clinical and antibody	55
Southwest Africa	July 1986	1	0	Tick bite	Unknown	Clinical, virus isolations, and antibody	55
Europe							
Bulgaria	Spring—summer, 1953—1965	717	17	Tick bite	Agricultural and related workers	Clincial, virus isolations, and antibody	4, 20, 22, 23
		42	41	Patient contact			
	1968—1973	129	16	Tick bite			
Middle East							
Pakistan	January—February 1976	14	29	Index case, unknown, and patient contact in hospital (10 cases) and home (3 cases)	Shepherd, medical workers, and others	Clinical, virus isolations, and antibody	20, 34, 35
Dubai[b]	November 1979	6	50	Index case, unknown, patient contact in hospital (5 cases), and other	Index case, unknown, and medical workers	Clinical, virus isolations, and antibody	36, 37
Sharjah[b]	February 1980	1	0	Unknown, tick bite likely	Storekeeper	Clinical	36
Iraq	September 1979—September 1980	55	64	Tick bite, patient contact suspected	Agricultural and related workers	Clinical, virus isolations, and antibody	38—41

[a] Evidence of CCHF based on viral isolations and antibody was obtained during and after 1967 and 1968; serological evidence of CCHF viral infections was demonstrated during the late 1960s by CF tests for 72% of sera collected from 96 patients during outbreaks in Bulgaria and southern U.S.S.R. between 1954 and 1967.

[b] United Arab Emirates.

walpindi, Pakistan.[34] The index case, a shepherd, and 14 secondary and tertiary cases were diagnosed, of which 4 had fatal outcomes. In May 1976, a soldier was admitted to the military hospital in Quetta, Pakistan, and died of CCHF.[20,35] About 20 other CCHF cases were diagnosed among personnel during the same month. Two of the first five patients died. However, the criteria for confirmation, as well as the actual number of cases during this outbreak, were not presented.

An outbreak of CCHF was documented during September 1979 in Dubai of the United Arab Emirates[36,37] and in Iraq.[38-41] In Dubai, a CCHF patient who had lived close to a cattle market died in the El Rashad Hospital on November 9, 1979. Subsequently, five secondary cases, including two fatalities, were diagnosed among hospital staff who had attended this patient. One additional, apparently unassociated CCHF case, an Afghan storekeeper from the Emirate of Sarajah, was admitted to the same hospital during February 1980, but survived after suffering severely. The first indication of the CCHF outbreak in Iraq was a female patient who died on September 9, 1979 in Yarmouk Hospital in Baghdad. Subsequently, a physician and a nurse of the hospital contracted CCHF and died on September 19, 1979. Apparently, an additional 53 non-nosocomial cases were recognized primarily during September to November 1979, and April to June 1980. Of the 55 patients, 35 (64%) died. The affected persons resided in Baghdad and in various localities along the Tigris and Euphrates River where they had frequent close association with sheep, goats, and (occasionally) cattle.

While sporadic cases of CCHF have been documented in Zaire and Uganda,[13-15,42,55] and more recently in Mauritania[43] and Burkina Faso[44] of Africa, a total of 19 outbreaks of CCHF occurred between February 1981 and May 1986 in the Republic of South Africa. CCHF was first diagnosed in the latter country during 1981 as the cause of death of a schoolboy.[45-47] Apparently, infection was acquired from the bite of an infected tick. Subsequently, two human cases, one each in November 1983 and February 1984, occurred after contact with livestock.[48] During May 1984, five cases, including one fatality, were diagnosed among farm workers who apparently contracted CCHF while butchering a dead cow.[49] In late August 1984, a patient who apparently acquired CCHF viral infection from either livestock or tick bite, died in Tygerberg Hospital near Capetown. Subsequently, seven medical workers associated with the latter patient in the hospital developed CCHF, and one died.[50-54] Another single case was diagnosed during November 1984, and was attributed to contact with either an infected ostrich or a tick. An additional eight outbreaks occurred during 1985 and five occurred during 1986.[55] Of 14 cases, 7 had fatal outcomes. Cases were caused by bite of infected ticks, contact with infected domestic animals and patients, and other unknown sources of infection. Also during 1986, one human case was recognized in Tanzania, one in Southwest Africa, and one fatal case in Zaire.[55] These cases were diagnosed by investigators in the Republic of South Africa.

C. Social and Economic Impact

Any attempt to assess the public health importance of CCHF is hampered by the paucity of data required to understand the true incidence of this disease. Nevertheless, the estimated ratio of apparent to inapparent infections suggests that one of every five persons infected develops CCHF.[56] This exceptionally high incidence of disease and the fact that CCHF cases are almost always severe, with a high case fatality rate, have potentially detrimental social and economic implications (Table 1).

Studies have not been conducted on the impact of CCHF on the socioeconomics of affected communities, but more apparent implications are the consequences that may arise from anxiety and fear and the high morbidity and mortality associated with contracting the disease. This could lead to disruption among workforces, particularly those at risk to CCHF, such as agricultural and animal husbandry workers. Similarly, anxiety and fear among medical workers may present problems in caring for and administering treatment to patients. In addition, the communicable nature of the disease and its severe and frequently protracted clinical course of illness, lasting from 3 to 6 weeks, may be excessively demanding of the medical personnel and require the use of expensive intensive care and patient isolation facilities.[20,34,36,48,49] Further socioeconomic inflictions and losses are implied by the extended period of convalescence which may endure for several months.[20] Finally, economic losses may result from restriction on the exportation of livestock to CCHF virus nonenzootic countries, and limited attempts to control CCHF have been an unaffordable expense.[20]

II. THE VIRUS

A. Antigenic Relationships

CCHF virus is a member of the genus *Nairovirus* (family Bunyaviridae).[57,58] Members of this genus possess tripartite ribonucleic acid (RNA) genomes of approximately 31 to 49 (L), 1.5 to 1.9 (M), and 0.6 to 0.7 (S) $\times$ 10^6 daltons. An antigenic relationship among members of the genus was demonstrated by cross-immune precipitation.[59] Morphologically, negative stained, electron microscopic observations revealed that the surface units of CCHF virions are smaller than representative viruses of other genera of the family Bunyaviridae.[60]

The Nairoviruses are organized into related antigenic serogroups: CCHF virus group — CCHF, Hazara, Khasan; Dera Ghazi Khan virus group — Abu Hammad, Dera Ghazi Khan, Kao Shuan, Pathum Thani, Pretoria; Hughes virus group — Hughes, Punta Salinas, Soldado, Zirqa; Nairobi sheep disease virus group — Dugbe, Nairobi sheep disease, Ganjam; Qalyub virus group — Bandia, Qalyub; Sakhalin virus group — Avalon, Clo Mor, Paramushir, Sakhalin, Taggert.[61] The antigenic relationship among these groups was demonstrated by complement-fixation (CF) hemagglutination-inhibition (HI), indirect fluorescent antibody (IFA), and neutralization (N) tests.[57]

B. Host Range

1. Vertebrates

CCHF virus has been isolated from domestic and wild vertebrates, including man, cattle,[14,16] goats,[16,62] sheep,[31] hares (*Lepus europaeus*),[63] hedgehogs (*Erinaceus (Atelerix) albiventris*),[16,17] and a multimammate mouse (*Mastomys* spp.).[64] Serological evidence of CCHF viral infections also has been demonstrated in these and other species of wild vertebrates (Table 2) and in humans and domestic animals (Table 3). In addition, experimental studies indicated that a limited number of wild and domestic animals were susceptible to CCHF viral infection (Section V.B.2).

2. Invertebrates

Among invertebrates, CCHF viral infection has been demonstrated only in ticks, including viral isolations from numerous species/subspecies of seven genera of the family Ixodidae, and two species of the family Argasidae (Table 4). As described in Section V.B.1, infection has been confirmed by experimental studies for only a few species of the genus *Ixodes*. Limited attempts to experimentally infect *Aedes aegypti* mosquitoes were unsuccessful.[20]

C. Strain Variation

In spite of the wide geographic distribution of CCHF virus and the diversity of invertebrate and vertebrate hosts, kinetic N test comparisons failed to demonstrate significant differences among CCHF viral strains.[79] Also, earlier studies employing modified agar gel diffusion-precipitation (AGDP), mouse neutralization, cell-culture interference, and CF tests demonstrated that there were no apparent antigenic differences among strains from several different geographic locations in the U.S.S.R. and Africa.[11,12,19] Similar results were obtained for CCHF viral strains that were isolated during different years in the U.S.S.R.[80] More

Table 2
SEROLOGICAL EVIDENCE OF CCHF VIRAL INFECTION IN WILD VERTEBRATES

Location	Vertebrate species	% antibody positive (no. sera tested)	Serological tests	Ref.
U.S.S.R.	Hare *(Lepus capensis)*	Positive[a]	AGDP[b]	20
	Common red fox *(Vulpes vulpes)*	40 (05)	AGDP	20
	Hare *(L. europaeus)*	25 (20)	IHI[c]	20
	Great gerbil *(Rhombomys o. opimus)*	Positive	CF[d]	65
	Long-clawed ground squirrel *(Spermophilopsis l. leptodactylus)*	Positive	Not stated	20
	Long-eared hedgehog *(Hemiechinus auritus)*	Positive	AGDP	20
	Pallas cat *(Felis manual)*	Positive	Not stated	20
	Red-tailed Libyan jird *(Meriones libycus)*	Positive	CF	65
Pakistan	Black rat *(Rattus rattus)*			
	Gerbil *(M. hurrianae)*			
	Naked-soled gerbil *(Tatera indica)*	6 (157)	CF	66
	Norway rat *(R. norvegicus)*			
Iran	Common noctule *(Nyctalus n. noctula)*	Positive	AGDP	20
	House mouse *(Mus musculus bactrianus)*	Positive	AGDP	20
	Large mouse-eared bat *(Myotis blythi omari)*	Positive	AGDP	20
	Swinhoe's jird *(Meriones crassus swinhoei)*	Positive	AGDP	20
	Williams' Jerboa *(Allactaga euphratica williamsi)*	Positive	AGDP	20
Egypt	Black rat *(R. rattus)*	4 (72)	CF	67
	Field rat *(Arvicanthis n. niloticus)*	5 (113)	CF	67
	Norway rat *(R. norvegicus)*	7 (176)	CF	67
Mauritania	Field rat *(A. n. niloticus)*	16 (43)	IFA[e]	68
	Multimammate mouse *(Mastomys erythroleucus)*	27 (11)	IFA	68
Senegal	Genet *(Genetta g. senegalensis)*	Positive	AGDP, CF	20
	Multimammate rat *[Praomys (Mastomys) natalensis]*	Positive	AGDP, CF	20
Republic of South Africa	Eland antelope *(Taurotragus oryx)*	100 (3)	RPHI[f]	47
	Ground squirrel *(Xerus inauris)*	25 (4)	RPHI	47
	Hare *(L. capensis)*	25 (44)	RPHI	47
	Hare *(L. saxatillus)*	15 (117)	RPHI	47
	Hare *(L. species)*	40 (10)	IFA, RPHI	47

[a] Data not presented.
[b] Agar gel diffusion-precipitation.
[c] Indirect hemagglutination inhibition.
[d] Complement-fixation.
[e] Indirect fluorescence antibody.
[f] Reverse passive hemagglutination inhibition.

detailed molecular comparisons among CCHF viral strains have been hindered by the need for biocontainment when working with the agent and difficulties in producing adequate concentrations of the virus.

D. Methods for Assay

CCHF virus has been isolated most frequently from humans, wild and domestic animals, and ticks by intracranial inoculation of suckling mice.[4,20,34,39,43,44,47] The virus replicates and causes illness and death in mice. Several vertebrate cell lines support replication of the virus, including primary chicken embryo and passaged human embryo, primary green monkey kidney cells, CF-1 continuous *Cercopithicus aethiops* cells CER cells, hamster cells of

Table 3
SUMMARY OF CCHF VIRAL ANTIBODY PREVALENCE IN HUMANS AND DOMESTIC ANIMALS, SELECTED SEROSURVEYS

Location	Species	% antibody positive (no. sera tested)	Serological tests	Ref.
Africa				
Burkina Faso and Benin	Human	00.10 (1595)	IFA	73
Egypt	Human	00.02 (433)	CF[a]	20
	Camel	09.00 (34)	CF	20
	Cattle	14.00 (36)	CF	20
	Sheep	18.00 (66)	CF	20
Kenya	Human	00.20 (741)	IFA[b]	69
	Cattle	05.00 (343)	AGDP[c]	20
Mauritania	Human	00.02 (59)	IFA	68
	Cattle	32.00 (25)	IFA	68
Nigeria	Human	10.00 (250)	MNT[d]	70
	Cattle	26.00 (1164)		71
Senegal	Human	00.00 (159)	AGDP	20
	Cattle	15.00 (26)	AGDP, CF	20
	Cattle	09.00 (93)	AGDP, CF	20
	Cattle	12.00 } (247)	AGDP, CF	20
	Sheep	06.00 }		
	Cattle	06.00 } (263)	AGDP, CF	20
	Sheep	01.00 }		
	Goats	01.00 (70)	AGDP, CF	
Uganda	Human	00.20 (813)	AGDP, CF	72
	Cattle	37.00 (104)	AGDP	72
Republic of South Africa	Human	07.00 (74)	IFA, RPHI[e]	47
	Human	02.00 (1109)	IFA	48
	Cattle	27.00 (6128)	IFA	48
	Cattle	64.00 (170)	RPHI	47
	Cattle	28.00 (8667)	RPHI	55
	Sheep	27.00 (270)	RPHI	47
Zimbabwe	Human	04.00 (486)	IFA	74
	Cattle	45.00 (763)	RPHI	55
Eurasia				
China	Human	01.00 (135)	CF	31
	Sheep	30.00 (125)	CF	31
India	Human	02.00 (367)	AGDP	20
	Human	03.00 (152)	IHI[f]	20
	Cattle	02.00 (58)	CF	20
U.S.S.R	Goats	03.00 (30)	CF	20
Crimean Oblast	Cattle	01.00 (143)	CF	20
	Cattle	01.00 (299)	AGDP	20
Kalmyx ASSR	Human	00.00 (360)	AGDP	20
	Cattle	00.30 (861)	AGDP	20
	Cattle	00.40 (7966)	AGDP	20
Kirgizia SSR	Human	00.00 (782)	CF	20
	Human	01.00 (2355)	CF	20
Krasnodar Region	Human	00.00 (1035)	AGDP	20
	Human	03.00 (450)	AGDP	20
	Cattle	01.00 (2567)	AGDP	20
Rostov Oblast	Cattle	08.00 (336)	AGDP	20
Stavropol Region	Cattle	00.30 (350)	AGDP	20
	Sheep	00.50 (2748)	AGDP	20
Turkmen SSR	Human	03.00 (28)	CF	20
Europe				
Bulgaria	Human	05.00 (3012)	AGDP, CF	20

Table 3 (continued)
SUMMARY OF CCHF VIRAL ANTIBODY PREVALENCE IN HUMANS AND DOMESTIC ANIMALS, SELECTED SEROSURVEYS

Location	Species	% antibody positive (no. sera tested)	Serological tests	Ref.
	Human	09.00 (580)	AGDP	20
	Human	08.00 (5398)	AGDP	20
	Cattle	47.00 (DNP)[g]	AGDP	20
	Goats	Positive[h]	AGDP	20
	Horses	Positive	AGDP	20
	Sheep	28.00 (DNP)	AGDP	20
Greece	Human	06.00 (65)	IFA, HI[i]	75
	Goats	33.00 (422)	AGDP	20
	Sheep	12.00 (294)	AGDP	
Hungary	Human	03.00 (587)	AGDP	20
	Cattle	00.08 (687)	AGDP	
	Sheep	31.00 (48)	AGDP	
Portugal	Human	01.00 (190)	IFA	76
Turkey	Human	02.00 (1100)	HI	20
Middle East				
Afghanistan	Cattle	06.00 (230)	AGDP	20
	Goat	09.00 (233)	AGDP	20
Iran	Human	14.00 (351)	AGDP	20, 129
	Human	04.00 (100)	HI	20, 129
	Cattle	18.00 (130)	AGDP	20, 129
	Sheep	38.00 (728)	AGDP	20, 129
	Goats	36.00 (135)	AGDP	20, 129
Iraq	Human	29.00 (233)[j]	CF	77
	Human	07.00 (733)[k]	CF	77
	Human	11.00 (166)[l]	CF	77
	Human	29.00 (257)[m]	CF	77
	Human	00.07 (251)[n]	CF	77
	Sheep	00.60 (533)	CF	38
	Goat	00.52 (435)	CF	38
	Cattle	00.33 (343)	CF	38
Kuwait	Human	04.00 (502)	CF, IFA	78

[a] Complement-fixation.
[b] Indirect fluorescent antibody.
[c] Agar gel diffusion-precipitation.
[d] Mouse neutralization test.
[e] Reverse passive hemagglutination inhibition.
[f] Indirect hemagglutination inhibition.
[g] Data not presented.
[h] Antibody detected, but data not presented.
[i] Hemagglutination inhibition.
[j] Animal breeders.
[k] Abattoir workers.
[l] Hospital workers.
[m] Contact with CCHF patients.
[n] Noncontact wtih CCHF patients and random sample population.

dubious parentage, LLC-MK_2 cells, and other primary and secondary cell lines.[20,34,47,39] Many of these cell lines produce CCHF viral plaques in cell cultures under semisolid overlay, but in general fail to cause discernible cytopathogenic effect (CPE) under fluid overlay. Strains of the virus apparently differ in their ability to produce plaques. Thus, indirect methods such as interference[81] or fluorescent antibody visualization,[47,49] have been developed and employed to detect CCHF viral antigen(s) in cell lines that support viral replication.

Table 4
TICK SPECIES/SUBSPECIES FROM WHICH CCHF VIRUS HAS BEEN ISOLATED, AND SOME OF THE ASSOCIATED VERTEBRATE HOSTS

Family (species/subspecies of ticks)	No. of vertebrate hosts[a]	Vertebrate hosts: Larvae and nymphs	Vertebrate hosts: Adult
Family Argasidae			
Argas (Persicargas) persicus (Oken)	5—20	Wild and domestic birds and mammals	Wild and domestic birds and mammals
Ornithodoros lahorensis Neumann	Multiple		
Family Ixodidae			
Amblyomma (Theileriella) variegatum (Fabricius)	3	Small mammals and birds	Cattle
Boophilus annulatus (Say)	1	Cattle	Cattle
B. decoloratus (Koch)	1	Cattle	Cattle
B. microplus (Canestrini)	1	Cattle	Cattle
Dermacentor (D.) daghestanicus Olenev	3	Hedgehogs, rodents, and hares	All domestic mammals, deer, and man
D. (D.) marginatus (Sulzer)	3	Insectivores, rodents, hares, and small carnivores	Livestock and other domestic and wild herbivores
Haemaphysalis (Aboimisalis) punctata Canestrini and Fanzago	3	Wild birds and hares	All domestic mammals and large wild mammals
Hyalomma (H.) anatolicum anatolicum Koch	2	Domestic mammals	Domestic mammals
H. (Hyalomma) asiaticum asiaticum Schulze	3	Hedgehogs, hares, rodents, and carnivores	Camels and other domestic herbivores, pigs, and gazelles
H. (H.) detritum Schulze	2	Cattle and horses	Cattle and horses
H. (H.) dromedarii Koch	2, 3	Hares and other small mammals	Camels and cattle, sheep and goats
H. (H.) impeltatum Schulze and Schlottke	3	Gerbils, jerboas, jirds, hedgehogs, hares, lizards, and birds	Domestic herbivores, gazelle, antelope, dog, wild pig, camel, and cattle
H. (H.) marginatum impressum Koch	3	Hedgehogs and rodents	Cattle, horses, camel, sheep, and dog
H. (H.) marginatum marginatum Koch	2	Hedgehogs, hares, and birds	Cattle, goats, sheep, horses, and camels
H. (H.) nitidum Schulze	3	Hares	Cattle, horses, goats, bushpig, Cape buffalo, DeFassa waterbuck, and Roan antelope
H. (H.) marginuatum rufipes Koch	2	Birds and hares	Cattle
H. (H.) marginatum turanicum Pomerantsev	2	Birds and hares	Cattle

Table 4 (continued)
TICK SPECIES/SUBSPECIES FROM WHICH CCHF VIRUS HAS BEEN ISOLATED, AND SOME OF THE ASSOCIATED VERTEBRATE HOSTS

Family (species/subspecies of ticks)	No. of vertebrate hosts[a]	Vertebrate hosts: Larvae and nymphs	Vertebrate hosts: Adult
H. (H.) truncatum Koch	2, 3	Hares, birds, rodents, and domestic animals	Cattle, goats, sheep, horses, camels, pigs, dogs, and numerous large wild mammals, tortoises, and birds
Ixodes (Ix.) ricinus Latrielle	3	Small mammals and birds (larvae) and large mammals and birds (nymphs)	Large mammals
Rhipicephalus (R.) appendiculatus Neumann	3	Hares and cattle	Cattle
R. (Digineus) bursa Canestrini and Fanzago	2	Wild and domestic mammals	Wild and domestic mammals
R. (Evertsi) evertsi Neumann	2	Hares and large herbivores	Large herbivores
R. (Lamellicauda) pulchellus Gerstacker	3	Small and large wild and domestic herbivores	Large herbivores
R. (R.) pumilio Schulze	3	Hedgehogs, hares, and birds	Hares, hedgehogs, and large-sized rodents, domestic and wild mammals
R. (R.) rossicus Yakimov and Kohl-Yakimova	3	Hedgehogs, rodents, and hares	Wild and domestic mammals from the size of hares to camels
R. (R.) sanguineus Latrielle	3	Insectivores and rodents	Wild and domestic ungulates and carnivores
R. (R.) turanicus Pomerantsev and Matikashvili	3	Small mammals	Large mammals

[a] The number of similar or dissimilar vertebrate hosts parasitized during and after the development of a particular tick species/subspecies.

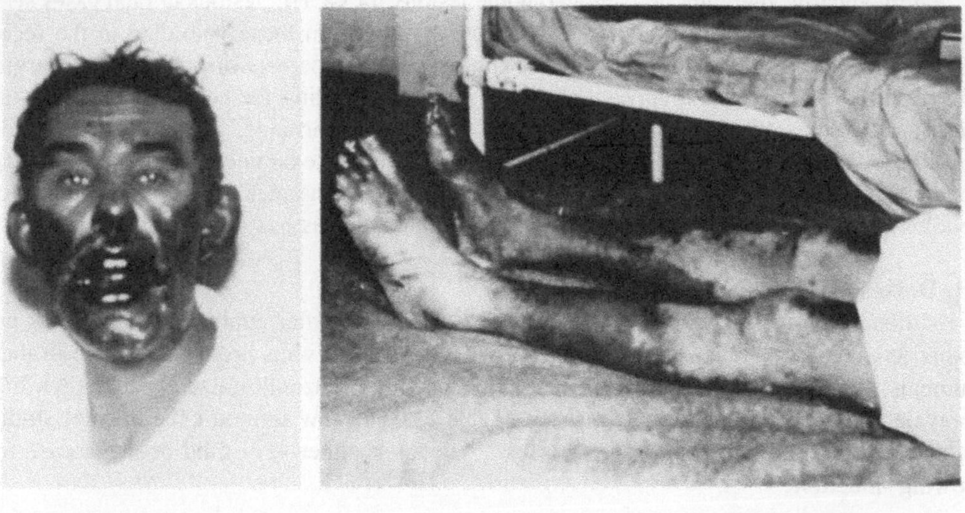

A B

FIGURE 1. Ecchymoses on the face (A) and legs (B) of a CCHF patient. (Courtesy of Dr. D. I. H. Simpson.)

The fluorescent antibody assay has been used by Soviet workers to detect CCHF viral infections in experimentally infected ticks[82] and in ticks collected in the field.[83] Enzyme immunoassays (EIA) also appear to be effective for detecting viral antigen in infected ticks.[84]

III. DISEASE ASSOCIATIONS

A. Humans

Humans are the only natural host of CCHF virus for which disease has been confirmed.

Early speculation that CCHF was less severe among African patients than those in Eurasia has not been confirmed. In fact, there have been severe and fatal cases of CCHF in the Republic of South Africa,[47-50,55] one fatal case each in Uganda[13,14] and Zaire,[55] and a severe case each in Mauritania[43] and Burkina Faso.[24] The clinical course of illness was similar to that described for the classical CHF disease syndrome.[20,85] Apparently, the hypothesis that CCHF was mainly a febrile illness in Africa was based on initial observations of predominantly febrile cases in Zaire and Uganda.[13,14]

The onset of clinical disease follows exposure to CCHF virus by 3 to 12 days. Among the 31 CCHF cases in the Republic of South Africa, reliable data for 21 patients indicated that the onset followed exposure to tick bite by 3.2 days, to blood or tissue of livestock by 5.0 days, and to blood of human cases by 5.6 days.[55]

A characteristic clinical feature of CCHF is a sudden onset with severe headache, dizziness, photophobia, and stiffness.[20,36,38,39,48-50,55] Fever with chills occurs at about the same time. Leg and back pains and general myalgia are said to be intense. Petechial or ecchymotic rashes/lesions (Figure 1) are commonly described in severe cases and appear after 3 to 6 days. Bleeding in the form of melena, hematemesis, and epistaxis is common and follows onset by 4 to 5 days. Those patients who develop severe illness are reported to enter hepatorenal failure from about day 5 postonset and become progressively drowsy, stuporous, and comatose. Of 31 CCHF patients who died in the Republic of South Africa, 11 developed multiple organ failure including cerebral, liver, and kidney with cardiac and pulmonary insufficiency.

Recovery from CCHF begins around day 9 or 10 with the abatement of the rash and a general improvement of the patient. Convalescence may be extended and accompanied by weakness, confusion, and asthenia; hair loss and local neuralgia have been observed for some patients.

Pathologically, there are no pathognomonic lesions for CCHF; lesions in fatal cases seem to be primarily of vascular origin.[37,52,55] Liver samples from those who died in the recent South African outbreaks have shown varying degrees of necrosis of hepatocytes, varying from 25 to 75% cellular involvement. Other organs, including the central nervous system and the kidneys, have also shown congestion, focal hemorrhage, and necrosis. Cellular infiltration appeared minimal in those pathologic specimens examined. Except for the large outbreaks in Eurasia, CCHF has occurred sporadically or, in areas where clinical and morphological pathology facilities were limited; the pathogenesis has not been well defined.

B. Domestic Animals

Although humans have been infected with CCHF virus after contact with livestock and other animals,[20,49,55] there is no definitive evidence that the virus produces disease in these animals. Antibody surveys among livestock in endemic (enzootic) areas have shown high prevalence among both cattle and sheep (Table 3). The few limited experimental studies suggested that calves, lambs,[86] and horses[87,88] develop either very mild or no disease following infection.

C. Wildlife

CCHF viral infection has been demonstrated more commonly among smaller wildlife species that act as hosts for the immature stages of the tick vectors (Table 2). Virus and serological surveys suggested that hares and hedgehogs are more commonly infected.[20] These vertebrates have been infected in the laboratory, but there has not been any reported evidence of disease.[80-91] Similarly, clinical signs of disease were not observed for the ground squirrel Little Suslik *(Citellus pygmaeus)*,[92] or African primates[10] following experimental infections with CCHF virus.

D. Applicable Diagnostic Procedures

Diagnostic procedures for CCHF viral infections depend upon the isolation of the virus, detection of viral antigen, or the demonstration of an increase in specific antibody titers. Isolation of virus in suckling mice or cell cultures (Section II.D) and the detection of viral antigens in tissues or blood by direct and/or indirect assays can provide a definitive diagnosis.[93-101] Because antibody may be present in individuals who have had remote clinical or inapparent infections, demonstration of a rise in CCHF virus-specific antibody titers for paired specimens and demonstration of specific IgM antibody are the serological methods of choice.

The IFA,[96] indirect hemagglutination (IHA),[93,97] and indirect hemagglutination inhibition assay (IHIA) seem to be the most reliable of the serological diagnostic tests for CCHF viral infections. The CF and AGDP tests are also useful, but suffer from a lack of sensitivity. The EIA test for CCHF viral antibody seems to offer a great deal of promise, but its sensitivity and specificity remain to be defined.[84]

E. Adverse Effects of Virus on Vector

Only a few controlled experimental studies have been conducted in an attempt to demonstrate the vector and reservoir potential of selected tick species for CCHF virus (Section V.B.1). However, none has been designed to address the possibility that CCHF viral infection may have a detrimental effect on this arthropod's potential role in the ecology and epidemiology of CCHF virus.

IV. EPIDEMIOLOGY

A. Geographic Distribution

The geographic range of CCHF virus is the most extensive among the tickborne viruses

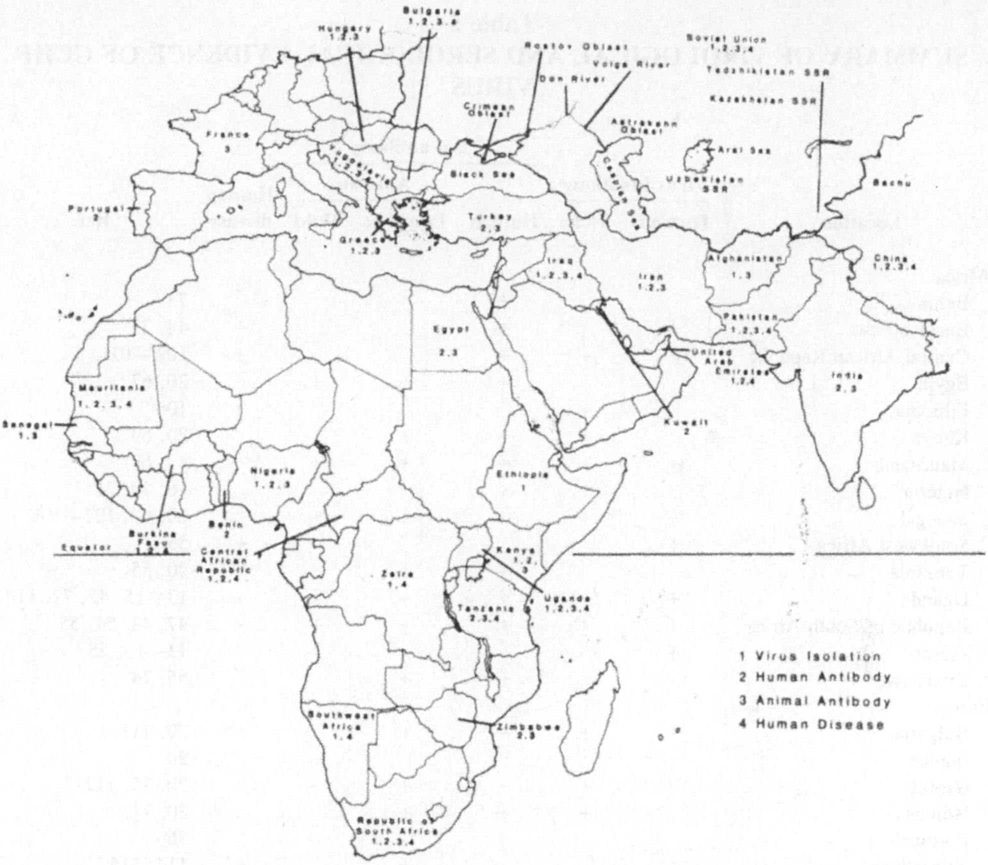

FIGURE 2. The geographic distribution of CCHF virus by country. Specific locations of recognized enzootic foci are indicated either in the text or in the references.

associated with human disease, and with the exception of dengue viruses, the most widespread of all the medically important arboviruses (Figure 2). Sporadically distributed enzootic foci originally described throughout southern Eurasia have since been recognized in western China and other countries in southern Europe. A similar focal distribution pattern extends southward and spans a vast portion of the Middle East region, possibly including India and a large portion of Africa extending into the Southern Hemisphere. Evidence of CCHF virus enzootic foci for most countries is based on virus isolations from either humans or ticks, and/or antibody detection in humans and domestic animals (Table 5). However, the evidence for Benin, Egypt, France, India, Portugal, and Turkey is based on very limited serological data, particularly for France, where serological reactors have been detected in only 2 of 19 bats.[20]

Historically, the recognition of CCHF virus enzootic foci has been characterized by an unpredictable and sudden occurrence of human CCHF cases in presumably nonenzootic areas. While this phenomenon is not understood, evidence indicates that CCHF virus persists in silent cycles involving ticks and nonhuman vertebrate hosts (Section IV.E). Also, it may be that new enzootic foci may be established by infected ticks introduced by parasitized vertebrates, particularly birds and livestock that can disperse ticks within and outside of enzootic foci (Section VI.H).

B. Incidence

Numerous reports alluded to in a previous review described the incidence of CCHF during

Table 5
SUMMARY OF VIROLOGICAL AND SEROLOGICAL EVIDENCE OF CCHF VIRUS

	Virus isolations[a]		Viral antibody				
				Animals		Human	
Location	Human	Ticks	Human	Domestic	Wild	disease	Ref.
Africa							
Benin			+[b]				73
Burkina Faso	+		+			+	44, 73
Central African Republic	+	+	+			+[c]	102—105
Egypt			+	+	+		20, 67
Ethiopia		+					106
Kenya		+	+	+			20, 69
Mauritania	+	+	+	+	+	+	43, 68
Nigeria		+	+	+			16, 70, 71
Senegal		+		+	+		20, 62, 107—109
Southwest Africa	+					+	55
Tanzania			+	+		+	20, 55
Uganda	+	+	+	+		+	13—15, 42, 72, 110
Republic of South Africa	+	+	+	+	+	+	47, 48, 54, 55
Zaire	+					+	13—15, 55
Zimbabwe			+	+			55, 74
Europe							
Bulgaria	+	+	+	+		+	20, 111
France					+		20
Greece		+	+	+			20, 75, 112
Hungary		+	+	+			20, 113
Portugal			+				76
Yugoslavia		+	+	+		+	114, 115
Middle East							
Afghanistan		+		+			20, 116
Iran		+	+	+	+		20, 117, 118, 129
Iraq	+		+	+		+	38, 39, 40, 77
Kuwait			+				78
Pakistan	+	+	+	+	+	+	18, 34, 66
United Arab Emirates	+		+			+	36
Eurasia							
China	+	+	+	+		+	31
India			+	+			20
Turkey			+	+			20, 119
Union of Soviet Socialist Republics							
Armenian SSR		+	+	+		+	20, 111
Astrakhan Oblast	+	+		+	+	+	20, 111
Azerbaijan SSR		+		+			20, 111
Crimean Oblast	+	+		+		+	111
Daghestan ASSR				+			20
Kalmyx ASSR				+			20, 111
Kazakhstan SSR	+	+	+	+		+	20, 111
Kirgizia SSR	+	+		+		+	20, 111
Krasnodar Region (RSFSR)		+	+	+		+	20, 111
Moldavian Oblast	+	+				+	20, 111
Rostov Oblast (RSFSR)	+	+	+	+	+	+	20, 111
Stavropol Region (RSFSR)			+	+		+	20, 111
Tadzhikistan SSR	+	+	+	+	+	+	20, 111, 120, 121

Table 5 (continued)
SUMMARY OF VIROLOGICAL AND SEROLOGICAL EVIDENCE OF CCHF VIRUS

Location	Virus isolations[a]		Viral antibody			Human disease	Ref.
	Human	Ticks	Human	Animals: Domestic	Animals: Wild		
Turkmen SSR		+	+	+	+	+	20, 65, 111, 122
Uzbekistan SSR	+	+	+	+	+	+	20, 111
Voroshilovgrad and Kherson Oblasts			+			+	20, 111

[a] CCHF virus-specific fluorescence observed in salivary glands of this tick; all other species/subspecies yielded isolations of CCHF virus.

[b] The + indicates either clinical, virological, or serological evidence of CCHF virus.

[c] CCHF virus isolated from laboratory worker who experienced a mild illness; naturally acquired disease not reported for the Central African Republic.

epidemics in Eurasia as merely "sporadic".[20] Apparently, there was no systematic nationwide surveillance system, and estimates of the number of cases were based on different methods that varied from region to region. Despite the lack of definitive data, morbidity data reveal a characteristic pattern of the incidence of CCHF. Excluding nosocomial, household, and other contagion-related cases, this pattern was characterized by the sporadic occurrence and widely scattered nature of cases, both spatially and temporally.

A summary of the estimated number of CCHF cases and mortality rates reported during epidemics and outbreaks is presented in Table 1. Since the Crimean epidemic, CCHF has been recognized in the Oblast sporadically, but not annually, over the subsequent 25 years.[63] Apparently, cases were not reported in the Crimean Oblast between 1970 and 1974, but CCHF virus was isolated from ticks.[63] The series of epidemics from 1953 to 1963 in the Astrakhan Oblast was followed by a low number of cases through 1968.[21] As an example of the sporadic incidence and scattered distribution pattern of CCHF cases, only 3 cases were recorded in the latter Oblast during 1953; 9 in 1954; 11 in 1955; 1 annually in 1956 to 1961; 32 in 1962; and 44 in 1963. Only single cases occurred in 45 of 72 agricultural settlements, but there were 2 to 5 cases during 1, 2, 3, or 4 seasons in 1 to 9 other settlements. During the CCHF epidemic in Rostov Oblast between 1963 and 1969, the reported morbidity rate was 13.5/100,000 persons.[25] A few cases were registered during 1970 and 1971, and none was registered in 1972 and 1973. However, a recent report stated that since 1975, cases of CCHF have been increasing, and that between 1963 and 1976, 12.5% of all CCHF cases known in the world occurred in the Rostov Oblast.[56] Among the epidemics that have occurred in Eurasia, the highest morbidity rate reported was 0.71% during a series of outbreaks between 1953 and 1965 in Bulgaria.[22]

The bitterly cold winter of 1969—1970 in the steppes of Ukrainian SSR and southwestern Oblasts of the RFSFR (Volga and Don River areas) drastically reduced the population density of the CCHF virus vector, *H. m. marginatum*, as well as the incidence of CCHF cases in Eurasia.[20] Virtually no data have been reported on the incidence of CCHF or other aspects of this disease since the mid-1970s, despite continuous surveillance in this and other areas of the U.S.S.R.

Estimates of CCHF cases in the Central Asian Republics of the U.S.S.R. have not been readily discernible, but there have been numerous anecdotal reports.[20] Since the last reported cases in 1969 in Tadzhikistan SSR, 10 isolates of CCHF virus were obtained from blood samples taken from 1093 febrile patients in hospitals from 1976 through 1981,[120] and 9 CCHF cases were reported between 1976 and 1980.[121] Also, during 1980, two CCHF cases were recorded in Dushanbe and Parkhar.[95] Apparently, no other cases have been reported from the Central Asian U.S.S.R. Republics.

Elsewhere, the incidence of CCHF has also been characterized by sporadic outbreaks and episodes, including cases acquired from tick bites and by contagion. A substantial portion of these cases, excluding most of the 55 cases reported during the outbreak in Iraq, was due to contact with patients, domestic animals, and occasional laboratory accidents (Table 1).

Although nosocomial cases occurred during the CCHF outbreak in Iraq, apparently most cases were attributed to tick bite. In addition to the 55 cases, mild and/or inapparent infections[77] were suggested by a serosurvey of persons associated with patients as well as others (Table 3). The infection rate reported for medical workers was 11% (18/166); 15% (76/508) for relatives and other patient contacts, 29% (67/233) for animal husbandry workers; and 7% (57/773) for abattoir workers. Only 2 of 100 noncontacts and none of 151 randomly sampled persons had evidence of CCHF viral infection. It is notable that a mortality rate of 64% for the outbreak in Iraq was among the highest reported for tickborne-associated cases. However, confirmation of CCHF viral infection has been reported for only a few of these cases.[38-41]

The number of CCHF cases during nosocomial, household, and animal contact-related episodes has ranged from 1 to 13, excluding the index cases. The case attack rate, based on available data, was 59% (10/17) for medical workers and 22% (2/9) for patient relatives during a nosocomial outbreak in Pakistan.[34] The number of possible contacts in the village of the index case was 60, but only 1 relative contracted CCHF, and antibody was not demonstrated in 6 of the index patient's family members. In the Republic of South Africa, 13% (7/55) of medical workers considered at-risk developed CCHF during a nosocomial outbreak.[53] Of an additional 452 patient contacts, apparently none of 217 that were followed contracted CCHF.[53]

C. Seasonal Distribution

Ecological and epidemiological observations revealed evidence of epidemic CCHF virus transmission from early spring until late autumn during outbreaks in Bulgaria and the Eurasian regions of the U.S.S.R. (Table 1). Epidemic curves depicted a gradual increase in human cases from the onset, usually during April, May, or June, until case rates peaked either in June, July, or August.[20] Subsequently, the number of CCHF cases decreased rapidly, but a few scattered cases were recognized as late as October and November, with an occasional human infection occurring during the winter in the semiarid regions of Central Asia. The seasonal distribution and fluctuation in the number of CCHF cases coincided with the seasonal population dynamics and activity period of the adult stage of the major vector, or tick species (Section VI.D).

While a spring-summer periodicity is characteristic of CCHF virus epidemic transmission in the more northern temperate portion of its geographical range, the temporal distribution pattern has not been clearly defined elsewhere (Table 1). In general, this can be attributed to the infrequent recognition of CCHF human cases within a particular geographic location, and the occurrence of predominantly point-like episodes. However, a seasonal periodicity was suggested for CCHF virus transmission during the epidemic in Iraq.[38] Among the 55 suspected and/or confirmed CCHF cases recognized between August 1979 and September 1980, 49% (27/55) of the cases were reported during September, October, and November 1979 and 35% (19/55) were reported during April, May, and June of the following year.

In the Republic of South Africa, CCHF cases have occurred more commonly during the spring and summer seasons of the Southern Hemisphere, but cases have been recorded every month of the year except June (Figure 3). However, all cases recognized during September were acquired by contact with a patient hospitalized during late August 1985.[50] The pattern closely parallels the seasonal feeding activity period, and the peak population density of the

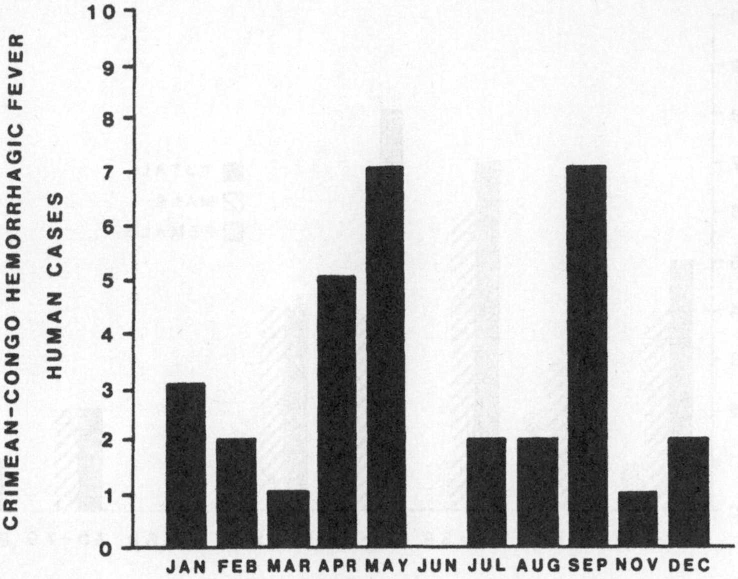

FIGURE 3. The monthly distribution of CCHF cases, Republic of South Africa, February 1981 to May 1986.

suspected outbreak CCHF virus tick vectors[123] (Section VI.D). In Nigeria, CCHF virus was isolated from wild and domestic animals and ticks throughout the year, but most isolates were obtained during October, November, and December, or the late rainy and early dry seasons.[124] Thus, the data clearly demonstrate the potential for the occurrence of CCHF virus transmission to humans throughout the year in milder climatic regions where ticks may remain active.

D. Risk Factors

CCHF has been recognized more commonly as a disease among agriculture workers, yet any person in contact with CCHF virus must be considered at-risk of infection. Agricultural practices, particularly those associated with large domestic animals, are important risk factors for CCHF contracted from the bite of infected ticks (Table 1). Also, exposure by contagion such as crushing infected ticks and butchering infected animals has been a frequent source of CCHF viral infection among the latter workers. Similarly, numerous contagion-acquired cases have been documented among medical workers and others who care for CCHF patients, as well as laboratory workers who handle materials containing virus.

Evidence based mainly on overt illness indicated that all ages and both sexes appeared equally susceptible to CCHF viral infection. An unequal distribution of cases among males and females is not uncommon, and can be attributed to specific occupational activities that allow for differential exposure to the sources of CCHF viral infection.[20,34,36,38,47-50] Similarly, this is the most likely explanation for the predominance of cases among the adult working cohort, even though cases have been documented for all ages. This age and sex distribution pattern (Figure 4) and the sex distribution of cases in relation to the sources of CCHF viral infection (Figure 5) are illustrated by data summarized for CCHF outbreaks in the Republic of South Africa.[45-50,55]

Whether humans come in contact and are bitten by infected ticks is dependent on an extremely complex and dynamic ecological and biological system that is not very well understood. However, it is becoming increasingly evident that the greatest risk is in areas where adults of one or two species of ticks of the genus *Hyalomma* are the predominant

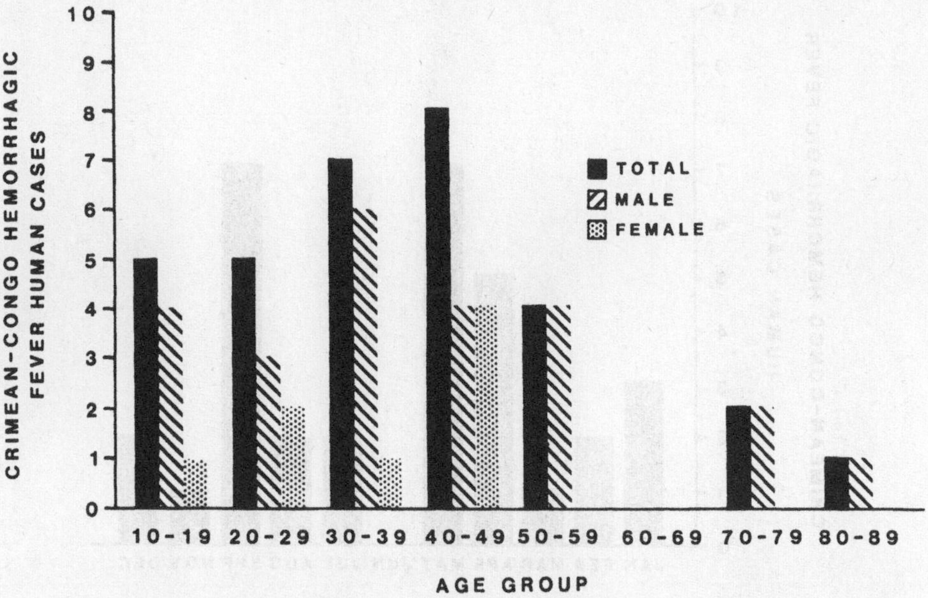

FIGURE 4. CCHF cases according to the age and sex of patients, Republic of South Africa, February 1981 to May 1986.

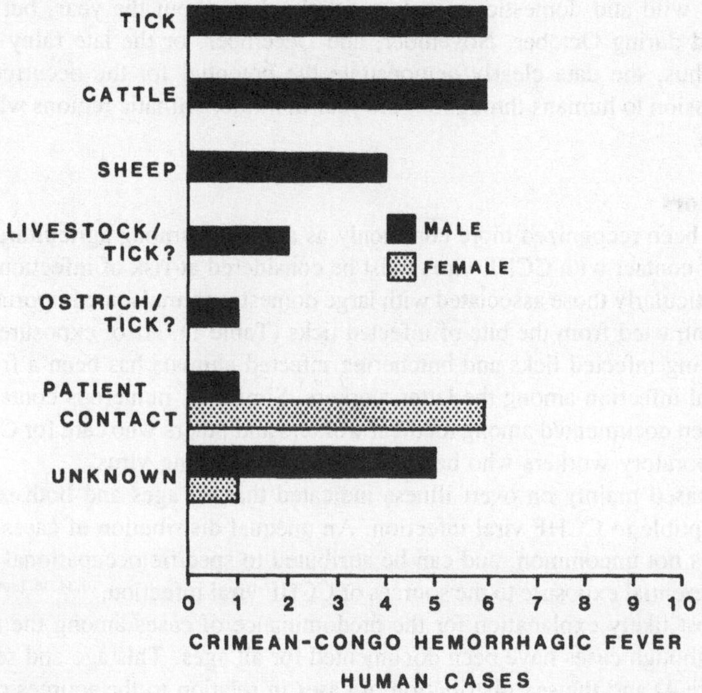

FIGURE 5. Sources of CCHF human viral infections, Republic of South Africa, February 1981 to May 1986.

parasites of humans and large domestic animals.[20] The degree of risk posed by infected ticks is, in general, unknown but incidence rates for CCHF during an epidemic in Rostov suggested that the risk of human infection was directly proportional to the rate of attachment of *H. m. marginatum* on humans.[56] The observed incidence rates during the latter epidemic were comparable to those derived by a mathematical model that depicted a low risk for human cases, but the predicted ratio of apparent to inapparent infection was 0.21, or 1 CCHF case for every 5 persons infected regardless of age.

Estimated human case attack rates during nosocomial outbreaks indicate that the risk of contracting CCHF can be extremely high, especially during the hemorrhagic phase of illness. For example, five of six medical workers who were in contact with an index case contracted CCHF during a nosocomial outbreak in Pakistan.[34] CCHF cases were also diagnosed among 5 of 11 similar workers, and 2 of 9 relatives who were in contact with secondary cases, for an overall estimate of 12 cases among the 26 at-risk persons. Estimates of the degree of risk posed by patients during a nosocomial episode in the Republic of South Africa revealed that 33% (3/9) of medical workers who had contact with patients through accidental needle pricks developed CCHF, and 8.7% (4/46) contracted this disease by other contacts with the patients' blood.[53] Of the seven cases, six were contacts of the index case, and one was in contact with secondary cases. Not including the 7 cases, a total of 452 persons were considered at-risk during the outbreak. None of 217 of the 452 who were followed developed CCHF, but 117 had suggestive symptoms. The possibility that some of these reflected mild CCHF viral infections cannot be excluded. As stated above, antibody to CCHF virus considered to represent mild and/or inapparent CCHF viral infection was demonstrated in medical workers who attended patients, as well as in patients' relatives during an epidemic of CCHF in Iraq.[77] Thus, a more accurate estimate of the degree of risk for humans in contact with CCHF patients may be obtained on the basis of both clinical and serological data.

The risk for human CCHF viral infections posed by infected animals has been described previously,[20] and more recently illustrated by human cases contracted from infected cattle and sheep (Figure 5) in the Republic of South Africa.[48-50] Among laboratory workers, it is notable that six cases of CCHF were contracted between 1958 and 1977 in Uganda by persons who handled virus-infected materials.[13,14,42] Also, one laboratory-associated case occurred in Zaire, one in the Central African Republic,[104] and several cases have been reported in the U.S.S.R., including one fatal case.[20] Aerosol transmission was suspected as the source of infection for at least four cases in the U.S.S.R.

Aside from the risk of acquiring CCHF from patients, evidence obtained in epidemics in Eurasia suggested that contagion-associated disease may be more severe than disease contracted by tick bite.[20] Mortality rates during epidemics in Bulgaria were 17% for nonnosocomial cases, presumably acquired by tick bite, and 41% for nosocomial cases.[22] Epidemics in Tadzhikistan resulted in a 25% case fatality rate for tick bite-related cases, and 50% for persons infected by contact with patients.[30] However, limited data for CCHF cases in the Republic of South Africa indicated that 22% (4/18) of the better-documented, contagion-associated cases were fatal; 17% (1/6) of tick bite cases were also fatal.[55]

E. Serologic Epidemiology

A substantial amount of ecological and epidemiological data pertaining to CCHF virus has been derived from serosurveys. However, the degree of reliability and the relevance of results to the total populations sampled have been difficult to ascertain. Sample sizes have varied and seldom have been defined as to the proportion of the total population or to what extent the data were based regarding random or nonrandomized samples. Comparative assessments of the relative prevalence of CCHF viral infections within and among different populations have been hampered by these discrepancies and by variation in the type of serological tests.[20,125]

In many serosurveys, the AGDP and CF tests have been widely employed (Table 3), but the results are not readily interpretable due to problems related to sensitivity of these techniques and possibly the low and transient nature of antibody produced by CCHF viral infections. As illustrated by a study in Rostov Oblasts, CF and AGDP antibody had declined below detectable levels in 33 and 42%, respectively, of convalescing CCHF patients after 1 year.[126] No relationship was noted between the duration of detectable antibody and age, but antibody persisted longer for persons who experienced severe CCHF than for those who had a mild clinical course. The duration of detectable antibody as measured by the AGDP and CF tests appears to be considerably more transient in domestic animals, raising difficulties in interpretation of antibody prevalence data.[127,128]

Variation in serological test results was also demonstrated by a study in Iran.[129] CCHF viral antibody was detected in 55% (30/55) of human sera by the AGDP test. Of these same sera, 9% (5/55) were positive by HI, 0/38 by CF, and 0/15 by N test. Of the latter 15 negative sera, 11 were positive by the AGDP test. In contrast, 67% (70/105), 65% (68/105), 22% (23/105), and 68% (71/105) of sheep sera were seropositive by the AGDP, HI, CF, and N tests, respectively.

Serological techniques such as the HI test routinely employed for most arboviruses have not been used routinely because not all viral strains yield agglutinating antigens.[20] N tests have not been considered acceptable because of nonspecific, antiviral activity associated with serum of both human and lower vertebrates. However, nonspecific factors were eliminated by acetone-ether treatment of human sera. Also, the lack of and/or variation among CCHF viral strains to replicate and produce plaques in cell culture has precluded extensive use of the plaque-reduction neutralization test (PRNT). More recent results revealed that CCHF virus, including strains from Africa, Pakistan, and the U.S.S.R., consistently produced distinct plaques in 4 days in a human adenocarcinoma, adrenal cortex cell line (SW-13).[130] Furthermore, no appreciable nonspecific antiviral activity was exhibited by a limited number of human and domestic animal sera employing this cell line in PRNTs.

The IFA and the reverse passive hemagglutination inhibition (RPHI) tests have proved effective for serosurveys.[47] For example, preliminary results indicated that the IF and RPHI tests yielded comparable estimates of the prevalence of CCHF viral antibody among humans, sheep, and hare sera, but the RPHI was more sensitive than the IF for detecting antibody in cattle sera. Compared to the ID test, both the RPHI and IF tests were more sensitive. The extent of difference in sensitivity varied in relation to the period of lapsed time since the animal was infected. For recent infection, both the RPHI and ID tests were considered suitable for detecting antibody, but the sensitivity of the ID test decreased markedly with time compared to the RPHI test.[33] Preliminary results suggest that the EIA may prove to be a valuable test.[84] However, the sensitivity and specificity of this as well as other techniques mentioned above need further assessment. Antibody induced by other antigenically related Nairoviruses,[57] such as Hazara virus, and the possibility of viral strain differences among CCHF viruses could lead to misinterpretation of data.

CCHF viral antibody prevalence rates determined by various techniques during selected serosurveys among humans and domestic animal populations are presented in Table 3. Antibody has been demonstrated among human populations in many countries of Asia, Europe, the Middle East, and Africa. Where concurrent human and domestic animal surveys have been conducted in the same areas, antibody was usually present in both, but the prevalence was markedly higher in domestic animals. This is likely to reflect differential exposure to the source of infection. However, data must be interpreted with caution, as illustrated by the variation in test results for Iran[129] and other serosurveys reviewed previously.[20]

The significance of the prevalence of CCHF viral antibody among humans and other vertebrates in areas where CCHF cases have not been recognized is difficult to interpret

(Table 3). Overt infections may not have occurred or occurred and were not detected and/or were possibly misdiagnosed. Alternatively, serological reactors may reflect nonspecific, antiviral activity or prior infection with another *Nairovirus*.

An illustration of the occurrence of CCHF virus and viral antibody in the absence of confirmed human cases is derived from studies conducted between 1964 and 1973 in Nigeria.[131] From 1964 to 1970, 168 arboviruses representing 15 different viruses were isolated predominantly from children between 1 and 4 years of age. Most of the 12,131 blood specimens that yielded the viral isolates by assays in 2-day-old mice were from this age group as well. Nonetheless, CCHF virus was not isolated from patients, although 48 strains of the virus were obtained during a comparable time period from domestic cattle and goats, African hedgehogs, a culicoid midge, and 6 species/subspecies of ticks.[16,17,124] CCHF viral isolates included 27 strains from tick species commonly implicated as the vectors associated with human infections in Africa; however, most were from the cattle tick *Boophilus decoloratus*. Despite the failure to isolate CCHF virus from humans in Nigeria, serological evidence for prior CCHF viral infections was detected by mouse neutralization tests in 16% (22/141) of males and 26% (28/109) of females.[70] The ages ranged from less than 1 year to 43 years; however, only 4 blood specimens were tested from older persons. Among the 0- to 10-year age group, antibody with $\log_{10}$ neutralization indexes of 1.0 to 2.8 was detected in 12% (10/84) of males and 30% (24/79) of females in the age group yielding most of the above viral isolates. As of 1984, human CCHF cases had not been reported in Nigeria.[132]

Retrospective evidence of CCHF viral infections based on the prevalence of antibody in cattle and wild vertebrates also demonstrated that CCHF virus had been enzootic in the Republic of South Africa long before human disease was recognized.[47] In this regard, it is notable that the second and third human cases in that country were not recognized as CCHF on the basis of clinical evidence, but were confirmed by retrospective laboratory studies.[48]

V. TRANSMISSION CYCLES

A. Evidence from Field Studies

1. Vectors

Until the development of techniques required to isolate and identify CCHF virus in 1967 and 1968, a few tick species, mainly of the genus *Hyalomma*, had been circumstantially implicated as vectors during CCHF epidemics in Eurasia.[20] A vector role was suspected on the basis of a temporal and spatial association between the seasonal distribution, population density, and adult activity period of *Hyalomma* ticks and the occurrence of CCHF cases. Of greater significance was the observation that patients revealed evidence of being bitten by *Hyalomma* or other tick species, or that they had crushed ticks with their hands. These observations supported a tickborne route of transmission for CCHF virus, but it was not until the late 1960s that this virus was isolated from adult *Hyalomma*, as well as from several other tick species.[20,133] Viral isolates were also obtained from field-collected eggs and unfed immature stages of *H. m. marginatum*, thus demonstrating evidence of transovarial and transstadial transmission, respectively.[134-136] These observations, and data derived from laboratory studies (Section V.B.1), fully incriminated *H. m. marginatum* as a vector and possible reservoir of CCHF virus. The proposed maintenance and transmission cycle for CCHF virus by this tick species is depicted in Figure 6.

While only a few species of ticks have been incriminated as vectors of CCHF virus, an enormous number of species/subspecies have been implicated primarily by viral isolations (Table 4). In 1973, only 6 years after the first isolations of CCHF virus, a total of 10 species/subspecies had yielded isolates of this virus.[137] A remarkable and especially important epidemiological feature that emerged was not only the large number of implicated vector tick species, but the association of CCHF virus with ticks in a variety of different ecological

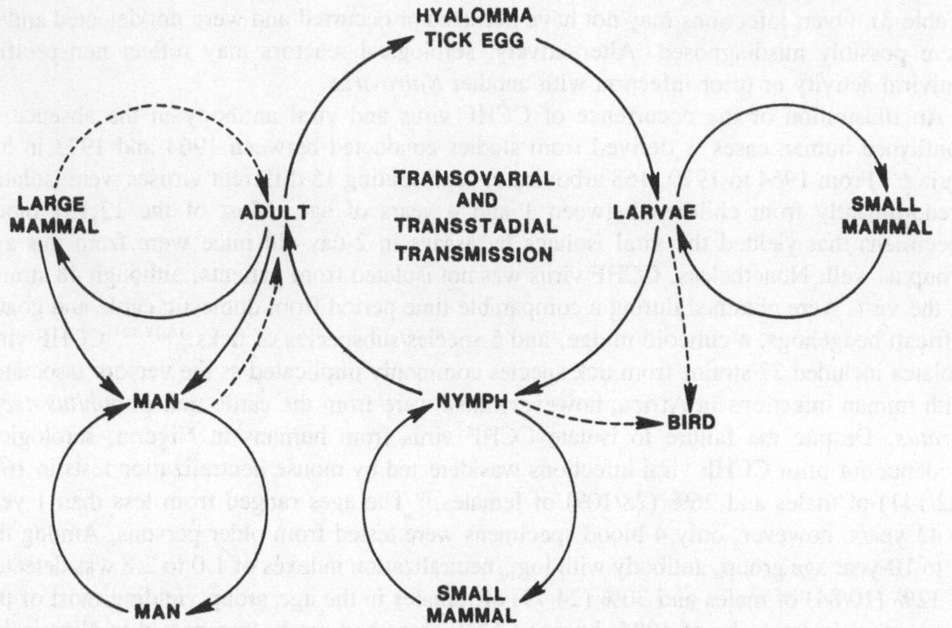

FIGURE 6. CCHF viral maintenance and transmission cycle involving *Hyalomma marginatum marginatum* and associated vertebrate hosts. (——) Viral transmission cycle demonstrated by field and laboratory observations. (----) Either unconfirmed or unknown.

biotypes in the Palearctic, Oriental, and Ethiopian faunal regions. For example, approximately 30 strains of this virus were obtained from ticks collected between 1967 and 1968 in Eurasia,[133] 27 strains from ticks collected between 1964 and 1968 in Nigeria,[16] 26 strains from ticks collected between 1969 and 1974 in Senegal,[107-109] and 2 strains from ticks collected during 1965 in Pakistan.[18] As of 1979, CCHF virus had been isolated from a total of 26 species/subspecies of ticks.[20] Since then, CCHF viral isolations have been reported from *Ornithodoros lahorensis* in Iran,[117,118] *Rhipicephalus appendiculatus* in Uganda,[42] and *R. e. evertsi* in the Republic of South Africa.[47] A summary of the species/subspecies of ticks from which CCHF virus has been isolated in different countries is presented in Table 6.

Among the total 29 species/subspecies of ticks associated with CCHF virus, most are either two- or three-host ticks of the family Ixodidae. One-host ticks are represented by three *Boophilus* species of the latter family, and two multihost species, *Argas persicus* and *O. lahorensis*, of the family Argasidae have yielded isolations of CCHF virus.

The significance and relative importance of the large number of species/subspecies of ticks in the maintenance and transmission cycle of CCHF virus are not readily understood. Viral isolation alone from the vast majority of the tick species/subspecies does not incriminate them as vectors. Numerous reports referred to in Table 1 presented CCHF viral isolation rates for different tick species, but difficulties in interpretation arise from the lack of uniformity of trapping and collecting methods, and the irregularity and lack of continuity in studies. However, the viral isolation data are valuable for identifying potential vectors that can be further evaluated by field and laboratory studies.

An especially important biological feature of ticks in general as potential vectors/reservoirs of arboviruses is their ability to transmit arboviruses transovarially.[138] Evidence of this phenomenon for CCHF virus in nature is based mainly on limited isolations from eggs of *H. m. marginatum* and *Dermacentor marginatus*.[136] Also, immature and adult ticks can be infected with CCHF virus as a result of transstadial transmission and by feeding on viremic

Table 6
TICK SPECIES AND SUBSPECIES FROM WHICH CCHF VIRUS HAS BEEN ISOLATED BY COUNTRY

Location	Species and subspecies	Ref.
Africa		
Central African Republic	*Amblyomma variegatum*	102—104
	Boophilus annulatus	
	Hyalomma nitidum	
Ethiopia	*H. impeltatum*	106
Kenya	*Rhipicephalus pulchellus*	20
Mauritania	*H. rufipes*	68
Nigeria	*A. variegatum*	16
	B. decoloratus	
	H. anatolicum	
	H. impeltatum	
	H. rufipes	
	H. truncatum	
Senegal	*A. variegatum*	62, 107—109
	B. decoloratus	
	H. impeltatum	
	H. impressum	
	H. rufipes	
	H. truncatum	
Uganda	*A. variegatum*	42, 110
	R. appendiculatus	
Republic of South Africa	*H. rufipes*	47
	H. truncatum	
	R. evertsi	
Europe		
Bulgaria	*B. annulatus*	20, 111
	Dermacentor marginatus	
	H. marginatum	
	Ix. ricinus	
	R. bursa	
	R. sanguineus	
Greece	*R. bursa*	112
Hungary	*Ix. ricinus*	113
Yugoslavia	*Ix. ricinus*	114
	H. marginatum	
Middle East		
Afghanistan	*A. variegatum*	116
	H marginatum	
Pakistan	*B. microplus*	18
	H. anatolicum	
Iran	*Ornithodorus lahorensis*	117, 118
Asia, Eurasia		
China	*H. asiaticum*	31
U.S.S.R.	*Argas persicus*	20, 111, 122
	B. annulatus	
	D. daghestanicus[a]	
	D. marginatus	
	Haemaphysalis punctata	
	Hyalomma anatolicum	
	H. asiaticum	
	H. detritum	
	H. dromedarii	
	H. marginatum	
	H. turanicum	

Table 6 (continued)
TICK SPECIES AND SUBSPECIES FROM WHICH CCHF VIRUS HAS BEEN ISOLATED BY COUNTRY

Location	Species and subspecies	Ref.
	Ix. ricinus	
	R. bursa	
	R. pumilio	
	R. rossicus	
	R. sanguineus	
	R. turanicus	

[a] CCHF virus-specific fluorescence observed in salivary glands of this tick; all other species/subspecies yielded isolations of CCHF virus.

vertebrates. CCHF virus has been isolated from field-collected, unfed *H. m. marginatum* nymphs and adults[134] and from *H. a. anatolicum*.[135,136] In addition, virus has been isolated from and antibody has been demonstrated in vertebrate hosts of immature ticks.

Hematophagous arthropods other than ticks have not been implicated as vectors of CCHF virus. Attempts to isolate the virus from more than 25,000 mosquitoes (6 species) collected between 1967 and 1969 in a CCHF enzootic foci in Astrakhan were unsuccessful.[20] Sentinel laboratory mammals also failed to reveal evidence of infection during the same period in the latter foci. As mentioned in Section I.A, an isolate of CCHF virus was obtained from 1 of 377 pools of *Culicoides* (biting gnats) collected in Nigeria.[16] Apparently, the source of virus was vertebrate blood that had been ingested recently by this insect and, therefore, was not considered evidence of infection.

While a certain amount of progress has been made, the role of most tick species/subspecies as vectors of CCHF virus can only be postulated. This can be attributed not only to the enormous number of suspected vectors, but to the extremely complex and diverse ecological and biological features of ticks. In addition, technological difficulties, and the human health risks posed by working with CCHF virus, undoubtedly, have hindered progress in understanding the relative importance of ticks as vectors of this virus. The comprehensive review alluded to earlier provides an excellent overall coverage of the current understanding of the vector status of most species/subspecies of ticks.[20]

2. *Vertebrate Hosts*

Vertebrates are essential as a source of blood for the development and perpetuation of ticks as a species. A wide variety of vertebrate species are parasitized by tick species associated with CCHF virus (Table 4). However, the qualitative and quantitative roles, if any, of vertebrates in the maintenance and transmission cycle of this virus are poorly understood. Existing data have unequivocally demonstrated that humans and domestic animals are infected by CCHF virus. A variety of species of small mammals has also been implicated as hosts, but the evidence is based mainly on limited serosurveys (Table 2). While several species have been considered, the number of a particular species studied has, for the most part, been insignificant. Furthermore, the data must be interpreted in view of the problems inherent to the serological tests (Section IV.E).

The role, if any, of humans in the perpetuation of the natural cycle of CCHF virus is unknown (Figure 6). Whether CCHF viral infections of humans produce a sufficient viremia to infect feeding ticks has not been determined, but a human-to-human transmission cycle can be initiated by contact with blood of CCHF patients and infected domestic animals (Section IV.D).

Evidence of CCHF viral infection among domestic animals, particularly livestock, has been demonstrated by viral isolations (Section II.B.1) and serological evidence of infection (Table 3). Antibody prevalence among livestock has varied according to the time after infection,[127,128] the serological tests, and the geographical location, but generally has revealed substantial evidence of CCHF viral infection. Despite the well-documented occurrence of infection, the significance of viremic livestock (Section V.B.2) as a source of infection for ticks is not clear (Figure 6). Even if proved to be a dead-end host for the virus, livestock may play a significant role in the ecology of CCHF virus if they are an essential source of blood feeding in the cycle of development of the vector tick species.

Among the vertebrate species known to be susceptible to CCHF virus, small mammals appear to have the greatest potential for contributing to maintenance and transmission cycles. Evidence of infection has been demonstrated by the isolation of CCHF virus from hares (*L. europaeus*) in the U.S.S.R., hedgehogs (*E. a. albiventris*) in Nigeria,[16,17] and a multimammate mouse (*Mastomys* sp.) in the Central African Republic.[64] Serological evidence of infection has been demonstrated in these as well as several other vertebrate species (Table 2), but only a moderate amount of data is available for hares, and data are extremely limited for other vertebrate species. Hares were implicated as important hosts of CCHF virus during outbreaks of virus in Eurasia[20] and, more recently, were considered hosts in the Republic of South Africa.[47] While further studies are needed, field and experimental data (Section V.B.2) indicate that hares and hedgehogs may serve as amplifying hosts.

Although large wild vertebrates are utilized as a source of blood by adult ticks, they have not been investigated as possible hosts for CCHF virus. Studies are especially warranted in Africa where all the tick species from which CCHF virus has been isolated feed as adults on large, wild vertebrates.[20] Data from preliminary studies in progress in the Republic of South Africa suggested that Eland antelopes were infected with CCHF virus.[47]

Several avian species are commonly parasitized by immature stages of tick species associated with CCHF virus, but no apparent role has been demonstrated for this vertebrate other than as a source of blood for ticks.[20] Attempts to detect evidence of CCHF viral infection have been largely unsuccessful, but like other vertebrate species, definitive studies are lacking. Thus, the more apparent significance of avians at this stage in the evolution of knowledge on the ecology and epidemiology of CCHF virus is their historically documented role as a dispersal mechanism, and as a source of blood for ticks.

B. Evidence from Experimental Infection Studies

1. Vectors

Only a very few experimental data have been reported on the role of ticks as vectors/reservoirs of CCHF virus. Furthermore, virtually no attempts have been made to quantitatively assess the relative importance of individual tick species. Limited data revealed that *H. m. marginatum*,[89,90,92,139] *Rhipicephalus rossicus*,[92] and *D. marginatus*[92] became infected with CCHF virus after feeding upon viremic vertebrates. Similarly, CCHF viral infection was demonstrated in *H. m. rufipes*[140-142] and *Amblyomma variegatum*[142] after parenteral inoculation. Data were not presented for *H. a. anatolicum*, but all other species transmitted CCHF virus transstadially and transovarially. In addition, viral infection was demonstrated in vertebrates fed upon by one or more developmental stages of *H. m. marginatum, R. rossicus, D. marginatus, H. m. rufipes*, and *A. variegatum*. CCHF viral infection was demonstrated for almost 2 years in various stages in some of these species.

2. Vertebrate Hosts

Although CCHF virus has an extensive geographical distribution and has serious disease consequences, there are comparatively few experimental data of the possible role of vertebrates as virus-amplifying hosts. Early Soviet studies found many infected nymphal ticks on rooks (*Corvus f. frugilegus*).[20] After experimental inoculation of rooks and rock doves (*Columba livea*) with CCHF virus, the birds remained healthy, and evidence of a viremia or an immune response was not demonstrable.[143,144] Antibody prevalence data from ecological studies also supported this evidence,[145] but field and experimental data are limited and poorly documented for the possible role of avians in the maintenance and transmission cycle of CCHF virus.[20]

The European hedgehog, *E. europaeus,* and the long-eared hedgehog, *Hemiechinus auritus*, have been evaluated as potential amplifying hosts of CCHF virus.[89] The long-eared (but not the European) hedgehog developed a viremia. CF and AGDP antibody could not be demonstrated in either species during a 1-month observation period following inoculation. However, virus was recovered from nymphal *H. m. marginatum* which had fed as larvae on the infected *H. auritus*.

Both experimentally infected hares (*L. europaeus*) and long-eared hedgehogs were used to provide CCHF virus-infectious blood meals for *H. m. marginatum* which transmitted virus both transstadially and transovarially.[90] Other studies confirmed the presence of virus in the blood of infected hares.[91] Viral titers in hare blood peaked at 3.6 dex and persisted as long as 15 days.

The little suslik (*Citellus pygmaeus*), experimentally infected with CCHF virus, provided an infective blood meal to *H. m. marginatum*.[92] Blood was positive for virus on days 2 through 7.

Two calves were infected with a Nigerian strain of CCHF virus, and both had demonstrable viremias.[16] Calves inoculated with a Rostov strain of CCHF did not transmit virus to *H. m. marginatum, D. marginatus,* or *R. rossicus*.[86] At least one of two 2-month-old calves developed a viremia, but virus was not demonstrable in the blood of two 6-month-old calves. All inoculated calves had demonstrable CF, AGDP, N, and indirect HI antibodies.

Two horses were inoculated with high doses of CCHF virus but "no pronounced infection was recorded in either", although antibody responses were measured by CF and AGDP.[87,88]

Humans were experimentally infected in early Soviet studies and they developed the disease.[4-6] Viremia levels in humans are relatively high[55] and it would seem probable that ticks could be infected from human blood meals, but, as indicated in Section V.A.2, the role of humans as a source of CCHF viral infection for ticks is not understood.

C. Maintenance / Overwintering

In general, data derived from limited observations suggest that ticks are the principal reservoirs of CCHF virus. As a host, the virus may be transmitted transovarially and transstadially, but the efficiency as maintenance mechanism is not understood. Data for European tickborne encephalitis virus-infected *Ixodes ricinus*[147] and for Russian spring-summer encephalitis virus-infected *Ix. persulcatus*[148] indicated that these viruses were transmitted transovarially to less than 10% of their progeny. This was not considered sufficient for indefinite maintenance, and therefore implied that viral amplifications by infected vertebrates were required to sustain these viruses. If applicable to CCHF virus, it is likely that small mammals serve as the principal virus-amplifying hosts in the proposed maintenance cycle involving *Hyalomma* ticks as vectors/reservoirs (Figure 6). Large wild and domestic mammals cannot be excluded, but limited experimental data suggest that ticks are not readily infected by feeding on large domestic mammals during the viremic phase of CCHF viral infections (Section V.B.2).

Either transovarially/transstadially acquired CCHF viral infections or infection acquired from viremic vertebrates appears to allow vector tick species to maintain CCHF virus during the winter in temperate regions (Figure 6). Evidence that ticks served as overwintering hosts in Eurasia included the isolation of CCHF virus from unfed nymphs and adult female *H. m. marginatum* collected in the field during the spring in the Crimean, Rostov, and Astrakhan

Oblasts[20,145] and from unfed adult *H. a. anatolicum* taken during the spring in Tadzhikistan.[20,146] Experimental data also showed that *H. m. marginatum, R. rossicus,* and *D. marginatus* remained infected with CCHF virus under simulated winter temperature.[92] The persistence of CCHF virus during adverse climatic conditions in subtropical and tropical regions is likely to rely on similar mechanisms of infection of the vector tick species, but the climatic conditions may be permissive for a continuous transmission cycle involving ticks and vertebrates.[123]

VI. ECOLOGICAL DYNAMICS

A. Macro- and Microenvironment

The primary faunal common denominator responsible for the enzootic distribution of CCHF virus in the Palearctic, Oriental, and Ethiopian regions appears to be *Hyalomma* ticks.[20] The only recognized exception is in the deciduous forest of Moldavia where *Ix. ricinus* replaces *Hyalomma* ticks. The geographical range of these ticks disappears in the fauna of Burma, immediately east of India, coinciding with apparent absence of CCHF virus activity. Within enzootic foci, *Hyalomma* ticks appear to be restricted to lowlands, foothills, low mountain belts with arid to semiarid climates, or at least long dry seasons. This can include deserts, semideserts, steppes (Eurasia), tropical savannah and grasslands, arid grasslands, and possibly desert regions (Africa). However, the most intense viral activity is in restricted areas where the variety and population densities of ticks and their hosts are greatest. These areas may be along river floodplains with rich grasslands (Astrakhan Oblast); shrub and tree vegetation (Uzbekistan); desert or semidesert environs; sparse saxaul forests scattered in desert sands (Kazakhstan); and forests and thickets of rough steppe lands (Rostov Oblast). If confirmed, the detection of CCHF viral antibody in tropical forested areas in India represents a new ecological habitat, unique among the currently recognized distribution of CCHF virus.

B. Climate and Weather

Enzootic foci of CCHF virus occur primarily in areas that are characterized by warm summers and relatively mild winters.[20] These areas range from the arid deserts and semideserts of Eurasia and North Africa to the wet Central African forests of Zaire and Uganda to the semiarid high-altitude areas of East and South Africa.

In Eurasia, foci are characterized by common physiogeographical conditions within the limits of the 2800 to 5000 sum of effective annual temperatures above 10°C in the transitional atmospheric humidity zone between the forest-steppe and desert. The distribution of enzootic foci in the lowland desert and semidesert area of southern Tadzhikistan appears to be confined to areas with sum annual temperatures ranging from 3000 to 5700°C, but may exceed a sum of 6000°C in warm years. In this focus annual rains, which fall mostly in the spring, ranged from 150 to 300 mm at 500 m to 900 mm.

In Eurasia, the primary tick vector, *H. m. marginatum*, and CCHF virus activity seem to require a relatively mild winter (lowest monthly mean temperatures above −20°C).[20] For example, in the Osh Oblast semidesert foci in Kirgiz SSR where the annual sum of effective temperatures is 4000°C, winters are relatively mild (January mean, 9 to −3°C). In the semidesert-steppe transitional zone of the Astrakhan Oblast, temperatures rarely drop to −20°C along the Volga Delta. However, in the winter of 1968—1969, temperatures dropped to −30°C and remained at −20°C or lower for more than 2 months, and the ground froze to a depth of 1 m. The adult tick abundance index per cow fell from approximately 20 in 1968 to less than 0.1 in May 1969, and no CCHF cases were reported during the summer of 1969. However, CCHF virus remained enzootic in two cold-adapted tick species, *R. rossicus* and *D. marginatus*.

It is not clear whether climatological parameters described for Eurasia can be applied to other areas. Certainly the environmental factors pertaining to the enzootic forests of Moldavia and the poorly described known foci in the Middle East and most of Africa need to be examined more closely before conclusions can be drawn.

C. Vector Oviposition

Species/subspecies associated with CCHF virus, or the hard ticks of the family Ixodidae, generally oviposit only once in a lifetime. Eggs are laid without regard to location at one time after detaching from the host. Usually the eggs are very numerous, sometimes over 10,000. For example, a single *H. m. marginatum* female laid from 4300 to 15,500 eggs 6 days after detaching from the host,[149] and an average of 6867 eggs were laid by *H. m. rufipes*.[150] Oviposition may be delayed or prevented during cold periods. At a temperature below 15°C, *H. asiaticum kozlovi* females did not oviposit.[151]

Soft tick species of the family Argasidae (*Argus, Ornithodorus*) oviposit at intervals in small batches, usually totaling only a few hundred, in niches where females seek shelter. Female *A. persicus* may oviposit up to six or seven times in a lifetime, each oviposition usually after a single blood meal.[152] From 195 to 646 eggs are laid after the first blood meal. Subsequent blood meals yield decreasing numbers of eggs. Oviposition starts 3 to 10 days after feeding in summer, but is delayed for weeks or months in the winter. Eggs of *A. persicus* have a very high degree of tolerance to fluctuating climatic factors, and *O. moubata* is known to envelop each egg during oviposition with a waxy, waterproof coating that enables the eggs to withstand very dry conditions.

D. Vector Density, Fecundity, and Longevity

Epidemic transmission of CCHF virus coincides with an increased population density and feeding of the vector tick species.[20] For example, during the outbreaks in Bulgaria, the principal vector, adult *H. m. marginatum*, first appeared, attained maximum population density, and declined about 1 month prior to the occurrence of human cases in April, peak numbers in June, and decreasing numbers of cases thereafter, respectively. Similarly, the seasonal dynamics of *H. a. anatolium*, the species implicated as a vector in the Central Asia Republics of the U.S.S.R., and the seasonal distribution of CCHF cases were closely interrelated.

Vector density also exhibits distinct seasonal variation in tropical areas. In Zambia and Zimbabwe, peak population densities of *H. m. rufipes* and *H. truncatum* adults occur during the warm seasons.[153,154] These species have been implicated as vectors of CCHF virus in the Republic of South Africa where they exhibit peak abundance during summer and winter.[123] In Kenya, immature *Hyalomma* spp. parasitized hares during most of the year, but peak population densities were attained just before the long rainy season and declined markedly during the rainy period.[155] While a seasonal periodicity in density is characteristic of ticks, abundance and activity can be affected by temperature, humidity, predators, fire, flooding, and host availability.

The effective fecundity of a vector population in a particular year is a product of the total number of eggs oviposited by the individual females of a generation and the number of generations completed during that year. As discussed in Section VI.C, the number of eggs produced by individual females of different species varies greatly, but in general, the Argasidae lay many fewer eggs than do the prolific Ixodidae.

The minimum time required for a tick generation may be 4 or 5 months; however, climatic conditions and host availability may drastically increase generation time and decrease the overall fecundity of a population. In the western Transvaal of South Africa, both *H. m. rufipes* and *H. truncatum* probably complete two generations per year.[123] However, *H. m. rufipes* only completes one generation per year in southeastern Zimbabwe[156] *H. m. margin-*

atum and *H. truncatum* and other *Hyalomma* sp. in Eurasia complete only one generation per year because of the more severe winter climate.[20]

Ticks can survive for months or years prior to oviposition without blood feeding if poor climate or host unavailability dictates. This adaptive behavior tends to maximize the chances of taking a blood meal and of completing a gonotrophic cycle. The longevity of a tick, due to its extended developmental cycle, may also reduce its chances of survival; however, this can maximize blood feeding required for viral transmission. *H. m. marginatum* adults held in optimal laboratory conditions survived more than 800 days without a blood meal.[157]

E. Biting Activity and Host Preference

Studies of the biting activity of larval, nymphal, and adult ticks usually examine the seasonal incidence of questing ticks on vegetation or attached ticks on their hosts. Section VI.D describes the seasonal biting activities of some CCHF virus vectors.

Host preference data are of crucial epidemiological importance to understanding CCHF viral maintenance and activity. The number of hosts parasitized by an individual tick during its lifetime, the potential variety of hosts, and the degree of host specificity, are biological parameters that must be understood. The number of vertebrate hosts and some of the species parasitized by tick species associated with CCHF virus are presented in Table 4.

One-, two-, three-, and multihost ticks have yielded isolations of CCHF virus (Table 4). All *Boophilus* species are one-host ticks, i.e., the developmental stages are completed on the same vertebrate host.[20] *Boophilus* ticks seldom attack man. Their role in the ecology/epidemiology of CCHF virus is not clear, but appears to be potentially important as an enzootic vector.

The two-host ticks can be regarded as two subgroups on the basis of their feeding characteristics.[20] The first group includes those that utilize similar vertebrate host species for development of the immature and adult stages. Unlike the one-host ticks, the immature stages complete their development on a host and detach as blood-engorged nymphs. The nymphs molt and reattach as adults on a similar vertebrate host species. This pattern is characteristic of *R. d. bursa*, *R. e. evertsi*,[158,159] *H. detrium*, *H. a. anatolicum*, and possibly *H. truncatum*. The second group includes those species that rely on two dissimilar host species for their development; i.e., the vertebrate host species utilized by the immature stages differs from that of the adult stage. Several of the *H. marginatum* complex species exhibit this feeding pattern. Of these two subgroups, all of the *Hyalomma* ticks — including *H. detrium*, *H. a. anatolicum*, *H. m. marginatum*, *H. m. rufipes*, and *H. m. turanicum* — are considered to be especially important as epidemic/outbreak vectors as well as CCHF viral enzootic vectors and possibly reservoirs.

The three-host ticks include those species for which three different vertebrates typically serve as host for the three developmental stages.[20] Species associated with CCHF virus in Eurasia include *I. ricinus*, *H. punctata*, *D. marginatus*, *D. daghestanicus*, *H. a. asiaticum*, *R. pumilio*, *R. rossicus*, *R. sanguineus*, and *R. turanicus*. CCHF virus has been isolated from six three-host ticks in Africa, including *H. impeltatum*, *H. nitidum*, *H. truncatum*, *A. variegatum*, *R. pulchellus*, and *R. appendiculatus*. Existing data suggest that these three-host ticks are involved primarily as enzootic vectors of CCHF virus. However, in Moldavia, CCHF viral transmission to humans was attributed to *Ix. ricinus*, and possibly *D. marginatus*, and *H. a. punctata*. The principal vector, *H. m. marginatum*, in adjacent regions was rarely found in Moldavia. Apparently, this is the only instance during which CCHF virus was transmitted to humans in the absence of one or more *Hyalomma* tick species/subspecies as the predominant ticks parasitizing humans and domestic animals.

Argasid ticks typically feed once as larvae, two to four times as nymphs, and several times as adults.[20] From 5 to 20 hosts may be utilized as a source of blood during the life of these ticks. No particular vector or reservoir role has been postulated for the two species, *A. persicus* and *O. lahorensis*, which have yielded isolations of CCHF virus.

Despite the few tick species/subspecies that have been linked to the transmission of CCHF virus to man, almost all species associated with this virus may feed on humans under certain circumstances.[20] For example, the especially strong zoophilic *Boophilus* ticks have been documented to bite man, apparently after being dislodged from a cow.

F. Vertebrate Host Density and Immunological Background

The general requirement for adequate numbers of vertebrates to support the population of vector tick species is a fundamental requirement for the perpetuation of CCHF virus. However, the relative role of transovarial transmission and horizontal entry of the virus into the tick population has not been adequately studied. If horizontal transmission of the virus between vertebrates and the various stages of the tick is essential to the maintenance of the virus, the density of susceptible vertebrate species is a major factor in determining the prevalence of ticks capable of transmitting virus to susceptible hosts. On the other hand, if vertical transmission of this virus occurs with the same efficiency as demonstrated for experimentally infected familial "lines" of mosquitoes,[160] the role of vertebrates may be secondary in regard to viral amplification, but essential as a source of blood for sustaining the vector population.

The necessity of high-population densities of both susceptible vertebrate species and arthropod vectors has been demonstrated for epidemic conditions with mosquitoborne viruses.[161] The nature of tickborne disease epidemiology differs because ticks are relatively immobile, and enjoy an extended life cycle and longevity of each stage, which is not so ephemeral as that found among the diptera. The virus might survive in areas for fairly extensive periods of time by relying on vertical transmission of the virus among ticks and survival of already infected ticks which are awaiting vertebrate hosts upon which to feed. Hence, with few exceptions, CCHF disease in humans seems to occur irregularly with low numbers of cases. Occasionally, environmental and climatic conditions may favor increased densities of vertebrates and their associated tick species; if humans are unfortunate enough to be undertaking major projects in these areas, larger numbers of cases may occur, as illustrated by the epidemics in the Crimea.

Another characteristic of multiple-host ticks, which relates to their vector potential, is reliance on smaller mammals or ground-dwelling birds for larval and nymphal feeding. These vertebrates have rapid population turnovers, and a major proportion of the population each year will be comprised of immatures and juveniles, i.e., animals with no previous exposure to either the virus or the tick.

The immune status of the vertebrate hosts may play a twofold role in the maintenance and transmission of CCHF virus. First, if horizontal transmission of virus is a critical factor, flourishing populations of young and susceptible animal hosts may amplify the prevalence of virus among the susceptible tick population. Second, immune mechanisms of vertebrates have been shown to reduce the feeding success of attaching ticks;[162] thus, this may hinder the ability of the tick population to expand and to serve effectively as a vector and/or reservoir of CCHF virus.

G. Vector Competence

As stated in Section V.B.1, virtually no quantitative data have been reported on the vector competence of ticks for CCHF virus. However, the biological features of ticks in general are especially suitable and unique among hematophagous arthropods in regard to their potential ability to serve as efficient vectors and reservoirs.[137,138,163-165] Under favorable climatic and ecological conditions, ticks are long-lived, extremely prolific, imbibe large quantities of blood from a wide variety of animal species, and can adapt to many different ecological environments. All stages are parasitic, including both males and females, and the characteristic feeding features, particularly of ixodid ticks, allow for extended periods of attachment to their hosts. This provides a mechanism for dispersal over long distances, and allows for colonization of new areas. More importantly, as potential vectors, the extended period of attachment to a host is likely to increase the probability of ingesting an infective blood meal and of viral transmission to the host. Arboviral infections persist throughout the lives of ticks, and as described in Section V.A, vector potential and efficiency are likely to be enhanced by the different transmission routes, including the possibility of transmission from male to female ticks during copulation.[166]

H. Movements, Migrations of Vectors and Hosts

Various avian species are responsible for both intra- and intercontinental dissemination of ticks associated with CCHF virus.[20] Dispersal of ticks may be restricted to short distances during local postbreeding flights or long distances during migration flights.

Studies of the tick parasites found on birds migrating through Egypt between 1955 and 1973 provided important information on the intercontinental dispersion of ticks.[20] Autumnal surveys of birds migrating from Eurasia to Africa revealed tick species which were characteristic of the fauna of Europe and nearby Asian areas of the Palearctic Faunal Region. More than 90% of immature ticks found on birds migrating south between 1959 and 1981 were species which had been associated with CCHF virus. Spring collections of birds migrating north from subsaharan Africa to Eurasia were parasitized almost exclusively by *H. m. rufipes*.

The movements of domestic animals to new pasture lands, markets, and abattoirs, and the migrations of wild mammals may also contribute to the dissemination of CCHF virus from enzootic foci.[20] Both infected animals and their ticks may serve as a source of CCHF virus in new areas. In Tadzhikistan, apparently infected ticks were introduced into the cooler central region and into the Pamir Mountains with cattle and sheep driven from the south to summer pastures.[30] Numerous CCHF virus-infected ticks collected in Nigeria were taken from domestic animals which had been driven through a variety of ecological zones enroute to the abattoir.[16]

I. Human Element in Disease Ecology

In enzootic foci, humans become infected sporadically with CCHF virus when they inadvertently interrupt ongoing viral circulation between the tick vector and the natural host. Unusual or new human activities can alter the natural vector-host cycle and create unnatural conditions predisposing to an outbreak.[20,167] The enemy occupation of the Crimean Oblast during World War II (1941 to 1944), which disrupted normal agricultural and hare hunting activities, dramatically increased hare and vector densities.[168] This environmental disturbance apparently led to the CCHF epizootic of 1944—1945. Changes in the rural economy of an area and alterations in local agricultural and animal husbandry practices can also increase tick densities or increase human and cattle contact with the tick vector. The 1963 CCHF outbreak in Rostov Oblast was attributed to increased cattle contact with a vector after moving cattle to pastures infested by numerous *H. m. marginatum*.[169]

VII. SURVEILLANCE

A. Clinical Hosts

Since humans are the only vertebrates with remarkable disease consequences, a clinical case diagnostic surveillance system may serve effectively as an indicator of enzootic foci of CCHF virus, especially since disease appears to be severe in a high proportion of infected humans. In areas where the disease has been documented previously, clinical awareness of CCHF cases is likely to be high and a reporting system in place. Since the first confirmed CCHF case in the Republic of South Africa in 1981,[45-47] additional cases have been recognized,[48-50,55] including other countries in Africa.[43,44,55] The assembly of interested physicians, epidemiologists, and microbiologists provides an environment in which the diagnosis, epidemiological impact, and control of CCHF may be facilitated.

B. Wild Vertebrates

The involvement of small, wild vertebrates as hosts for the larval and nymphal stages of ticks incriminated or implicated as vectors of CCHF virus and serological evidence indicate that lagomorphs and insectivores can serve effectively as indicators of CCHF viral activity. The capture of these animals and determination of the species of ticks associated with them may therefore be a reliable indicator of CCHF viral enzootic foci.

C. Vectors

The frequent association of CCHF virus with *Hyalomma* ticks has been emphasized in other sections of this chapter. The capture of ticks and the isolation and characterization of the virus can provide information about the presence of CCHF, the associations of the stages of the tick with vertebrate hosts, and the geographical range of the vector. Methods employed successfully for collecting ticks include carbon dioxide-baited sticky traps and removal from parasitized animals and from materials such as towels, bed sheets, and other fabrics after dragging through tick habitats. However, caution should be exercised in handling, shipping, and isolation attempts as laboratory infections have been documented and the severe/fatal course of the disease dictates the use of maximum precautions and high containment facilities.

D. Sentinels

Serosurveys of domestic animals, particularly livestock, seem to be one of the most economical approaches as a surveillance system for CCHF virus. Livestock species such as sheep, goats, and cattle have all exhibited high antibody prevalence to the virus in areas where human disease has been documented (Table 3). As demonstrated in the Republic of South Africa,[47,48,55] the use of existing livestock disease surveillance and control programs to obtain serum already acquired by livestock authorities is efficient, economical, and probably offers the greatest hope of obtaining epidemiologically significant samples from livestock populations. If wildlife species are being hunted in an environment where wildlife management authorities are involved, blood samples for serological analysis might be obtained during the hunting season.

VIII. INVESTIGATION OF EPIDEMICS

A characteristic epidemiological feature of CCHF is the relatively low incidence and sporadic occurrence of human cases. Primary exposure of non-nosocomial cases is dependent upon the density and host-seeking activity and behavior of infected ticks. Transmission from infected domestic animals and nosocomial transmission are likely to occur and contribute to the overall morbidity, but are also dependent, initially, on infected ticks as the primary source of CCHF virus.

The previous sections of this chapter have emphasized that although CCHF is an important and widespread disease, there are many questions that remain concerning the fundamental ecology of the virus. Because CCHF virus depends upon complex ecological factors for transmission and maintenance, an adequate investigation of a focus which has resulted in human disease will involve a multidisciplinary approach. This approach will necessarily involve entomologists, veterinarians, microbiologists, mammalogists, and epidemiologists, as well as physicians. Special consideration should be given to the dire consequences of human CCHF viral infection and adequate biological containment facilities should be available.

The relative role of the numerous tick species as vectors and reservoirs of CCHF virus is a question which has far-reaching implications in understanding its ecology and epidemiology, and in the ultimate control of the virus during enzootic and epidemic transmission. Therefore, studies should be carried out to determine the vector species and the environmental and biological features that allow for epidemic transmission of CCHF virus. Observations derived from field studies, such as viral isolation data for ticks, should be further investigated by controlled experimental studies to determine the relative importance of different species as potential epidemic vectors.

Investigation of the vertebrate hosts which may contribute to the amplification of CCHF virus should be considered by controlled and field investigations. Determination of the prevalence and distribution of antibody among vertebrate populations will identify potential hosts for experimental studies and ultimately elucidate the role of individual species. The role of CCHF virus in disease of domestic animals also requires more investigation; natural foci of infection may provide this opportunity.

If personnel and resources are available, population-based sampling schemes would be useful in defining the ratio of apparent to inapparent infections, incidence, attack rates, risk factors, virulence, and pathogenicity of the CCHF viral strains circulating in the immediate area of the outbreak. Otherwise, case control studies may also be employed to answer some of these questions.

Patients who have been infected provide an important resource in the validation of newer diagnostic tools. Sera and body fluids from confirmed cases provide means of validating emerging technologies for serology and detection of CCHF virus/viral antigens. Serial collections of serum from patients recovered from CCHF provide useful means of determining the duration of detectable antibody with newer serological techniques and the validity of these tests for determining the prevalence of CCHF viral infection. Collection of these materials should be kept in mind when there are available cases.

Epidemiological studies of the human population should not be limited to a short-term study of the immediate conditions surrounding an "epidemic", but should include an in-depth analysis of the interaction of the vector, the nonhuman vertebrate hosts, the environment, and human intrusion into enzootic foci (Section X).

IX. PREVENTION AND CONTROL

A. Vector Control

The risk of CCHF viral infection of humans in enzootic areas may be reduced by implementing efficient tick control, repellent, interception, and avoidance measures. Control measures may involve environmental modification (see Section IX.C), control of vertebrate hosts (see Section IX.B), and application of acaricides. Commercially available repellents against ticks may be applied to exposed skin or impregnated on clothing. Diethyl toluamide (Deet) has been used for these purposes; however, permethrin appears to be an even more effective repellent for certain tick species.[170] Ticks can be intercepted before biting by carefully inspecting one's clothing and body for crawling ticks. A useful method of preventing tick bite is to wear clothing that prevents ticks from crawling under pant leg cuffs, wristbands, or neckbands of shirts. More information about tick seasonality and host-seeking behavior needs to be accumulated to better determine questing behavior.

Vector control on domestic animals can best be accomplished by direct chemical use. Acaricide treatment of cattle with Sevin during the period of adult attachment was found to be the most efficient control measure for *H. m. marginatum* in Astrakhan Oblast.[171] Application of acaricides only to specific body regions where adults are known to attach can increase treatment efficiency.

Attempts to control ticks on cattle and in pastures by aerial spraying of acaricides (insecticides) were made during epidemics of CCHF in the Rostov Oblast.[172] Results employing DDT dust and chlorophos were inconsistent for reducing tick population, and it was concluded that these measures were not effective for preventing CCHF, regardless of the effect on the population density of tick vectors.

B. Control of Vertebrate Hosts

The density of tick vectors of CCHF may be reduced by controlling the primary vertebrate hosts of the immature ticks.[20] Suppression of rodent populations apparently reduced the numbers of *D. marginatus* in Europe and *H. a. asiaticum* in the Asian deserts and semideserts. *Hyalomma* ticks were reduced in Europe by controlling hares and hedgehogs. Control of birds could also limit the dispersion of tick vectors. *H. m. marginatum* and *H. a. punctata* densities in CCHF foci could be substantially reduced by control of the rook, *C. f. frugilegus*.

C. Environmental Modification

Environmental modification has been shown to be effective in controlling vector population density and CCHF viral activity.[20] Clearing areas around resorts, woodlots, and pathsides of shelters where ticks may survive has been stressed as a useful measure to reduce *H. m. marginatum* population density and the chances of human contact with the ticks. The recommended strategy for controlling *H. a. anatolicum* was the removal of vegetation on hillslopes, floodplain meadows, and abandoned alfalfa fields in Kazakhstan.[173] Environmental modification which increases larval tick exposure to sunlight and to winter temperature causes mortality. Irrigation and plowing were also considered effective in reducing the density of *H. a. anatolicum*.

D. Epidemiological Consideration of the Use of Vaccines

The first consideration regarding the possible use of a vaccine for CCHF is the likely target population. Previous sections have dealt with the enzootic nature of the disease and the risk groups predominantly attacked by the vectors and, therefore, at risk to infection. These include primarily agriculturally related workers who enter the habitat of the local tick vector or who, by virtue of animal husbandry practices, are exposed to either tick-infested or virus-infected animals. Secondarily, there are those health care providers who may be exposed to virus-contaminated body fluids during the course of treatment of CCHF patients. Another target population is laboratory workers who handle clinical specimens or who are engaged in the production of viral materials or who conduct scientific investigations.

The potential target population for any CCHF viral vaccine is probably not large and therefore the impetus to develop a universally acceptable vaccine does not appear warranted. On the other hand, the disease is serious and certain geographic foci have had sufficiently high incidence to make targeted vaccination desirable.

Vaccination of domestic animal populations does not, superficially, seem to be warranted; domestic animals do not suffer serious disease consequences from infection. However, should these animals contribute to the amplification of the virus in the community, animal vaccination may be one strategy for overall reduction of human disease risk.

Serological comparisons of CCHF viral strains have not indicated that antigenic variation would be a major hindrance to the development of effective vaccine (see Section II.C).

A suckling-mouse-brain-produced, killed vaccine has been employed in the Rostov area.[174,175] A total of 1500 persons received the vaccine and were said to show a high frequency of detectable antibody by the N test. A CCHF vaccine was also given to 583 human volunteers in Bulgaria. Antibody was induced in 96.6% (CF test) and 82.1% (AGDP test).[176]

X. FUTURE RESEARCH

A. Field Studies

A priority area for field research is the need to understand dynamics of the enzootic maintenance and transmission cycle of CCHF virus. In most enzootic areas, with the exception of the U.S.S.R., there are virtually no qualitative or quantitative data pertaining to the environment, vertebrate hosts and tick vectors, and their interactions that allow persistence of this virus in enzootic foci. An understanding of these parameters will require longitudinal, in-depth ecological studies to define and quantify climatic and vegetative features of the macro- and microhabitat, and the numerous biological and ecological characteristics of the vector(s) and vertebrate hosts. Unless these parameters are understood, including the recognition and the relative role of each of the components of the basic cycle, the critical information required to design, evaluate, and implement disease surveillance and control strategies is not likely to be available.

Epidemiological studies should be conducted, preferably in parallel with investigations on the enzootic cycle of CCHF virus, within a particular foci. As mentioned in Section VIII stratified serosurveys and monitoring of clinical cases in well-defined human populations and through outpatient and hospital-based sampling will enhance our understanding of the basic epidemiological features of CCHF. Data gained from the latter studies should be analyzed and interpreted in relation to the environmental and biological determinants of the enzootic maintenance and transmission cycle. The ultimate outcome of these and other field investigations should provide a better understanding of the factors that predispose humans to CCHF viral infection. A mathematical model should be considered as a basis for implementing field research on this extremely complex and diverse ecological system.

B. Experimental Studies

An understanding of the ecology and epidemiology of arthropod-borne viral diseases relies in part on data generated by relevant and carefully controlled experimental investigations. Although data are limited, CCHF may prove to be unique in view of the large amount of research that may be required to understand its ecology and epidemiology. The significance of the ever-increasing list of tick species/subspecies as potential virus vectors and reservoirs clearly indicates the need for an immense research effort. The progress of such studies is not likely to be rapid in view of the extended developmental cycle of ticks and the fact that only a few institutions are equipped with the high-level biocontainment facilities required to work with CCHF virus. Nevertheless, it is important that the task be accepted as a challenge, essential to the clarification of the ecology and epidemiology of this virus.

Specific gaps in knowledge related to the role of ticks as vectors/reservoirs were alluded to in this chapter. Briefly, studies should focus on the development of experimental models to evaluate the vector and reservoir competence of ticks for CCHF virus. Assuming that parameters such as the relative threshold level of infection and efficiency of vertical, transstadial, and horizontal transmission can be determined and employed to identify potential vector species, further in-depth studies should be limited to the latter. These should determine the influence of environmental conditions, the immune response of the vertebrate host, and the dynamics and pathways of viral replication on vector and reservoir potential. Biological characteristics of the candidate species, such as longevity and fecundity of CCHF-virus-infected ticks, remain to be defined, as well as the possibility that CCHF virus may be transmitted from infected male to female ticks during copulation.

The relative importance of the numerous species of wild and domestic animals associated with CCHF virus needs to be determined. Susceptibility to infection, the magnitude and duration of viremia, and the immune response need to be defined for each species. However, controlled experimental studies may not be possible for species that cannot be maintained in captivity. In addition, an important aspect related to the use of wild animals is the possibility of prior infection and/or maternally acquired antibody in young animals. This may be avoided for some species by colonization and maintenance of animals in tick-free environments.

While major progress has been made in the development of new virological and immunological techniques, further studies are needed to ascertain the sensitivity and specificity, particularly of seroepidemiological techniques. More recent studies in the Republic of South Africa demonstrated the relative utility of a variety of serological tests. Whether these can be standardized and employed universally requires further study.

Effective treatment strategies for CCHF human viral infection are not available. Inhibition of viral replication should be addressed by evaluating antiviral drugs and interferon or interferon inducers as well as virus-specific immunoglobulins in cell culture and in animal models.

Available data concerning the pathogenetic/pathophysiological mechanism(s) of CCHF have been derived primarily from empirical studies on patients. Attempts to identify a suitable laboratory animal for this purpose have not been successful. Histopathological and pathophysiological observations should be made on patients in conjunction with definitive virus and immunology studies.

Although techniques for conducting molecular virology and biochemistry studies on most arboviruses are available, the extent to which these are applicable to CCHF virus will depend on whether satisfactory yields of virus can be propagated. Such studies are imperative for exploring the possibility of variations among viral strains and for consideration of vaccine development employing genetic engineering and/or other molecular level approaches. Also, virus characterization studies are needed to determine the relationship to other Nairoviruses and the potential for reassortment of genome segments among these viruses. Thus, further investigations are warranted to assess different vertebrate cell lines as well as lines derived from ticks for replicating adequate yields of CCHF virus.

ACKNOWLEDGMENTS

The authors gratefully acknowledge Mrs. Betty Hoffman and Ms. Joan Fry for typing the manuscript, Ms. Kathy Kenyon for editorial comments, and Drs. Charles L. Bailey and James W. LeDuc for their critical review. We also thank Dr. David I. Simpson for providing photographs of CCHF patients and Dr. Robert Swanepoel and and staff for sharing data derived from studies on CCHF in the Republic of South Africa and other African countries.

REFERENCES

1. **Chumakov, M. P.,** A new tick-borne virus disease — Crimean hemorrhagic fever, in *Crimean Hemorrhagic Fever (Acute Infectious Capillary Toxicosis)*, Sokolov, A. A., Chumakov, M. P., and Kolachev, A. A., Eds., Izd. Otd. Primorskoi Armii, Simferopol., Moscow, 1945, 13 (in Russian).
2. **Kolachev, A. A.,** Data on the clinical aspects and treatment of the so-called acute infectious capillary toxicosis, *Voen. Med. Zh.*, 6, 21, 1945 (in Russian).
3. **Chumakov, M. P.,** A new virus disease — Crimean hemorrhagic fever, *Nov. Med.*, 4, 9, 1947 (in Russian; in English, NAMRU3-T900).
4. **Chumakov, M. P.,** On 30 years of investigation of Crimean hemorrhagic fever, *Tr. Inst. Polio. Virusn. Entsefalitov Akad. Med. Nauk SSSR*, 22, 5, 1974 (in Russian; in English, NAMRU3-T950).
5. **Chumakov, M. P.,** Crimean hemorrhagic fever, *Entsikl. Slovar Voenn. Med.*, 3, 268, 1948 (in Russian).
6. **Chumakov, M. P.,** Some results of investigation of the etiology and immunology of Crimean hemorrhagic fever, *Tr. Inst. Polio. Virusn. Entsefalitov Akad. Med. Nauk SSSR*, 19, 7, 1971 (in Russian; in English, NAMRU3-T953).

7. **Chumakov, M. P., Butenko, A. M., Shalunova, N. V., Mary'yanova, L. I., Smirnova, W. E., Bashkirtsev, V. N., Zavodova, T. I., Rubin, S. G., Tkachenko, E. A., Karmy-Sheva, V. Ya., Reingold, V. N., Popov, G. V., and Savinov, A. P.,** New data on the virus causing Crimean hemorrhagic fever (CHF), *Vopr. Virusol.*, 13, 377, 1968 (in Russian; in English, NAMRU3-T596).
8. **Butenko, A. M., Chumakov, M. P., Bashkirtsev, V. N., Zavodova, T. I., Tkachenko, E. A., Rubin, S. G., and Stolbov, D. N.,** Isolation and investigation of Astrakhan strain ("Drosdov") of Crimean hemorrhagic fever virus and data on serodiagnosis of this infection, *Mater. 15. Nauchn. Sess. Inst. Polio. Virus. Entsefalitov* (October 1983), 3, 88, 1968 (in Russian; in English, NAMRU3-T866).
9. **Chumakov, M. P., Belyaeva, A. P., Voroshilova, M. K., Mart'yanova, A. M., Smirnova, S. E., Bashkirtsev, V. N., Zavodova, T. I., Rubin, S. G., Tkachenko, E. A., Karmysheva, V. Ya., Reingold, V. N., Popov, G. V., Kiroy, I., Stolbov, D. N., and Perelatov, V. D.,** Progress in studying the etiology, immunology and laboratory diagnosis of Crimean hemorrhagic fever in the U.S.S.R. and Bulgaria, *Mater. 15. Nauchn. Sess. Inst. Polio. Virus. Entsefalitov* (Moscow, October 1968), 3, 100, 1968 (in Russian; in English, NAMRU3-T613).
10. **Butenko, A. M., Chumakov, M. P., Smirnova, S. E., Vasilenko, S. M., Zavadova, T. I., Tkachenko, E. A., Zarubina, L. V., Bashkirtsev, V. N., Zgurskaya, G. N., and Vyshnivetskaya, L. K.,** Isolation of Crimean hemorrhagic fever virus from blood of patients and corpse material (from 1968 to 1969 investigation data) in Rostov, Astrakhan Oblast and Bulgaria, *Mater. 3. Oblast. Naunchn. Prakt. Konf.* (Rostov-On-Don, May 1970), p. 6, 1970 (in Russian; in English, NAMRU3-T522).
11. **Casals, J.,** Antigenic similarity between the virus causing Crimean hemorrhagic fever and Congo virus, *Proc. Soc. Exp. Biol. Med.*, 131, 233, 1969.
12. **Chumakov, M. P., Smirnova, S. E., and Tkachenko, E. A.,** Antigenic relationships between the Soviet strains of Crimean hemorrhagic fever virus and the Afro-Asian Congo virus strains, *Mater. 16. Nauchn. Sess. Inst. Polio. Virus. Entsefalitov* (Moscow, October 1969), 2, 152, 1969 (in Russian; in English, NAMRU3-T614).
13. **Simpson, D. I. H., Knight, E. M., Courtois, Gh., Williams, M. C., Weinbren, M. P., and Kibukamusoke, J. W.,** Congo virus: a hitherto undescribed virus occurring in Africa. I. Human isolations — clinical notes, *East Afr. Med. J.*, 44, 87, 1967.
14. **Woodall, J. P., Williams, M. C., Simpson, D. I. H., Ardoin, P., Lule, M., and West, R.,** The Congo group of agents, *Rep. East Afr. Virus Res. Inst.*, 14, 34, 1965.
15. **Woodall, J. P., Williams, M. C., and Simpson, D. I. H.,** Congo virus: a hitherto undescribed virus occurring in Africa. II. Identification studies, *East Afr. Med. J.*, 44, 93, 1967.
16. **Causey, O. R., Kemp, G. E., Madbouly, M. H., and David-West, T. S.,** Congo virus from domestic livestock, African hedgehogs, and arthropods in Nigeria, *Am. J. Trop. Med. Hyg.*, 19, 846, 1970.
17. **Kemp, G. E., Causey, O. R., Setzer, H. W., and Moore, D. L.,** Isolation of viruses from wild mammals in West Africa, 1966—1970, *J. Wildl. Dis.*, 10, 279, 1974.
18. **Begum, F., Wisseman, C. L., Jr., and Casals, J.,** Tick-borne viruses of West Pakistan. IV. Viruses similar to or identical with Crimean hemorrhagic fever (Congo-Semunya). Wad Medani and Pak *Argas* 461 isolated from ticks of the Changa Manga Forest, Lahore District, and Hunza, Gilgit Agency, West Pakistan, *Am. J. Epidemiol.*, 92, 197, 1970.
19. **Casals, J., Henderson, B. F., Hoogstraal, H., Johnson, K. M., and Shelokov, A.,** A review of Soviet viral hemorrhagic fevers, 1969, *J. Infect. Dis.*, 122, 437, 1970.
20. **Hoogstraal, H.,** The epidemiology of tick-borne Crimean-Congo hemorrhagic fever in Asia, Europe, and Africa, *J. Med. Entomol.*, 15, 307, 1979.
21. **Chumakov, M. P., Birulya, N. B., Butenko, A. M., Vasyuta, Yu. S., Egorova, P. S., Zalutskaya, L. I., Zimina, Yu. V., Leshchinskaya, E. V., Povalishina, T. P., and Stolbov, D. N.,** On the question of epidemiology of diseases of Crimean hemorrhagic fever in Astrakhan *Oblast, Mater. 11. Nauchn. Sess. Inst. Polio. Virus. Entsefalitov.*, p. 263, 1964 (in Russian; in English, NAMRU3-T165).
22. **Donchev, D., Kebedzhiev, G., and Rusakiev, M.,** Hemorrhagic fever in Bulgaria, *Bulg. Akad. Nauk Microbiol. Inst., 1. Kongr. Mikrobiol.* (1965), p. 777, 1967 (in Russian; in English, NAMRU3-T465).
23. **Vasilenko, S. M., Katsavov, G., Kirov, I., Radev, M., and Arnaudov, G.,** Etiological diagnosis of Crimean hemorrhagic fever in Bulgaria, *Tezisy 17. Nauchn. Sess. Inst. Posvyashch. Aktual. Probl. Virus. Profilakt. Virus. Zabolev.* (Moscow, October 1972), p. 337, 1972 (in Russian; in English, NAMRU3-T1049).
24. **Badalov, M. E., Koimchidi, E. K., Semenov, M. Ya., and Karinskaya, G. A.,** Crimean hemorrhagic fever in Rostov Region, *Tr. Inst. Polio. Virusn. Entsefalitov Akad. Med. Nauk SSSR*, 19, 167, 1971 (in Russian; in English, NAMRU3-T923).
25. **Perelatov, V. D. and Vostokova, K. K.,** Epidemiology of Crimean hemorrhagic fever in Rostov Region, *Tr. Inst. Polio. Virusn. Entsefalitov Akad. Med. Nauk SSSR*, 19, 174, 1971 (in Russian; in English, NAMRU3-T924).
26. **Meliev, A.,** A contribution to epidemiology of hemorrhagic fever in Uzbekistan, *Zh. Mikrobiol. Epidemiol. Immunobiol.*, 44, 93, 1967 (in Russian; in English, NAMRU3-T413).

27. **Meliev, A.,** Data on investigating hemorrhagic fever in Uzbek SSR (Avtoref. Diss. Soisk. Uchen. Step. Kand. Med. Nauk.), *Inst. Epidem. Mikrobiol. imeni N. F. Gamaleya. Akad. Med. Nauk. SSSR.* Moscow, p. 18, 1967 (in Russian; in English, NAMRU3-T1179).
28. **Dobritsa, P. G.,** Epidemiology and prophylaxis of hemorrhagic fever in Chimkent Region of the southern Kazakhstan, *Tr. Inst. Polio. Virusn. Entsefalitov Akad. Med. Nauk SSSR,* 7, 262, 1965 (in Russian; in English, NAMRU3-T196).
29. **Temirbekov, Zh. T., Dobritsa, P. G., Kontaruk, V. M., Vainshtein, E. K., Marushchak, O. N., Dobritsa, M. A., and Shvets, M. Ya.,** Investigation of Crimean hemorrhagic fever in Chimkent Region of the Kazakh SSR, *Tr. Inst. Polio. Virusn. Entsefalitov Akad. Med. Nauk SSSR,* 19, 160, 1971 (in Russian; in English, NAMRU3-T949).
30. **Pak, T. P. and Mikhailova, L. I.,** *Crimean Hemorrhagic Fever in Tadzhikistan,* Izd. "Irfron," Dushanbe 1973 (in Russian; in English, NAMRU3-T1000).
31. **Yu-Chen, Y., Ling-Xiong, K., Ling, L., Yu-Qin, Z., Feng, L., Bao-Jian, C., and Shou-Yi, G.,** Characteristics of Crimean-Congo hemorrhagic fever virus (Xinjiang strain) in China, *Am. J. Trop. Med. Hyg.,* 34, 1179, 1985.
32. **Heneberg, D., Heneberg, N., Celina, D., Filipovic, D., Markovic, Z., Zubi, D., Zivcovic, B., Simic, M., Zonjic, S., and Pantelic, M.,** Crimean hemorrhagic fever in Yugoslavia, *Vojnosanit. Pregl.,* 25, 181, 1968.
33. **Obradovic, M. and Gligic, A.,** Specific antibodies in the sera of patients formerly affected by Crimean-Congo hemorrhagic fever, *Zbl. Bakt., Abt. I.,* 9, 267, 1980.
34. **Burney, M. I., Ghafoor, A., Saleen, M., Webb, P. A., and Casals, J.,** Nosocomial outbreak of viral hemorrhagic fever caused by Crimean hemorrhagic fever — Congo virus in Pakistan, January 1976, *Am. J. Trop. Med. Hyg.,* 29, 941, 1980.
35. **Hayes, C. C. and Burney, M. I.,** Arboviruses of public health importance in Pakistan, *J. Pak. Med. Assoc.,* 31, 16, 1981.
36. **Suleiman, M. N. E. H., Muscat-Baron, J. M., Harries, J. R., Satti, A. G. O., Platt, G. S., Bowen, E. T. W., and Simpson, D. I. H.,** Congo/Crimean haemorrhagic fever in Dubai. An outbreak at the Rashid Hospital, *Lancet,* 2, 939, 1980.
37. **Baskerville, A., Satti, A. G. O., Murphy, F. A., and Simpson, D. I. H.,** Congo-Crimean haemorrhagic fever in Dubai: histopathological studies, *J. Clin. Pathol.,* 34, 871, 1981.
38. **Casals, J.,** Report on a Visit to Iraq to Investigate a Viral Haemorrhagic Fever of Man Caused by Crimean Haemorrhagic Fever-Congo Virus, World Health Organization, Geneva, 1981.
39. **Al-Tikriti, S. K., Al-Ani, F., Jurji, F. J., Tantawi, H., Al-Moslih, M., Al-Janabi, N., Mahmud, M. I. A., Al-Bana, A., Habib, H., Al-Munthri, H., Al-Janabi, Sh., Al-Jawahry, K., Yonan, M., Hassan, F., and Simpson, D. I. H.,** Congo/Crimean haemorrhagic fever in Iraq, *Bull. WHO,* 59, 85, 1981.
40. **Tantawi, H. H., Al-Moslih, M. I., Hassan, F. K., and Al-Ani, F. S.,** *Crimean-Congo Haemorrhagic Fever,* Vol. 1, Al-Muthanna House for Printing and Publishing, Baghdad, 1980.
41. **Tantawi, H. H., Al-Moslih, M. I., Al-Janabi, N. Y., Al-Bana, A. S., Mahmud, M. I. A., Jurji, F., Yonan, M. S., Al-Ani, F., and Al-Tikriti, S. K.,** Crimean-Congo haemorrhagic fever virus in Iraq; isolation, identification and electron microscopy (short communication), *Acta Virol. (Prague),* 24, 464, 1980.
42. **Kalunda, M., Lule, M., Sekyalo, E., Mukuye, A., and Mujomba, E.,** Virus isolation and identification, *Rep. East Afr. Virus Res. Inst.,* 27, 17, 1983.
43. **Saluzzo, J. F., Aubry, P., McCormick, J., and Digoutte, J. P.,** Haemorrhagic fever caused by Crimean-Congo haemorrhagic fever virus in Mauritania, *Trans. R. Soc. Trop. Med. Hyg.,* 79, 268, 1985.
44. **Saluzzo, J. F., Digoutte, J. P., Cornet, M., Baudon, D., Roux, J., and Robert, V.,** Isolation of Crimean-Congo haemorrhagic fever and Rift Valley fever viruses in Upper Volta, *Lancet,* 1, 1179, 1984.
45. **Gear, J. H. S.,** The hemorrhagic fevers of southern African with special references to studies in the South African Institute for Medical Research, *Yale J. Biol. Med.,* 55, 207, 1982.
46. **Gear, J. H. S., Thomson, P. D., Hopp, M., Andronikou, S., Cohn, R. J., Ledger, J., and Berkowitz, F. E.,** Congo-Crimean haemorrhagic fever in South Africa. Report of a fatal case in the Transvaal, *S. Afr. Med. J.,* 62, 576, 1982.
47. **Swanepoel, R., Struthers, J. K., Shepherd, A. J., McGillivray, G. M., Nel, M. J., and Jupp, P. G.,** Crimean-Congo hemorrhagic fever in South Africa, *Am. J. Trop. Med. Hyg.,* 32, 1407, 1983.
48. **Swanepoel, R., Shepherd, A. J., Leman, P. A., and Shepherd, S. P.,** Investigations following initial recognition of Crimean-Congo haemorrhagic fever in South Africa and the diagnosis of 2 further cases, *S. Afr. Med. J.,* 68, 638, 1985.
49. **Swanepoel, R., Shepherd, A. J., Leman, P. A., Shepherd, S. P., and Miller, G. B.,** A common-source outbreak of Crimean-Congo haemorrhagic fever on a dairy farm, *S. Afr. Med. J.,* 68, 635, 1985.

50. **Van Eeden, P. J., Joubert, J. R., Van De Wal, B. W., King, J. B., DeKock, A., and Groenewald, J. H.,** A nosocomial outbreak of Crimean-Congo haemorrhagic fever at Tygerberg Hospital. I. Clinical features, *S. Afr. Med. J.*, 68, 711, 1985.
51. **Van Eeden, P. J., Van Eeden, S. F., Joubert, J. R., King, J. B., Van De Wal, B. W., and Michell, W. L.,** A nosocomial outbreak of Crimean-Congo haemorrhagic fever at Tygerberg Hospital. II. Management of patients, *S. Afr. Med. J.*, 68, 718, 1985.
52. **Joubert, J. R., King, J. B., Rossouw, D. J., and Cooper, R.,** A nosocomial outbreak of Crimean-Congo haemorrhagic fever at Tygerberg Hospital. III. Clinical pathology and pathogenesis, *S. Afr. Med. J.*, 68, 722, 1985.
53. **Van De Wal, B. W., Joubert, J. R., Van Eeden, P. J., and King, J. B.,** A nosocomial outbreak of Crimean-Congo haemorrhagic fever at Tygerberg Hospital. IV. Preventive and prophylactic measures, *S. Afr. Med. J.*, 68, 729, 1985.
54. **Shepherd, A. J., Swanepoel, R., Shepherd, S. P., Leman, P. A., Blackburn, N. K., and Hallett, A. F.,** A nosocomial outbreak of Crimean-Congo haemorrhagic fever at Tygerberg Hospital. V. Virological and serological observations, *S. Afr. Med. J.*, 68, 733, 1985.
55. **Swanepoel, R., Shepherd, A. J., LeMan, P. A., Shepherd, S. P., McGillivray, G. M., Erasmus, M. J., Searle, L. A., and Gill, D. E.,** Epidemiological and clinical features of Crimean-Congo hemorrhagic fever in southern Africa, *Am. J. Trop. Med. Hyg.*, in press.
56. **Goldfarb, L. G., Chumakov, M. P., Myskin, A. A., Kondratenko, V. F., and Reznikov, O. Yu,** An epidemiological model of Crimean hemorrhagic fever, *Am. J. Trop. Med. Hyg.*, 29, 260, 1980.
57. **Casals, J. and Tignor, G. H.,** The *Nairovirus* genus: serological relationships, *Intervirology*, 14, 144, 1980.
58. **Mathews, R. E. F.,** Classification and nomenclature of viruses, *Intervirology*, 17, 115, 1982.
59. **Clerx, J. P. M., Casals, J., and Bishop, D. H. L.,** Structural characteristics of Nairoviruses (Genus *Nairovirus*, Bunyaviridae), *J. Gen. Virol.*, 55, 165, 1981.
60. **Martin, M. L., Lindsey-Regnery, H., Sasso, D. R., McCormick, J. B., and Palmer, E.,** Distinction between Bunyaviridae genera by surface structure and comparison with Hantaan virus using negative strain electron microscopy, *Arch. Virol.*, 86, 17, 1985.
61. **Karabatsos, N.,** *International Catalogue of Arboviruses, Including Certain Other Viruses of Vertebrates*, 3rd ed., American Committee on Arthropod-borne Viruses, American Society of Tropical Medicine and Hygiene, Washington, D.C., 1985.
62. **Robin, Y.,** Centre regional O.M.S. de reference pour les arbovirus in Afrique de'Ouest, *Rapp. Inst. Pasteur (Dakar)*, 1973.
63. **Chumakov. M. P., Andreeva, S. K., Zavodova, T. I., Zgurskaya, G. N., and Kostetsky, N. V., Mart'yanova, L. I., Nitkitin, A. M., Sinyak, K. M., Smirnova, S. E., Turta, L. I., Ustinova, E. D., and Chunikhin, S. P.,** Problems of Crimean hemorrhagic fever virus ecology in natural foci of this infection in the Crimea, *Tr. Inst. Polio. Virusn. Entsefalitov Akad. Med. Nauk SSSR*, 22, 19, 1974 (in Russian; in English, NAMRU3-T1110).
64. **Digoutte, J. P. and Heme, G.,** Rapport annuel 1984, *Rapp. Inst. Pasteur (Dakar)*, 1, 1985.
65. **Ermakov, N. M. and Berdyev, A.,** Data on natural focality of diseases in Turkmenia, *Tezisy Dokl. 5. Vses. Akarol. Sovesch.* (Fruze, May 1985), 121, 1985 (in Russian; in English, NAMRU3-T1816).
66. **Darwish, M. A., Hoogstraal, H., Roberts, T. J., Ghazi, R., and Amer, T.,** A seroepidemiological survey for Bunyaviridae and certain other arboviruses in Pakistan, *Trans. R. Soc. Trop. Med. Hyg.*, 77, 446, 1983.
67. **Darwish, M. A. and Hoogstraal, H.,** Arboviruses infecting humans and lower animals in Egypt: a review of thirty years of research, *J. Egypt. Public Health Assoc.*, 56, 112, 1981.
68. **Saluzzo, J. F., Digoutte, J. P., Camicas, J. L., and Chauvancy, G.,** Crimean-Congo haemorrhagic fever and Rift Valley fever in south-eastern Mauritania, *Lancet*, 1, 116, 1985.
69. **Johnson, B. K., Ocheng, D., Gitau, L. G., Gichogo, A., Tukei, P. M., Ngindu, A., Langatt, A., Smith, D. H., Johnson, K. M., Kiley, M. P., Swanepoel, R., and Isaacson, M.,** Viral haemorrhagic fever surveillance in Kenya, 1980—1981, *Trop. Geogr. Med.*, 35, 43, 1983.
70. **David-West, T. S., Cooke, A. R., and David-West, A. S.,** Seroepidemiology of Congo virus (related to the virus of Crimean haemorrhagic fever) in Nigeria. Brief communications, *Bull. WHO*, 51, 543, 1974.
71. **Umoh, J. U., Ezeokoli, C. D., and Ogwu, D.,** Prevalence of antibodies to Crimean-haemorrhagic fever-Congo virus in cattle in northern Nigeria, *Int. J. Zoonoses*, 10, 151, 1983.
72. **Butenko, A. M. and Minja, T.,** Serological study of geographical distribution of CHF-Congo virus and some other tick-borne viruses in East Africa, *Public Health Med. Sci.*, 17, 1979.
73. **Gonzalez, J.-P., Baudon, D., and McCormick, J. B., Premieres etudes serologiques dans les populations humaines de Haute-Volta et du Benin sur les fievres hemorragiques Africaines d'origine virale,** *OCCGE Inf.*, 12, 113, 1984.
74. **Blackburn, N. K., Searle, L., and Taylor, P.,** Viral haemorrhagic fever antibodies in Zimbabwe schoolchildren, *Trans. R. Soc. Trop. Med. Hyg.*, 76, 803, 1982.

75. **Antoniadis, A. and Casals, J.,** Serological evidence of human infection with Congo-Crimean hemorrhagic fever virus in Greece (brief communication), *Am. J. Trop. Med. Hyg.*, 31, 1066, 1982.
76. **Filipe, A. R., Calisher, C. H., and Lazuick, J.,** Antibodies to Congo-Crimean haemorrhagic fever; Dhori, Thogoto and Bhanja viruses in Southern Portugal, *Acta Virol.*, 29, 324, 1985.
77. **Al Tikriti, S. K., Hassan, F. K., Moslih, I. M., Jurji, F., Mahmud, M. I. A., and Tantawi, H. H.,** Congo/Crimean haemorrhagic fever in Iraq: a seroepidemiological survey, *J. Trop. Med. Hyg.*, 84, 117, 1981.
78. **Al-Nakib, W., Lloyd, G., El-Mekki, A., Platt, G., Beeson, A., and Southee, T.,** Preliminary report on arbovirus-antibody prevalence among patients in Kuwait: evidence of Congo/Crimean virus infection, *Trans. R. Soc. Trop. Med. Hyg.*, 78, 474, 1984.
79. **Tignor, G. H., Smith, A. L., Casals, J., Ezeokoli, C. D., and Okoli, J.,** Close relationship of Crimean hemorrhagic fever-Congo (CHF-C) virus strains by neutralizing antibody assays, *Am. J. Trop. Med. Hyg.*, 29, 676, 1980.
80. **Kuchin, V. V., Karinskaya, G. A., and Badalov, N. E.,** Antigenic interrelationships of Crimean hemorrhagic fever virus strains isolated in different years, *Misc. Publ. Entomol. Soc. Am.*, 9, 139, 1974.
81. **Chumakov, M. P., Shalunova, N. V., Semashko, I. V., and Belyaeva, A. P.,** Use of interference phenomenon in tissue culture for detecting Crimean hemorrhagic fever virus (CHF), *Tr. Inst. Polio. Virusn. Entsefalitov Akad. Med. Nauk SSSR*, 7, 202, 1965 (in Russian; in English, NAMRU3-T832).
82. **Zgurskaya, G. N., Popov, G. V., Beregin, V. V., Smirnova, S. E., and Chumakov, M. P.,** Application of fluorescent antibody method (FAM) in detecting CHF virus in tick vectors, *Tezisy Dokl. Vop. Med. Virus., Inst. Virus. Imeni Ivanovsky, D. I. Akad. Med. Nauk SSSR* (19—21 October), 2, 135, 1971 (in Russian; in English, NAMRU3-T509).
83. **Zgurskaya, G. N., Smirnova, S. E., and Chumakov, M. P.,** Immunofluorescent antibody technique (FAT) application to detect Crimean hemorrhagic fever (CHF) in naturally infected ticks, *Tezisy 17. Nauchn. Sess. Inst. Posvyashch. Aktual. Probl. Virus. Profilakt virus Zabolev.* (Moscow, October 1972), p. 362, 1972 (in Russian; in Engish NAMRU3-T1069).
84. **Donets, M. A., Rezapkin, G. V., Ivanov, A. P., and Tkachenko, E. A.,** Immunosorbent assays for diagnosis of Crimean-Congo haemorrhagic fever (CCHF), *Am. J. Trop. Med. Hyg.*, 31, 156, 1982.
85. **Leshchinskaya, E. V.,** Clinical picture of Crimean hemorrhagic fever and its comparison with hemorrhagic fevers of other types *(Avtoref. Diss. Soisk, Uchen. Step. Dokt. Med. Nauk) Akademiya Meditsinskikh Nauk SSR*, Moscow, 1967 (in Russian; in English, NAMRU3-T1180).
86. **Zarubinsky, V. Ya., Kondratenko, V. V., Blagoveshchenskaya, N. M., Zarubina, L. V., and Kuchin, V. V.,** Susceptibility of calves and lambs to Crimean hemorrhagic fever virus, *Tezisy Dokl. 9 Vses. Konf. Prirod. Ochag. Bolez. Chelov. Zhivot* (Omsk, May 1976), p. 130, 1976 (in Russian; in English, NAMRU3-T1178).
87. **Milyutin, V. N., Butenko, A. M., Artyushenko, A. A., Bliznickenko, A. G., Zavodova, T. I., Zarubina, L. V., Novikova, E. M., Rubin, S. G., Chernyshev, N. I., and Chumakov, M. P.,** Experimental infection of horses with Crimean hemorrhagic fever virus. Report 1: clinical observations, *Mater. 16. Nauchn. Sess. Inst. Polio. Virus. Entsefalitov* (Moscow, October 1969), 2, 145, 1969 (in Russian; in English, NAMRU3-5851).
88. **Blagoveshchenskaya, N. M., Butenko, A. M., Vyshnivetskaya, L. K., Zavodova, T. I., Zarubina, L. V., Karinskaya, G. A., Kuchin, V. V., Milyutin, V. N., Novikova, E. M., Rubin, S. G., and Chumakov, M. P.,** Experimental infection of horses with Crimean hemorrhagic fever virus. Report 2: virological and serological observations, *Mater. 16 Nauchn. Sess. Inst. Polio. Virus. Entsefalitov* (Moxcow, October 1969), 2, 126, 1969 (in Russian; in English, NAMRU3-T840).
89. **Blagoveshchenskaya, N. M., Donets, M. A., Zarubina, L. V., Kondratenko, V. F., and Kuchin, V. V.,** Study of susceptibility to Crimean hemorrhagic fever (CHF) virus in European and long-eared hedgehogs, *Tezisy Konf. Vop. Med. Virus.* (Moscow, October 1975), p. 269, 1975 (in Russian; in English, NAMRU3-T985).
90. **Zgurskaya, G. N., Berezin, S. E., Smirnova, S. E., and Chumakov, M. P.,** Investigation of the question of Crimean hemorrhagic fever virus transmission and interepidemic survival in the tick *Hyalomma plumbeum plumbeum*, Panzer, *Tr. Inst. Polio. Virusn. Entsefalitov Akad. Med. Nauk SSSR*, 19, 217, 1971 (in Russian); in English, NAMRU3-T911).
91. **Zgurskaya, G. N., Berezin, V. V., and Smirnova, S. E.,** Threshold levels of blood infectiousness for *Hyalomma p. plumbeum* tick during viremia in hares and rabbits caused by CHF virus, *Tezisy Konf. Vop. Med. Virus.* (Moscow, October 1975), p. 291, 1975 (in Russian; in English, NAMRU3-T997).
92. **Kondratenko, V. F.,** Importance of ixodid ticks in transmission and preservation of Crimean hemorrhagic feveragent in infection foci, *Parazitologiya*, 10, 297, 1976 (in Russian; in English, NAMRU3-T1116).
93. **Gaidamovich, S. Ya., Klisenko, G. A., Shanoyan, N. K., Obukhova, V. R., and Mel'nikova, E. E.,** Indirect hemagglutination for diagnosis of Crimean hemorrhagic fever, *Intervirology*, 2, 181, 1974.

94. **Zgurskaya, G. N., Chumakov, M. P., and Smirnova, S. E.,** Titration of antibodies to CHF virus in drops of cell suspensions from infected tissue cultures by the indirect immuno-fluorescence method, *Tezisy Konf, Vop. Med. Virus* (Moscow, October 1975), 293, 1975 (in Russian; in English, NAMRU3-T998).
95. **Lapina, T. F. and Gaidamovich, S. Ya.,** The indirect hemagglutination test for dianosing West Nile fever and Crimean hemorrhagic fever, *Sborn. Trud. Inst. Virus. Imeni D. I. Ivanovsky, Akad. Med. Nauk SSSR,* p. 132, 1981 (in Russian; in English, NAMRU3-T1575).
96. **Johnson, K. M., Elliott, L. H., and Heymann, D. L.,** Preparation of polyvalent viral immunofluorescent intracellular antigens and use in human serosurveys, *J. Clin. Microbiol.,* 14, 527, 1981.
97. **Swanepoel, R., Struthers, J. K., and McGillivray, G. M.,** Reversed passive hemagglutination and inhibition with Rift Valley fever and Crimean-Congo hemorrhagic fever viruses, *Am. J. Trop. Med. Hyg.,* 32, 610, 1983.
98. **Klisenko, G. A., Gaidamovich, S. Ya., Zarubinsky, V. Ya., Lapina, T. F., and Meliev, A. M.,** Rapid diagnosis of Crimean hemorrhagic fever by the indirect hemagglutination test, *Vopr. Virusol.,* 29, 566, 1984.
99. **Nikolaev, V. P. and Tsar'kova, V. A.,** Rapid solid-phase enzyme immunoassay for detection of viral antigens, *Vopr. Virusol.,* 29, 720, 1984.
100. **Tsar'kova, V. A. and Nikolaev, V. P.,** Detection of antigens to Crimean-Congo hemorrhagic fever virus by solid-phase enzyme immunoassay and indirect hemagglutination test, *Vopr. Virusol.,* 29, 724, 1984 (in Russian, with English summary).
101. **Smirnova, S. E. and Karavanov, A. S.,** Detection of Crimean haemorrhagic fever virus antigen by solid phase enzyme immunosorbent assay, *Acta Virol.,* 29, 87, 1985.
102. **Sureau, P., Ed.,** Rapport sur le fonctionnement technique de l'Institut Pasteur de Bangui, *Ann. 1974, Bangui,* 1974.
103. **Robin, Y.,** Centre collaborateur O.M.S. de reference et de recherche pour les arbovirus, *Rapp. Inst. Pasteur (Dakar),* 1975.
104. **Robin, Y.,** Centre collaborateur O.M.S. de reference et de recherche pour les arbovirus, *Rapp. Inst. Pasteur (Dakar),* 1977.
105. **Saluzzo, J. F., Gonzalez, J. P., Herve, J. P., and Georges, A. J.,** Enquete serologique sur las prevalence de certains arbovirus dans la population humaine du sud-est de la Republique Centrafricaine en 1979, *Bull. Soc. Pathol. Exot.,* 74, 490, 1981.
106. **Wood, O. L., Lee, V. H., Ash, J. S., and Casals, J.,** Crimean-Congo hemorrhagic fever, Thogoto, Dugbe, and Jos viruses isolated from *Ixodid* ticks in Ethiopia, *Am. J. Trop. Med. Hyg.,* 27, 600, 1978.
107. **Robin, Y. and Le Gonidec, G.,** Activities du laboratorie des arborvirus, *Rapp. Fonct. Tech. Inst. Pasteur (Dakar),* p. 37, 1972.
108. **Robin, Y.,** Centre regional O.M.S. de reference pour les arbovirus en Afrique de l'Ouest, *Rapp. Inst. Pasteur (Dakar),* 1972.
109. **Robin, Y.,** Centre regional O.M.S., de reference pour les arbovirus en Afrique de l'Ouest, *Rapp. Inst. Pasteur (Dakar),* 1974.
110. **Kirya, B. G. and Lule, M.,** Congo virus (AMP10358), *Rep. East Afr. Virus Res. Inst.,* 20, 18, 1971.
111. **Pak, T. P. and Pashkov, V. A.,** Criteria for epidemiological assessment of a locality for Crimean hemorrhagic fever, *Sb. Tr. Ekol. Virusol.,* 2, 129, 1974 (in Russian; in English, NAMRU3-T782).
112. **Papadopoulos, O. and Koptopoulos, G.,** Isolation of Crimean-Congo hemorrhagic fever (CCHF) virus from *Rhipicephalus bursa* ticks in Greece, *Acta Microbiol. Hell.,* 23, 20, 1978 (in Greek, with English summary).
113. **Molar, E.,** Occurrence of tick-borne encephalitis and other arboviruses in Hungary, *Geogr. Med (Budapest),* 12, 78, 1982.
114. **Gligic, A., Stamatovic, L., Stojanovic, R., Obradovic, M., and Boskovic, R.,** The first isolation of Crimean haemorrhagic fever virus in Yugoslavia, *Vojnosanit. Pregl.,* 34, 318, 1977 (in Serbo-Croatian; in English, MRIID-644).
115. **Obradovic, M., Gligic, A., Stojanovic, R., Stamatovic, L., and Boskovic, R.,** A serological and arachnoentomological investigation of natural foci of Crimean haemorrhagic fever in certain localities in Yugoslavia, *Vojnosanit. Pregl.,* 35, 253, 1978 (in Serbo-Croatian; in English, MRIID 642).
116. **Voltsit, O. V.,** Review of arboviruses isolated from ixodid ticks in Afghanistan, Pakistan and India, *Sb. Nauch. Tr. Inst. Virusol. D.I. Ivanovsky Akad. Med. Nauk SSSR,* p. 111, 1982 (in Russian; in English, NAMRU3-T1659).
117. **Sureau, P. and Klein, J. M.,** Arbovirus en Iran, *Med. Trop. (Marseilles),* 40, 549, 1980.
118. **Sureau, P., Klein, J. M., Casals, J., Digoutte, J. P., Salaun, J. J., Piazak, N., and Calvo, M. A.,** Isolement des virus Thogoto, Wad Medani, Wanowrie et de la fievre hemorragique de Crimee-Congo en Iran, a partir de tiques d'animaux domestiques, *Ann. Inst. Pasteur Paris Virol.,* 131E, 185, 1980.
119. **Serter, D.,** Present status of arbovirus seroepidemiology in the Aegean region of Turkey, *Zbl. Bakt., Abt. I.,* 9, 155, 1980.

120. **Pak, T. P.**, Etiological structure of arbovirus infections in Tadzhik SSR, *Sborn. Nauch. Trud. Inst. Virus. D. I. Ivanovsky, Akad. Med. Nauk SSSR*, p. 127, 1982, (in Russian, with English summary).
121. **Pak, T. P.**, Clinical-epidemiological characteristics of arbovirus infections in Tadzhikistan, *Sb. Tr. Inst. Virusol. (D. I. Ivanovsky Akad. Med. Nauk SSSR*, p. 101, 1981 (in Russian, with English summary).
122. **Skvortsova, T. M., Gromashevsky, V. L., Sidorova, G. A., Khutoretskaya, N. V., Aristova, V. A., Kondrashina, N. G., Polyakova, A. N., Muradov, Sh. M., Belousov, E. M., and Kurchenko, F. P.**, Results of virological investigation of arthropod vectors in the territory of Turkmenia, *Sb. Nauchn. Tr. Inst. Virusol. D. I. Ivanovsky Akad. Med. Nauk SSSR*, p. 39, 1982 (in Russian; in English, NAMRU3-T1664).
123. **Rechav, Y.**, Seasonal activity and hosts of the vectors of Crimean-Congo haemorrhagic fever in South Africa, *S. Afr. Med. J.*, 69, 364, 1986.
124. **Fabiyi, A.**, Congo virus in Nigeria: isolation and pathogenetic studies, *9th Int. Congr. Trop. Med. Malaria (Athens)*, 1, 44, 1976.
125. **Casals, J.**, Crimean-Congo hemorrhagic fever, in *Ebola Virus Haemorrhagic Fever*, Pattyn, S. R., Ed., Elsevier, New York, 1978, 301.
126. **Karinskaya, G. A., Chumakov, M. P., Butenko, A. M., Badalov, M. E., and Rubin, S. G.**, Investigation of blood samples from animals in Rostov Oblast for antibodies to Crimean hemorrhagic fever virus, *Mater. 3. Oblast. Nauchn. Prakt. Konf.* (Rostov-On-Don, May 1970), p. 55, 1970 (in Russian; in English, NAMRU3-T530).
127. **Blagoveshchenskaya, N. M., Butenko, A. M., Vyshnivetskaya, L. K., Zarubina, L. V., Kuchin, V. V., Milyutin, V. N., Novikova, E. M., and Chumakov, M. P.**, Dynamics of antibodies to Crimean hemorrhagic fever virus in hyperimmunized horses, *Mater. 3 Oblast. Nauchn. Prakt. Konf.* (Rostov-On-Don, May 1970), p. 50, 1970 (in Russian; in English, NAMRU3-T529).
128. **Kuchin, V. V., Yanovich, T. D., Butenko, A. M., and Kirsanova, K. S.**, Serological examination for antibodies to Crimean hemorrhagic fever virus in domestic animals of Rostov Oblast, *Mater. 3. Oblast. Nauchn. Prakt. Konf.* (Rostov-On-Don, May 1970), p. 61, 1970 (in Russian; in English, NAMRU3-T531).
129. **Saidi, S., Casals, J., and Faghih, M. A.**, Crimean hemorrhagic fever-Congo (CHF-C) virus antibodies in man, and in domestic and small mammals, in Iran, *Am. J. Trop. Med. Hyg.*, 24, 353, 1975.
130. **Watts, D. M., Hasty, S. E., Smith, J. F., and Peters, C. J.**, unpublished observations, 1986.
131. **Moore, D. L., Causey, O. R., Carey, D. E., Reddy, S., Cooke, A. R., Akinkugbe, F. M., David-West, T. S., and Kemp, G. E.**, Arthropod-borne viral infection of man in Nigeria, 1964 to 1970, *Ann. Trop. Med. Parasitol.*, 69, 49, 1975.
132. **Durojaiye, O. A.**, Viral zoonoses in Nigeria. II. Non-rabies zoonoses, *Int. J. Zoonoses*, 11, 69, 1984.
133. **Chumakov, M. P.**, Some results of investigation of the etiology and immunology of Crimean hemorrhagic fever, *Tr. Inst. Polio. Virusn. Entsefalitov Akad. Med. Nauk SSSR*, 19, 7, 1971 (in Russian; in English, NAMRU3-T953).
134. **Chumakov, M. P., Butenko, A. M., Rubin, S. G., Berezin, V. V., Derbedeneva, M. P., Badalov, M. E., and Stolbov, D. N.**, Question on the ecology of Crimean hemorrhagic fever virus, *Mater. 5. Simp. Izuch. Roli Pereletn. Plitsepererab. Rasprostr. Arbovirus* (Novosibirsk, July 1969), p. 222, 1972 (in Russian; in English, NAMRU3-T877).
135. **Tsilinsky, Ya., Lebedev, A. D., Pak, T. P., Gromashevsky, V. L., Timofeev, E. M., Ershov, R. I., Tsirkin, Yu. M., and L'vov, D. K.**, Isolation of Crimean haemorrhagic fever (CHF) virus from *Hyalomma plumbeum* ticks in Tadzhikistan, *Mater. Simp. Itogi. 6. Simp. Izuch. Virus. Ekol. Svyazan. Ptits*, (Omsk, December 1971), p. 94, 1972 (in Russian; in English, NAMRU3-T665).
136. **Kondratenko, V. F., Blagoveshchemskaya, N. M., Butenko, A. M., Vyshnivetskaya, L. K., Zarubina, L. V., Milyutin, V. N., Kuchin, V. V., Novikova, E. M., Rabinovich, V. D., Shevchenko, S. F., and Chumakov, M. P.**, Results of virological investigation of ixodid ticks in Crimean hemorrhagic fever focus in Rostov Oblast, *Mater. 3. Oblast. Nauchn. Prakt. Konf.* (Rostov-on-Don, May 1970), p. 29, 1970 (in Russian; in English, NAMRU3-T524).
137. **Hoogstraal, H.**, Viruses and ticks, in *Viruses and Invertebrates*, Gibbs, A. J., Ed., North-Holland, Amsterdam, 1973, 349.
138. **Burgdorfer, W. and Varma, M. G. R.**, Trans-stadial and transovarial development of disease agents in arthropods, *Ann. Rev. Entomol.*, 12, 347, 1967.
139. **Levi, V. and Vasilenko, S.**, Study on the Crimean hemorrhagic fever (CHF) virus transmission mechanism in *Hyalomma pl. plumbeum* ticks, *Epidemiol. Mikrobiol. Infekts. Boles.*, 9, 182, 1972 (in Bulgarian).
140. **Lee, V. H. and Kemp, G. E.**, Congo virus: experimental infection of *Hyalomma rufipes* and transmission to a calf, *Bull. Entomol. Soc. Niger.*, 2, 133, 1970.
141. **Okorie, T. G. and Fabiyi, A.**, The replication of Congo virus in *Hyalomma rufipes* Koch following intracoelomic inoculation, *Vet. Parasitol.*, 7, 369, 1980.
142. **Okorie, T. G.**, Congo virus — the development in the ixodid ticks — *Hyalomma rufipes* and *Amblyomma variegatum* Fabricius, *Abstr. 16, Int. Congr. Entomol.* (Kyoto, August 1980, p. 331, 1980.

143. **Berezin, V. V., Chumakov, M. P., Reshetnikov, I. A., and Zgurskaya, G. N.,** Study of the role of birds in the ecology of Crimean hemorrhagic fever virus, *Mater. 6. Simp. Izuch. Virus. Ekol. Svyazan. Ptits.* (Omsk, 1971), p. 94, 1971 (in Russian; in English, NAMRU3-T721).
144. **Berezin, V. V., Chumakov, M. P., Rubin, S. G., Stolbov, D. N., Butenko, A. M., and Bashkirtsev, V. A.,** Contribution to the ecology of Crimean hemorrhagic fever virus in the lower Volga River, *Mater. 16. Nauchn. Sess. Inst. Polio. Virus. Entsefalitov* (Moscow, October 1969), 2, 120, 1969 (in Russian; in English, NAMRU3-T912).
145. **Chumakov, M. P.,** Investigations of arboviruses in the U.S.S.R. and the question of possible association through migratory birds between natural arbovirus infection foci in the U.S.S.R. and warm climate countries, *Mater. 5. Simp. Izuch. Roli Pereletn. Ptits. Rasp. Arbovirus* (Novoskibirsk), July 1969), p. 133, 1972.
146. **Pak, T. P., Daniyarov, O. A., Kostyukov, M. A., Bulychev, V. P., and Kuima, A. U.,** Ecology of Crimean hemorrhagic fever in Tadzhikistan, *Mater. Resp. Simp. Kamenyuki "Belovezh. Puscha"* (Minsk, September 1974), p. 93, 1974 (in Russian; in English, NAMRU3-T1127).
147. **Benda, R.,** The common tick: *Ixodes ricinus* L. as a reservoir and vector of tick-borne encephalitis. II. Experimental transmission of encephalitis to laboratory animals by ticks at various stages of development, *J. Hyg. Epidemiol. Microbiol. Immunol. (Prague),* 2, 331, 1958.
148. **Smorodintsev, A. A.,** Tick-borne spring-summer encephalitis, *Progr. Med. Virol.,* 1, 400, 1958.
149. **Nuttall, G. H. F.,** Observations on the biology of Ixodidae. I., *Parasitology,* 6, 68, 1913.
150. **Knight, M. M., Norval, R. A. I., and Rechav, Y.,** The life cycle of the tick *Hyalomma marginatum rufipes* Koch (Acarina:Ixodidae) under laboratory conditions, *J. Parasitol.,* 64, 143, 1978.
151. **Yao, W. B.,** Effect of temperature and humidity on the oviposition of *Hyalomma asiaticum kozloi, Acta Entomol. Sin.,* 28, 713, 1985.
152. **Hooker, W. A., Bishopp, F. C., and Wood, H. P.,** The life history and bionomics of some North American ticks, *Bull. U.S. Bur. Entomol.,* 106, 1912.
153. **Matson, B. A. and Norval, R. A. I.,** The seasonal occurrence of adult ixodid ticks on cattle on a Rhodesian highveld farm, *Rhod. Vet. J.,* 8, 2, 1977.
154. **Pegram, R. G., Perry, B. D., Musisi, F. L., and Mwanaumo, B.,** Ecology and phenology of ticks in Zambia: seasonal dynamics on cattle, *Exp. Appl. Acarol.,* 2, 25, 1986.
155. **Clifford, C. M., Flux, J. E. C., and Hoogstraal, H.,** Seasonal and regional abundance of ticks (Ixodidae) on hares (Leporidae) in Kenya, *J. Med. Entomol.,* 13, 40, 1976.
156. **Minshull, J. I.,** Seasonal occurrence, habitat distribution and host range of four ixodid tick species at Kyle recreational park in southeastern Zimbabwe, *Zimbabwe Vet. J.,* 12, 58, 1981.
157. **Nuttall, G. H. F.,** Observations on the biology of Ixodidae, *Parasitology,* 7, 408, 1915.
158. **Norval, R. A. I.,** The limiting effect of host availability for the immature stages on population growth in economically important ixodid ticks, *J. Parasitol.,* 65, 285, 1979.
159. **Baker, M. K. and Ducasse, F. B. W.,** Tick infestation of livestock in Natal. I. The predilection sites and seasonal variations of cattle ticks, *J. S. Afr. Vet. Med. Assoc.,* 38, 447, 1967.
160. **Tesh, R. B.,** Experimental studies on the transovarial transmission of Kunjin and San Angelo viruses in mosquitoes, *Am. J. Trop. Med. Hyg.,* 29, 657, 1980.
161. **Reeves, W. C.,** Factors that influence the probability of epidemics of western equine, St. Louis, and California encephalitis in California, *Calif. Vector Views,* 14, 13, 1967.
162. **Brown, S. J.,** Immunology of acquired resistance to ticks, *Parasitol. Today,* 6, 166, 1985.
163. **Smith, G. E.,** Ticks and viruses, *Symp. Zool. Soc. Lond.,* No. 6, p. 199, 1962.
164. **Sonenshine, D. E.,** Vector population dynamics in relation to tick-borne arboviruses: a review, *Phytopathology,* 64, 1060, 1974.
165. **Nuttall, P. A.,** Transmission of viruses to wildlife by ticks, in *Virus Vectors,* Symp. Soc. Gen. Microbiol., Harrap, K. A. and Mayo, M., Eds., University of Dundee, Dundee, Australia, 1986, 135.
166. **Plowright, W., Perry, C. T., and Greig, A.,** Sexual transmission of African swine fever virus in the tick, *Ornithodoros moubata porcinus,* Walton, *Res. Vet. Sci.,* 17, 106, 1974.
167. **Hoogstraal, H.,** Changing patterns of tickborne diseases in modern society, *Ann. Rev. Entomol.,* 26, 75, 1981.
168. **Grobov, A. G.,** Carriers, of Crimean haemorrhagic fever, *Med. Parazitol. Parazit. Bolezni,* 15, 59, 1946 (in Russian; in English, NAMRU3-T36).
169. **Yanovich, T. D.,** Reports of the Committee on Coordinated Study of Prophylactic Measures Against Crimean Hemorrhagic Fever in Rostov Oblast, *Mater. 3. Oblast. Nauchn. Prakt. Konf.* (Rostov-on-Don, May 1970), p. 3, 1970 (in Russian; in English, NAMRU3-T521).
170. **Mehr, Z. A., Rutledge, L. C., Morales, E. L., and Inase, J. L.,** Laboratory evaluation of commercial and experimental repellents against *Ornithodoros parkeri* (Acari:Argasidae), *J. Med. Entomol.,* 23, 136, 1986.
171. **Vashkov, V. I. and Poleshchuk, V. D.,** Measures for control of vectors of CHF — *Hyalomma plumbeum plumbeum* Panz. Ticks, *Tr. Inst. Polio. Virusn. Entsefalitov Akad. Med. Nauk SSSR,* 19, 239, 1971 (in Russian; in English, NAMRU3-T983).

172. **Kondratenkov, V. F., Shevchenko, S. F., Perelatov, V. D., Badalov, M. E., Ionov, S. S., Semenov, M. Ya., Romanova, V. A., Lobanov, V. V., and Tekut'ev, I. V.,** Two year experiment on application of chemical campaign method against ixodid ticks in Crimean hemorrhgic fever focus of Rostov Oblast. *Mater. 3. Oblast. Nauchn. Prakt. Konf.* (Rostov-on-Don, May 1970), p. 157, 1970 (in Russian; in English, NAMRU3-T550).

173. **Dobritsa, P. G., Abdulimov, M. A., Bakirova, M. N., and Mamontov, S. I.,** Investigation of Crimean hemorrhagic fever (CHF) in Chimkent, Oblast, Kazakh SSR. Report 2: prevention of CHF in Kazakhstan conditions, *Tr. Inst. Polio. Virusn. Entsefalitov Akad. Med. Nauk SSSR,* 19, 231, 1971 (in Russian; in English, NAMRU3-T977).

174. **Tkachenko, E. A., Butenko, A. M., Butenko, S. A., Zavodova, T. I., and Chumakov, M. P.,** Characteristics of prophylactic vaccine against Crimean hemorrhagic fever, *Mater. 3. Oblast. Nauchn. Prakt. Konf.* (Rostov-on-Don, May 1970), p. 136, 1970 (in Russian; in English, NAMRU3-T546).

175. **Tkachenko, E. A., Butenko, A. M., Badalov, M. E., Zavodov, T. I., and Chumakov, M. P.,** Investigation of the immunogenic activity of killed brain vaccine against Crimean hemorrhagic fever, *Tr. Inst. Polio. Virusn. Entsefalitov Akad. Med. Nauk SSSR,* 19, 119, 1971 (in Russian; in English, NAMRU3-T931).

176. **Vasilenko, S. M.,** Results of the investigation on etiology, epidemiologic features and specific prophylactic of Crimean hemorrhagic fever (CHF) in Bulgaria, *Abstr. Inv. Pap. 9. Int. Cong. Trop. Med. Malar.* (Athens, October 1973), 1, 32, 1973.

Chapter 23

DENGUE

D. J. Gubler

TABLE OF CONTENTS

I. HISTORICAL BACKGROUND

A. Discovery of Agent and Vectors

Dengue fever has been known clinically for over 200 years,[1] but the etiology of the disease was not discovered until 1944. The first dengue viruses were isolated from soldiers who became ill in Calcutta, India, New Guinea, and Hawaii.[2] The viruses from India, Hawaii, and one strain from New Guinea were antigenically similar, whereas three other strains from New Guinea appeared to be different. They were called dengue 1 (DEN-1) and dengue 2 (DEN-2) and designated as prototype viruses (DEN-1, Hawaii and DEN-2, New Guinea-C).[2] Two more serotypes — dengue 3 (DEN-3) and dengue 4 (DEN-4) — were subsequently isolated from patients with a hemorrhagic disease during an epidemic in Manila in 1956.[3] Although many dengue viruses have been isolated from different parts of the world since that time, all have fit antigenically into the four-serotype classification. They are closely related antigenically to other flaviviruses and ecologically to some flaviviruses (e.g., yellow fever) and togaviruses (e.g., chikungunya).[4,5]

Many early workers suspected that dengue viruses were transmitted by mosquitoes, but actual transmission was first documented by Graham[6] in 1903. In 1906, Bancroft[7] demonstrated that *Aedes aegypti*, allowed to blood feed on a person during the acute phase of illness, was able to transmit the agent to another person after an incubation period of 10 days.[7] Subsequent studies in the Philippines, Indonesia, and the Pacific showed that *Ae. albopictus* and *Ae. polynesiensis* were also efficient vectors for these viruses.[8-11]

B. History of Epidemics

The first epidemics of dengue-like illness reported in the medical literature occurred almost simultaneously in 1779 in Cairo and Batavia (Jakarta), Indonesia.[12] In 1780, an epidemic of a similar illness occurred in Philadelphia.[1] Since that time, repeated epidemics have occurred in most tropical and subtropical regions of the world where the mosquito vectors occurred.[8,12,13] In Asia and the Americas, large epidemics (pandemics) of dengue-like illness moved through the regions at 10- to 30-year intervals.[12] There is no documentation, however, that dengue viruses were responsible for all of these epidemics because reports were based only on clinical diagnosis. A review of clinical descriptions from early epidemics has suggested that some of the outbreaks may have been caused by chikungunya, an alphavirus transmitted to humans by *Ae. aegypti* in a cycle similar to dengue.[12] It is likely that this did occur to some extent, but recent data show that the dengue viruses, not chikungunya, were responsible for the majority of these epidemics in the past 30 years. With the advent of the jet airplane in the 1960s and the subsequent increase in air travel by humans, there has been a significant increase in the frequency of epidemic dengue in most tropical areas of the world.

C. Social and Economic Impact

Historically, dengue was described as an adult disease of the expatriate (colonial) communities in tropical endemic areas of Asia and was not commonly observed or recognized in the local population. In nonendemic tropical, semitropical, and temperate areas where *Ae. aegypti* was common, the disease occurred as major epidemics at irregular intervals. Since dengue fever was generally benign and seldom recognized to have a fatal outcome, the social impact of the disease was limited, except during large epidemics. With the emergence of epidemic dengue hemorrhagic fever (DHF) in the 1950s, however, the impact of the disease became more pronounced because it affected primarily children of the local populations in endemic areas of Asia. If undiagnosed and untreated, it was associated with fatality rates as high as 30 to 40%. In the past 30 years, conservative estimates are that there have been over 700,000 children hospitalized with suspected DHF in Southeast Asia alone, and there have been over 20,000 fatalities.[14] The disease is currently one of the leading causes of hospitalization and death among children in several Southeast Asian countries.[15] Moreover, in the past 15 years, epidemic DHF has moved out of Southeast Asia and into the Pacific and the Americas.[16,17]

In addition to the public health aspects, the economic impact of epidemic dengue and DHF is considerable. In Jakarta, Indonesia, from 1975 to 1978, the average duration of hospitalization of over 1300 children admitted for DHF was 3.5 days.[179] Considering that many of these children were seriously ill and required intensive medical care, and that over 700,000 have been hospitalized in the Southeast Asian region, DHF then becomes a very important economic problem. In Thailand, it is estimated that the 1980 epidemic of DHF alone cost $6.8 million in hospitalization and mosquito control.[14] Millions of dollars are spent each year in the Southeast Asian region to control the mosquito vectors of dengue, unfortunately without much effect.

Recent examples of the economic impact of epidemic dengue and DHF are Puerto Rico and Cuba. The estimated cost of the 1977 epidemic of DEN-2 and DEN-3 in Puerto Rico was between $6 and 16 million (U.S.) in medical costs, lost work, and control measures.[18] There was no attempt to measure the impact on the tourist industry, but the cost must have been equally great. Since Puerto Rico has had five epidemics of dengue in the past 8 years, it is estimated that this disease has cost the people of the island between $100 and 150 million (U.S.) since 1977.[19]

The 1981 DHF epidemic in Cuba had an even greater economic impact with an estimated cost of over $100 million (U.S.) in control measures and medical costs during a 4-month period from July through October.[20] The Cuban authorities continued the control efforts in an attempt to eradicate the mosquito vector, *Ae. aegypti*, and currently have succeeded in reducing house indexes below 1%.[21]

In the Caribbean Basin, tourism plays a major role in the economic health of many countries. When an epidemic occurs in these countries, it can be devastating to the tourist industry, which in turn affects all segments of society. Unfortunately, the impact of epidemic dengue on tourism is difficult to measure, but any decrease in tourism caused by epidemic disease is intolerable. This problem has been recognized by most governments in the American region, but instead of attempting to prevent the disease by improved surveillance and more effective mosquito control, cases of disease are simply not reported until epidemic transmission has been reached, and then it is too late to initiate effective control measures.

II. THE VIRUS

A. Antigenic Relationships

The dengue viruses belong to the family *Flaviviridae* and the genus *Flavivirus*.[4] There are four dengue serotypes, designated DEN-1, -2, -3, and -4.[4,5] while there are extensive cross reactions among the four dengue viruses in most serological tests, there is no lasting cross-protective immunity.[5] Thus, a person living in an endemic area can have as many as four dengue infections during his lifetime, one with each serotype.

The extensive serological crossing that occurs among dengue viruses and between dengue viruses and other flaviviruses can be very confusing epidemiologically. The hemagglutination-inhibition (HI) test, which is used for most seroepidemiological surveys, is nonspecific and most flaviviruses cross react in this test.[22] Thus, it is difficult to determine whether the test is detecting antibodies to dengue or to other falviviruses that may be endemic in the area. The neutralization (N) test, which is more specific and more sensitive, may be used to distinguish specific dengue antibody from that of other flaviviruses.

Although the four dengue serotypes are antigenically distinct, there is some evidence that serological subcomplexes may exist within the group. Thus, DEN-1 and DEN-3 have been shown to share some antigenic determinants by N test and by immunofluorescence using monoclonal antibodies.[23,24] More recently, a close genetic relationship has been demonstrated between DEN-1 and DEN-4 using cDNA hybridization probes.[25,26] These viruses shared approximately 70% of their genome, whereas DEN-2 showed low sequence homology with all other serotypes (23 to 36%). Surprisingly, DEN-2 showed high sequence homology (71%) with Edge Hill virus, an ecologically distinct flavivirus that had not been previously considered genetically close to the dengue viruses.

B. Host Range

There are only three known natural hosts for dengue viruses: humans, lower primates, and *Aedes* mosquitoes. Experimental laboratory evidence shows that several species of lower primates (chimpanzees, gibbons, and macaques) become infected and develop viremia titers sufficient to infect mosquitoes.[27-30,180] Onset of viremia after infection in these animals is similar to that in humans, but the magnitude and duration of viremia generally are lower and shorter, respectively. Thus, viremia in humans may last from 2 to 12 days (average 4 to 5 days), with titers ranging from undetectable to over 10^8 mosquito infectious dose 50 (MID_{50})/mℓ.[31-33] In lower primates, viremias are more transient, often lasting only 1 or 2 days if detectable, with titers seldom reaching 10^6 MID_{50}/mℓ.[27-30,180]

Humans are the only known host that develops clinical expression of dengue virus infection, ranging from clinically inapparent infection to severe hemorrhagic disease, shock, and death. Lower primates appear to be particularly well adapted to the viruses and show no overt signs of illness.

Dengue viruses do not readily infect any other vertebrate animals. Even baby mice, which are used for isolation and assay of many other arboviruses, generally show no signs of illness after intracerebral (i.c.) inoculation with most unpassaged strains of dengue viruses.

Only mosquitoes of the genus *Aedes* appear to be natural hosts for dengue viruses. Species of the subgenus *Stegomyia* are the most widespread and important mosquito hosts and include *Ae. (S.) aegypti, Ae. (S.) albopictus, Ae. (S.) scutellaris* spp., *Ae. (S.) africanus,* and *Ae. (S.) leuteocephalus.*[7-11,34] Species of the subgenus *Finlaya [Ae. (F.) niveus]* and the subgenus *Diceromyia [Ae. (D.) taylori* and *Ae. (D.) furcifer*] appear to be important mosquito hosts involved in dengue forest maintenance cycles in Asia and Africa.[34,35] Two other species, *Ae. (Gymnometopa) mediovittatus* and *Ae. (Protomacleaya) triseriatus*, have been shown to be excellent experimental hosts for dengue viruses.[36,37] Genera of mosquitoes other than *Aedes* have generally been found to be poor hosts for dengue viruses, although a recent report from China incriminates *Culex p. quinquefasciatus* as a vector.[38] Other workers have found, however, that *Culex* species are refractory to dengue virus infection and do not play a role in transmission.[8,9,39]

C. Strain Variation

1. Evidence from Field Studies

The possibility that strain variation exists among dengue viruses has only recently received much attention. The first evidence that such variation might occur in nature was observed in the South Pacific in the early 1970s. In late 1971 and early 1972, explosive epidemics of DEN-2 occurred in Fiji, Tahiti, New Caledonia, and Niue Island.[16,40-42] All of these epidemics were associated with explosive transmission and an illness characterized by severe classical dengue fever, frequent hemorrhagic manifestations, and fatalities. The Niue epidemic was especially severe with several cases of fatal primary dengue infection.[16,43] Also in 1972, another, more limited outbreak of DEN-2 with mild illness occurred in American Samoa.[180] Of more interest, however, was the fact that DEN-2 was introduced into Tonga in 1973, where the virus was maintained in a human population previously unexposed to DEN-2 for nearly 1 year without being recognized by the medical community.[31] The mild illness in Tonga was not diagnosed as dengue infection until after fever of unknown origin studies were initiated in the hospital outpatient clinic. In addition to the mild illness, dengue infections in Tonga were characterized by very low viremia titers that were undetectable in most patients tested. This is in contrast to very high viremia titers associated with DEN-2 infection during the Tahiti and New Caledonia epidemics.[180]

DEN-1 was introduced into the South Pacific in 1975, and a similar pattern of epidemics was observed. With this serotype, however, Tonga experienced an explosive epidemic associated with severe and fatal disease,[31,43] while Tahiti experienced only mild illness. Also of interest was the island of Ponape in the Central Pacific, where both DEN-1 and DEN-2 viruses were transmitted silently.[180] Thus, a serological survey in the spring of 1975 revealed that 52% of persons under the age of 25 years had DEN-1 or DEN-2 N antibody, even though the last reported dengue transmission in Ponape was during the DEN-1 pandemic 31 years earlier in 1943—1944.

Similar strain variation was observed with DEN-3 in Indonesia. An explosive epidemic of severe and fatal dengue hemorrhagic fever/dengue shock syndrome (DHF/DSS) in Bantul, Central Java, in 1976—1977 was associated with high viremia titers.[33] A year later, another less explosive outbreak of DEN-3 occurred in Sleman, Central Java, about 45 km to the north. In the latter outbreak, illness was milder, there were no deaths, and viremia titers were significantly lower.[44]

Whether there is a relationship between viremia and severity of illness in dengue infection is not known. On an individual patient basis, there appears to be no such relationship, but this is difficult to study because most patients have already passed the period of peak viremia by the time of hospitalization. On the other hand, the explosiveness of the epidemic might be related to the magnitude and duration of human viremia.[31,44] Thus, persons with higher viremia titers would infect more mosquitoes and increase the probability of virus transmission. Although limited and circumstantial, the data suggest that viremia may be a marker for epidemic strains of dengue viruses, as opposed to endemic strains that are maintained via silent or sporadic transmission in nature.[44]

2. Evidence from Laboratory Studies

In addition to the field evidence cited above, there is also laboratory evidence that strain variation occurs among dengue viruses. DEN-3 viruses, isolated during epidemics in Puerto Rico in 1963 and in Tahiti in 1965 and 1969, were biologically and antigenically very similar to each other, but very different from the prototype and other Asian strains of DEN-3.[45] N test data showed similar differences between strains of DEN-4 from the Caribbean and Asia.[46] Moreover, growth studies in mosquitoes showed that the Tahiti strains of DEN-3 replicated more slowly and to lower titers in *Ae. albopictus* than the prototype or Thailand strains of DEN-3.[47]

Recent developments in molecular virology have provided further proof of strain variation among dengue viruses. The most extensive work has been done with the oligonucleotide fingerprint technique. Multiple genetic topotypes have been identified for DEN-1, -2, and -3 serotypes, even though only limited numbers of virus strains have been studied.[48-51,181] DEN-2 has been studied in greatest detail and 12 topotypes from different geographic regions have been identified.[182] The data suggest that most viruses circulating in a particular geographic area are similar to each other. This is not always the case, however, since two topotypes of DEN-2 have been documented in Caribbean Basin countries.[51] Also, marked differences were observed between DEN-2 viruses isolated from forest mosquitoes and those isolated from humans or from *Ae. aegypti* mosquitoes in an urban setting in West Africa, suggesting that enzootic strains of dengue virus in West Africa are genetically distinct from epidemic or endemic strains.[51]

Oligonucleotide fingerprinting of dengue viruses, however, is slow, expensive, and labor intensive. Moreover, it can only be done in laboratories that are well equipped and have highly trained personnel. More recently, two new methods for studying strain variation among dengue viruses have been described. Specific short DNA probes have been constructed for rapid detection of DEN-2 genetic variants by RNA/DNA hybridization.[51] The probes, which are complementary to unique RNase T1 oligonucleotides, were constructed for five DEN-2 topotypes and show subtopotype, topotype, and type specificity.

Antigen signature analysis, using radioimmunoassays with monoclonal antibodies that recognize distinct DEN-2-type-specific and flavivirus cross-reactive epitopes, has also been used to study strain variation among DEN-2-viruses.[52] This method showed excellent agreement with the oligonucleotide fingerprint topotype classification, but there were a few differences. Of particular interest was a 1983 Mexico DEN-2-virus which shared 85% of large oligonucleotides with the Puerto Rico topotype, but was antigenically distinct by signature analysis. Such differences are to be expected since the oligonucleotide fingerprinting technique only examines 10 to 15% of the total viral genome.[52]

Both cDNA hybridization probes and antigen signature analysis have the advantage over oligonucleotide fingerprinting of being rapid and relatively simple. Thus, new strains of dengue virus can be compared with the predominant topotype present in an area and easily characterized within 1 to 2 days.[52] These techniques can thus provide important information necessary for developing active surveillance systems based on monitoring dengue virus transmission in an area.

Collectively, field and experimental data suggest that there are distinct strain differences among dengue viruses. The role that strain variation plays in distribution and spread of epidemic dengue and in determination of disease severity, however, is not yet known. Unfortunately, there are still no known biological or biochemical markers that can be used to identify virulent or epidemic strains of dengue viruses.

D. Methods for Assay

Dengue viruses have been among the most difficult arboviruses to isolate and propagate. As noted above, they do not grow well in any of the laboratory animals or mammalian cell cultures normally used in virology laboratories. Dengue viruses were first isolated by i.c. inoculation of baby mice.[2] This host is very insensitive, however, and several blind passages are usually required to allow adaptation before evidence of infection can be observed. The development of the direct plaque assay using LLC-MK_2 cells provided a somewhat faster and more sensitive method, but still had the disadvantage of requiring adaptation of the virus to the cell culture.[53]

The development of the mosquito inoculation technique provided, for the first time, a sensitive and relatively rapid method for isolation and assay of dengue viruses.[54] Mosquitoes are natural hosts for these viruses and, therefore, are highly susceptible to infection and allow rapid viral replication. The direct fluorescent antibody test provided a simple technique to determine dengue virus infection in the mosquito.[55] Recent variations of the mosquito inoculation technique include i.c. inoculation of adult and larval *Toxorhynchites* mosquitoes.[56,57] Both techniques have similar sensitivity to intrathoracic inoculation of adult mosquitoes, but are more difficult and labor intensive. Their main advantage is that viruses can be isolated in 2 to 3 days. It should be noted, however, that this can also be done using mosquito cell lines that are less labor intensive.

Mosquito cell lines that are highly susceptible to dengue virus infection are also now available. Three cell lines — C6-36 (*Ae. albopictus*), AP-61 (*Ae. pseudoscutellaris*), and TRA-284 (*T. amboinensis*) — are the most widely used and provide a relatively simple and economical method for dengue virus isolation and assay.[58-61] In two separate comparative studies of the three cell lines, it was observed that the TRA-284 cell line was the most susceptible to wild-type dengue viruses.[62,63] Even so, this line is about 15 to 20% less sensitive than the mosquito inoculation technique for routine isolation of dengue viruses.[183] Since most laboratories do not have facilities to rear mosquitoes, however, the method of choice should be the TRA-284 cell line adapted to serum-free medium.[62] Immunofluorescence should be used to detect virus infection in these cells because cytopathic effects (CPE) are not reliable to detect all dengue viruses. Using the TRA-284 cell line, dengue viruses have been isolated and identified in as short a period as 2 days.[183]

The plaque reduction neutralization (PRNT) and complement-fixation (CF) tests have been the methods of choice for dengue virus identification.[23,64] Both tests, however, are very laborious, and the PRNT is also very costly. The recent development of serotype-specific monoclonal antibodies for dengue viruses[65] has greatly facilitated identification of newly isolated viruses using an indirect fluorescent antibody (IFA) test.[24,66] This test can be used on infected cell cultures or on infected mosquito tissue.[67] In those laboratories that do not have immunofluorescence capability, dengue viruses can also be identified using monoclonal antibodies in an antigen capture enzyme-linked immunosorbent assay (ELISA).[68]

III. DISEASE ASSOCIATIONS

Dengue infection causes a spectrum of illness in humans ranging from clinically inapparent to severe and fatal hemorrhagic disease. The majority of patients, however, present as mild, nonspecific febrile illness or as classical dengue fever. The factors determining whether a person develops mild or severe illness are not well understood, but it is now clear that disease expression can be influenced by a number of factors, including strain of virus and immune status of the individual.[43,44,69] The genetic makeup of the infected individual may also influence the severity and type of illness.[17]

Classical dengue fever is generally observed in older children and adults, and is characterized by sudden onset of fever, frontal headache, retroocular pain, and myalgias. Rash, joint pains, nausea and vomiting, and lymphadenopathy are common. The acute illness, which lasts for 3 to 7 days, is usually benign and self limiting, but it can be very debilitating and convalescence may be prolonged for several weeks.

The hemorrhagic form of the disease, DHF/DSS, is most commonly observed in children under the age of 15 years, but it can also occur in adults.[15,31,43] It is characterized by acute onset of fever and a variety of nonspecific signs and symptoms that may last 2 to 7 days. During this stage of the illness, it is difficult to distinguish DHF from any number of other viral or even some protozoal infections. In children, upper respiratory symptoms caused by concurrent infection with other viruses are not uncommon. The critical stage in DHF occurs when the fever subsides to or below normal.[15,70] At that time, the patient's condition may deteriorate rapidly with signs of circulatory failure, hemorrhagic manifestations, shock, and death if proper management is not implemented.[15] Skin hemorrhages such as petechiae, easy bruising, bleeding at the sites of venipuncture, and purpura/ecchymoses are the most common hemorrhagic manifestations, but gastrointestinal (GI) hemorrhage may also occur after, but in some cases, before onset of shock.[15,71]

Based on data collected primarily from Thailand, the World Health Organization (WHO) has defined strict criteria for diagnosis of DHF/DSS.[15] According to those criteria, patients with DHF/DSS have four major clinical manifestations: high fever, hemorrhagic manifestations, hepatomegaly, and circulatory failure. Thrombocytopenia and hemoconcentration are constant features. The WHO has classified DHF into four grades according to severity of illness:[15]

Grade I	Fever accompanied by nonspecific constitutional symptoms, with a positive tourniquet test as the only hemorrhagic manifestation
Grade II	The same as grade I, but with spontaneous hemorrhagic manifestations
Grade III	Circulatory failure manifested by rapid, weak pulse, narrowing of pulse pressure (20 mmHg or less), or hypotension
Grade IV	Profound shock with undetectable pulse and blood pressure

All grades of illness are considered DHF, while only grades III and IV are considered DSS. The presence of thrombocytopenia and concurrent hemoconcentration differentiates grades I and II DHF from classical dengue fever with hemorrhagic manifestations. Thus, the term DHF/DSS, according to the WHO criteria, defines a physiological condition in which the patient has increased vascular permeability.[72]

There is some disagreement, however, as to exactly what constitutes a case of DHF. In Indonesia, hemorrhagic disease associated with dengue is somewhat different from that in Thailand. While approximately 60 to 70% of laboratory-confirmed cases of hemorrhagic dengue in Indonesia meet WHO criteria, the rest do not.[71,73] The most striking difference noted in Indonesia was severe upper GI bleeding that often occurred prior to the onset of shock, and with no evidence of hemoconcentration. These cases were, in general, more difficult to manage than those with classical DSS.[73] Others argue that it is capillary permeability, not hemorrhagic manifestations, that define a case of DHF, and that just because a patient bleeds to death, it cannot be called DHF unless there is evidence of vascular permeability.[72] Because of the variation in clinical presentation, it has been proposed that the definition of DHF be changed to include patients such as those described in Indonesia.[71]

Also of interest in Indonesia was the fact that hepatomegaly was not a constant feature of confirmed DHF.[73,74] Similar findings have been reported from earlier epidemics in the Philippines, Singapore, Vietnam, and Burma in Asia, and Cuba in the Americas.[17,75-78] It is possible that some of the variation in hepatomegaly rates might be due to lack of expertise in palpating livers, but this is highly unlikely because most of the examinations were carried out by pediatricians trained in tropical countries where malaria and other diseases affecting the liver are common. Thus, it would appear that with the exception of Thailand, hepatomegaly is not a consistent clinical finding in DHF in most areas of the world.

Historically, dengue virus infection has been associated with a variety of neurological disorders.[79] More recently, cases of encephalitic disease have been reported from Indonesia, Malaysia, Burma, Thailand, Dominican Republic, and Puerto Rico.[80-84,184] While many of these patients present clinically as viral encephalitis, there is no evidence that the virus crosses the blood-brain barrier and infects the central nervous system.[79] Instead, it appears that the neurological symptoms may be secondary to cerebral hemorrhage, edema, or other indirect affects of dengue virus infection.

Dengue viruses do not cause disease in any animal other than humans. Although lower primates readily become infected and develop viremia (see Section II.B), specific disease associated with dengue infection in these animals has never been described.

Diagnosis of classical DHF/DSS as described by the WHO can be facilitated by several clinical laboratory tests.[15] Thrombocytopenia with concurrent hemoconcentration will help differentiate DHF from other diseases. There may be leukopenia or leukocytosis, hypoalbuminemia, hyponatremia, and elevated transaminase and blood urea nitrogen levels. In severe cases of DSS, prothrombin time may be prolonged, along with a concomitant increase in fibrinogen degradation products and thrombocytopenia, suggesting disseminated intravascular coagulation.

In Thailand, a constant finding has been that thrombocytopenia usually precedes the rise in hematocrit, which indicates onset of plasma leakage from the vascular compartment.[70] Also, a chest X-ray can be useful in detecting pleural effusion, another indicator of increased vascular permeability. Both tests can be of considerable diagnostic value for patients with classical DSS, but are of less value when dealing with patients with severe upper GI hemorrhage such as those observed in Indonesia.[71]

The major pathophysiologic abnormality in classical DSS is an increase in vascular permeability which leads to leakage of plasma.[15,70,72] Post-mortem examination of patients in Thailand has shown that many have serous effusion in the pleural and abdominal cavities and a variable amount of hemorrhaging in most major organs.[85] Studies have not revealed destructive or inflammatory vascular lesions, but some swelling and occasional necrosis have been observed in endothelial cells, as well as some perivascular edema.[2]

Treatment for DHF is presently symptomatic and the prognosis of the disease depends upon early recognition, initiation of corrective fluid replacement, and management of shock.[15,70] Definitive diagnosis, however, can only be made in the laboratory by serological and/or virological methods. These include virus isolation, which was discussed in Section II.D, as well as serological diagnosis. This field has been evolving rapidly in recent years and will continue to do so with the development of new and more rapid diagnostic methods based on molecular technology.

Currently, serological diagnosis of dengue infection depends upon four tests: HI,[22] CF,[86] PRNT,[87] and IgM capture enzyme-linked immunosorbent assay (MAC-ELISA).[88] The HI is most widely used in dengue laboratories and is very sensitive in detecting low-level antibody, even as long as 40 or more years after infection.[89] In primary infections, HI antibody is first detectable about 4 to 5 days after onset of illness, and antibody produced during the first 14 days of illness may be relatively type specific in few cases. During the first dengue infection, convalescent HI antibody titers usually do not exceed 1:640, although there are exceptions and titers of 1:10,240 have been documented.[16] Second, third, and probably fourth dengue infections induce immediate anamnestic responses with HI titers often exceeding 1:10,240. Significant HI antibody titers in such patients persist for months after infection and therefore it is difficult to make a diagnosis on the basis of a single serum sample. Most workers, however, consider an HI titer of $\geq$1:1280 to be a presumptive diagnosis for current dengue infection.

In general, HI antibody is very nonspecific and cross reactions occur not only among the dengue virus serotypes, but also between dengue and other flaviviruses. It is usually not possible, therefore, to determine the infecting dengue serotype or even whether it is a dengue infection with the HI test. Diagnostic laboratories should combine HI serology with virus isolation in order to determine the infecting virus serotype.

The CF test is a little more difficult to perform than the HI test and, therefore, is not as widely used. CF antibody appears somewhat later than HI and generally wanes at a faster rate. CF antibody in patients with primary dengue infection is generally more specific than HI antibody and if the serum samples are properly timed, the infecting serotype can be determined in some cases. In late convalescent sera and in sera from cases with secondary infection, however, CF reactions (like HI reactions) show heterologous cross reactions.

The PRNT is the least used test for routine dengue serology because it is expensive and most difficult to perform. It is, however, the most sensitive and specific test used in the dengue diagnostic laboratory. N antibody appears at about the same time as HI antibody and persists for the life of the patient.[89] The PRNT is slightly more sensitive in detecting low levels of dengue antibody than the HI test. Thus, in a small percentage of cases, one may find low titers of N antibody, but undetectable dengue HI antibody in acute sera or convalescent survey sera. The infecting virus serotype can nearly always be determined by PRNT in primary dengue infections. With secondary infections, however, heterologous cross reactions occur between the four serotypes and it is usually not possible to tell which serotype caused the current or recent infection. Some patients, however, may show higher PRNT antibody titers against one serotype even though the current infection is caused by another serotype (original antigenic sin).[16,90] This test, therefore, can be very useful to determine the sequence of infections in cases of DHF/DSS.

The MAC-ELISA is a relatively new test in dengue serology, but one that has turned out to be extremely useful for rapid diagnosis and surveillance for dengue and DHF. Studies carried out in Puerto Rico have shown that detectable dengue IgM antibody appears at about the same time as dengue HI antibody. By day 5 after onset of illness, most patients, both primary and secondary infections, have detectable IgM antibody.[185] The persistence of IgM is not as well known, but preliminary data suggest that the majority of patients have lost their IgM antibody by day 60 after onset, although in some patients, it is still detectable as long as 90 days after infection. Thus, caution must be exercised in interpretation of MAC-ELISA results because the presence of dengue-specific IgM does not necessarily mean a current dengue infection.

Few data are available on the sensitivity of the dengue MAC-ELISA, but results collected to date in Puerto Rico show good correlation between HI and MAC-ELISA results.[185] Preliminary results suggest that the MAC-ELISA is relatively specific in distinguishing dengue from other flavivirus infections, but is very nonspecific in distinguishing the infecting dengue serotype.[186]

The obvious need for more rapid diagnostic methods for dengue infection which could be of use to the physician managing a case of suspected DHF/DSS has prompted investigators in several laboratories to develop tests to directly detect virus or antigen in the acute serum. Unfortunately, this has proved to be extremely difficult and most have not been successful. Recently, however, two tests, a multisite monoclonal radioimmunoassay (RIA) and an immunofiltration assay have been developed which show promise as rapid diagnostic techniques.[91,92] The RIA was constructed using antibodies with high binding avidity and detected virus in 47% of acute viremic sera from patients with DEN-1, -2, and -3 infections, but only 10% of viremic sera from patients infected with DEN-4. Virus was detected in sera of patients with primary dengue infection (54%) more frequently than in sera from patients with secondary infection (16%), suggesting that heterologous antibody was interfering with the reaction in the latter sera. The immunofiltration assay is based on the use of filters for nonspecific capture and immobilization of antigen, and detection of radiolabeled viral components or enzyme-linked reagents. While detailed studies with dengue have not yet been done, preliminary laboratory results show promise.[187]

IV. EPIDEMIOLOGY

A. Geographic Distribution

The first reported epidemics of dengue-like disease occurred on three separate continents almost simultaneously in 1779 and 1780.[12] Although there is some disagreement as to whether all of these epidemics were caused by dengue viruses, it is clear that dengue and other arboviruses with similar ecology had widespread distribution in the tropics as long as 200 years ago. For the next 175 years, major pandemics of dengue-like illness occurred in Asia and the Americas at variable intervals ranging from 10 to 30 years.[12,13] With the advent of modern diagnostic virology, and the isolation and identification of the four dengue virus serotypes, their distribution became better known. Asia historically has been the area of highest endemicity, with all four dengue serotypes circulating in the large urban centers of most countries.

During and shortly after World War II, *Ae. aegypti* became more widespread in Asia, and with the subsequent urbanization that occurred in most countries, the incidence of dengue infection increased dramatically. This increase coincided with the emergence of epidemic DHF in the 1950s. The advent of the jet airplane in the 1960s then provided the ideal mechanism for the transport of dengue viruses by humans who had visited endemic areas, became infected, and were incubating the virus infection. Thus began a trend of increased spread of dengue viruses throughout the tropical world. In addition to Asia, increased epidemic activity was observed in the Pacific Islands and the Caribbean Basin in the 1970s, and major epidemics of all four dengue serotypes were documented in both regions.[19,93] In the American region, all four viruses are probably now endemic.[19]

In the 1980s, increased dengue activity also spread to Africa, where little has been known about dengue historically. Until recently, only DEN-1 and -2 were known from Nigeria, West Africa, and epidemics of documented dengue were rare.[94,95] In recent years, however, all four dengue serotypes have been documented in Africa, and for the first time in history, epidemics of disease have been documented in East Africa.[96-99] It should be noted that many of the recent dengue outbreaks were first reported as being caused by malaria, underscoring the lack of adequate surveillance for this disease in Africa. Most of tropical Africa has *Ae. aegypti* mosquitoes; therefore, is at risk for epidemic dengue activity. In addition, past records show that *Ae. aegypti* was present in most of the port cities of North Africa.

The current known world distribution of dengue viruses, along with areas considered to be at risk for transmission because of the presence of *Ae. aegypti*, is shown in Figure 1. Because current data are not available for North Africa, that area was not included in Figure 1 as being at risk for dengue transmission.

B. Incidence

The incidence of dengue infection has increased dramatically in the past 30 years, first in Asia, then in the Pacific and Americas, and finally in Africa. Unfortunately, there are no accurate data on actual numbers of cases of either dengue fever or DHF. From Figure 1 it will be noted that most of the tropical world, with an estimated population of over 1.5 billion persons, is at risk of infection with dengue.[100] Epidemics in recent years have caused millions of cases, but surveillance systems report only a small fraction of the actual numbers. Table 1 shows the number of reported cases and incidence for four countries in Southeast Asia. It will be noted that Thailand and Vietnam have had epidemics of DHF every 3 to 4 years, but that the trend has been a consistent increase in both the total number of cases

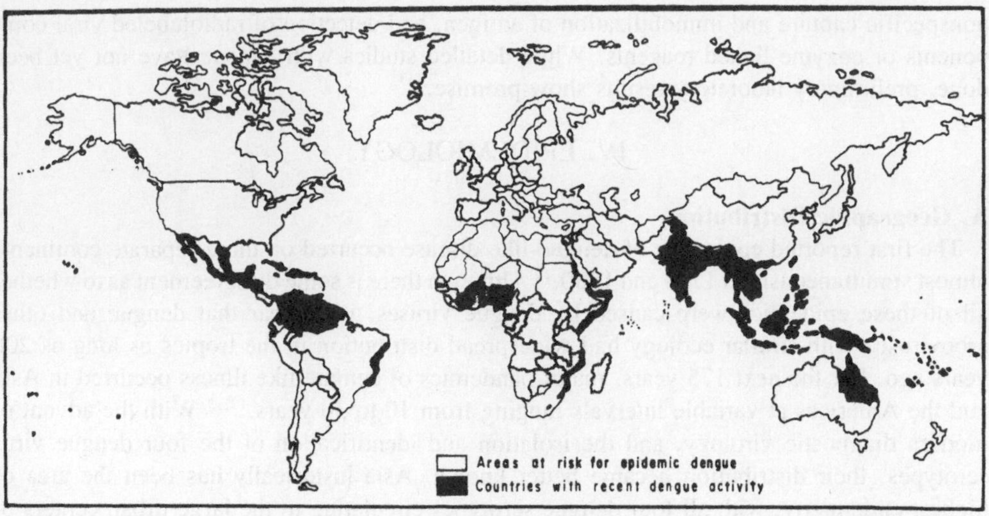

FIGURE 1 World distribution of dengue, 1986.

Table 1
REPORTED CASES AND INCIDENCE OF DHF IN THAILAND, INDONESIA, BURMA, AND VIETNAM, 1976 TO 1985

	Thailand		Indonesia		Burma		Vietnam	
Year	No. cases	Rate /100,000	No. cases	Rate /100,000	No. cases	Rate /100,000	No. cases	Rate /100,000
1976	9,616	21.4	4,548	3.0	3,158	9.0	25,722	52.3
1977	38,768	86.2	7,826	5.2	5,364	15.8	40,544	80.4
1978	12,547	27.9	6,989	4.7	2,029	6.0	43,684	84.5
1979	11,478	25.0	3,422	2.2	4,685	13.8	63,976	121.3
1980	43,382	94.3	5,007	3.2	2,026	5.8	95,146	177.1
1981	25,670	55.8	5,809	3.7	1,524	4.4	35,323	64.2
1982	22,250	48.4	4,665	3.0	1,706	4.9	39,806	71.0
1983	30,025	65.3	13,875	8.7	2,856	8.2	149,519	260.6
1984	69,101	147.0	12,710	7.9	2,323	6.6	25,122	43.3
1985	79,071	168.2	13,605	8.5	No data	—	No data	
Totals	341,908		78,456		25,671		518,842	

and incidence. In Indonesia and Burma, DHF is not as serious a health problem, but in Indonesia, there has been a marked increase in cases in recent years. This same trend has been observed in other countries of the region and in some, such as China, there have been major epidemics of dengue fever after an absence of the disease for over 35 years.

In the Americas, there has been a similar trend of increased incidence of dengue fever over the past 15 years caused mainly by increased frequency of epidemics (Table 2). In recent years, all four dengue virus serotypes have been involved and there has been an increase in the occurrence of hemorrhagic disease in the region. The first major epidemic of DHF occurred in Cuba in 1981, and the region as a whole appears to be at high risk for other such epidemics.

It is estimated that in Southeast Asia alone, there have been over 700,000 hospitalized cases of DHF in the past 30 years, with over 20,000 deaths.[14] It is clear that these figures are gross underestimates of the true incidence of DHF, since there have been over 14,400

Table 2
REPORTED CASES AND INCIDENCE RATES OF DENGUE AND DHF IN THE AMERICAS, 1976 TO 1985

	Dengue		DHF	
Year	No. cases	Rates per 100,000[a]	No. hospitalized	No. deaths
1976	193	0.12	—	—
1977	502,026	309.89	—	—
1978	89,539	55.27	—	—
1979	8,889	5.49	—	—
1980	54,555	33.68	—	—
1981	362,398	223.70	116,143	158
1982	30,244	18.67	3	1
1983	25,108	15.50	—	—
1984	31,270	19.30	8	4
1985	68,950	42.56	3	1
Totals	1,173,172		116,157	164

[a] Based on 1980 census data.

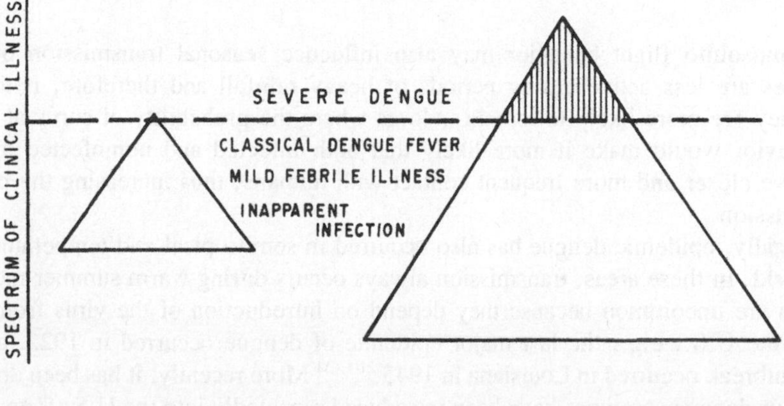

FIGURE 2. The iceberg concept of DHF: cases of severe dengue in relation to the total number of dengue infections.

deaths caused by DHF in the 4 countries listed in Table 1 just since 1976. Also, not all cases are hospitalized and there are many cases of dengue infection for every case of DHF.[14] This is best visualized as the iceberg concept of dengue/DHF (Figure 2). The iceberg (i.e., incidence) has increased dramatically in the past 30 years. Severe and fatal hemorrhagic disease represents only the tip of the iceberg, or that part protruding above the water. It is not known how many cases of dengue infection occur for every case of DHF, but this is not a constant quantitative relationship. Best esimates are that the ratio may vary from 1:100 to 1:500, and is probably influenced by the strain of the infecting dengue virus, the immune status of the host, and the genetic makeup of the host. Regardless of the other risk factors, however, increased incidence may be the most important factor in explaining the emergence of epidemic DHF (Figure 2). The two main factors responsible for increased incidence of dengue in recent years are (1) the almost complete lack of effective, routine mosquito control in most tropical countries, and (2) the jet airplane, which is responsible for the dramatic increase in the movement of dengue viruses between population centers of the tropics.

C. Seasonal Distribution

Dengue transmission occurs throughout the year in endemic tropical areas. In most countries, however, there is a distinct cyclical pattern, with increased transmission usually associated with the rainy season. Thus, in Thailand, peak DHF transmission occurs from June to October, while in Indonesia, peak transmission occurs from October through March. In both countries, these periods correspond to months with the highest rainfall and moderate to high temperatures. This is somewhat surprising because in both Thailand and Indonesia, the principal larval habitats for *Ae. aegypti* are clay water storage jars which are usually kept indoors and are not greatly influenced by rainfall. This would suggest that peak transmission has little to do with mosquito density in these areas. More likely, temperature and humidity conditions during the rainy season are more conducive to survival of the adult mosquito, thus increasing the probability that infected mosquitoes survive the extrinsic incubation period (EIP) and transmit the virus to other persons. The duration of the EIP is directly influenced by temperature, and at temperatures comparable to those observed in Thailand during the period of peak transmission, the EIP can be as short as 7 days for *Ae. aegypti*.[101] Also, in some rural and suburban areas, increased dengue transmission may be caused by other mosquito species such as *Ae. albopictus* which oviposit primarily outdoors and thus depend on rainfall.[102]

In the Americas, seasonal epidemic dengue activity is also associated with periods of highest rainfall and warmer temperatures in most countries. Because the majority of mosquito larval habitats are outdoors, however, increased transmission may also depend more upon mosquito population density. In certain Caribbean and Pacific Islands where rainfall is sporadic, water is often stored in metal drums and cement cisterns, thus providing continuous breeding.

Adult mosquito flight behavior may also influence seasonal transmission of dengue. Mosquitoes are less active during periods of heavy rainfall and therefore, if there is an option, they are more likely to remain indoors where the probability of survival is higher. This behavior would make it more likely that both infected and noninfected mosquitoes would have closer and more frequent contact with humans, thus increasing the probability of transmission.

Historically, epidemic dengue has also occurred in semitropical and temperature regions of the world. In these areas, transmission always occurs during warm summer months, and epidemics are uncommon because they depend on introduction of the virus from endemic areas. In the U.S., e.g., the last major epidemic of dengue occurred in 1922, although a smaller outbreak occurred in Louisiana in 1945.[103,104] More recently, it has been documented that all four dengue serotypes have been introduced repeatedly into the U.S.,[19] and although secondary transmission did occur in 1980, when DEN-1 moved up the east coast of Mexico and into Texas, a major epidemic never developed.[105] Most dengue virus introductions have occurred in the northeastern U.S. where *Ae. aegypti* does not occur regularly.

D. Risk Factors

In discussing risk factors, those which influence epidemic transmission must be separated from those which influence the severity of disease.

1. Epidemic Dengue

In permissive areas that are not endemic for dengue viruses, the mosquito vector and the human host are present, but epidemic transmission depends upon introduction of a suitable virus. Whether or not secondary dengue transmission occurs after virus introduction involves several interrelated factors including (1) the strain of virus, which may influence magnitude and duration of viremia in humans, (2) the susceptibility of the human population, (3) the density, behavior, and competence of the mosquito vector population, and (4) the introduction of the virus into an area where it has contact with the local mosquito population. Magnitude and duration of viremia, in conjunction with vector competence, density, and behavior of the mosquito population, determine how many mosquitoes become infected after the virus introduction. The susceptibility of the human population determines whether and how many indigenous cases of dengue will occur. For example, a person with high viremia may infect 100% of the mosquitoes taking a blood meal, and even though mosquito densities may be low, some may survive the EIP and transmit the virus. On the other hand, a person with low viremia may not infect any of the mosquitoes that take a blood meal, even though mosquito densities may be high.[31] Obviously, the susceptibility of the mosquito population is also important because a highly competent vector may become infected by feeding on a person with low viremia, while an incompetent vector may not become infected even if it feeds on a person with relatively high viremia. In the end, it is the number of mosquitoes that become infected and survive the EIP which determine whether secondary transmission occurs.

Adult mosquito behavior influences both survival and whether or not an infective mosquito that has survived the EIP actually feeds on another human, thus allowing transmission. In many of the large urban areas of Asia, all four dengue serotypes circulate simultaneously, although one serotype is usually dominant.[106] Epidemics may occur when levels of immunity to a certain serotype fall to a point where there are enough susceptible persons to support increased transmission. Evidence from Indonesia and the Pacific, however, suggests that the strain of dengue virus is also important.[31,43,44] Thus, like other animal viruses, dengue virus strains may vary in their epidemic potential and change genetically as they move through human and mosquito populations.

The selective pressures that act to change dengue viruses in nature are not known. One possibility is that incompetent mosquitoes that only become infected when feeding on persons with high viremia select only those virus strains that produce high viremia for propagation to the next generation of infections.[39,107] This would help explain why mosquitoes such as *Ae. aegypti*, which have relatively low susceptibility to oral infection with dengue viruses, are more frequently associated with explosive epidemic transmission.[107]

2. Severe and Fatal Disease

The risk factors associated with severe and fatal dengue infections are not well understood. Epidemiological studies in Thailand have suggested that an important risk factor for DSS is the presence of preexisting dengue antibody at subneutralizing levels.[14,69,72,100] This led to formulation of the "secondary infection" or "immune enhancement" hypothesis to explain DHF/DSS.[69] Briefly stated, the hypothesis implies that only those persons experiencing a second infection with a heterologous dengue serotype have DHF/DSS. This hypothesis is based on the belief that mononculear phagocytes are the primary site of dengue virus infection.[72] It is hypothesized that subneutralizing levels of heterologous dengue antibody complex with infective virus, thus enhancing phagocytosis. Once inside the cell, the complex breaks up and virus replication occurs. Disease severity is related to the number of infected monocytes that release vasoactive mediators into the bloodstream which may be responsible for the leaky capillary syndrome observed in DHF/DSS.

The association between secondary dengue infection and DSS has persisted in Thailand for over 20 years. More recently, well designed, prospective studies have suggested that the sequence of infecting virus serotypes is also important.[108] Thus, sequential infections in which the second infection was DEN-2 resulted in more severe disease. It should also be noted that the 1981 Cuban epidemic of DHF caused by DEN-2 was preceded in 1977 by a large DEN-1 epidemic.[109] Thus, the sequence of the epidemics (DEN-1 followed by DEN-2) occurring in a relatively short period (<5 years) agrees with data from Thailand.

Of interest is that infants under the age of 1 year also appear to be a significant risk group for DSS.[69,110] One possible explanation for these obvious cases of primary DSS is that as antibody passively acquired from the mother wanes below neutralizing titers, infection is enhanced in these infants.[69,111]

It should be noted that the question of preexisting dengue antibody as an important risk factor for DSS is still questioned by some.[112] Epidemiological and laboratory studies have demonstrated that the majority of cases in Southeast Asia occur in children who have had one or more previous dengue infections.[100] Age-stratified antibody prevalence studies, however, show that most children in the urban centers of Southeast Asia have had at least one dengue infection by the age of 6 years. The problem of determining the importance of secondary infection as a risk factor for DHF/DSS is that there are no good denominator data available in most endemic areas, and laboratory techniques required to determine whether a patient is experiencing a secondary or tertiary infection are not generally available in most laboratories. To confuse the issue even more, cases of primary DSS have been documented in older children and adults in most DHF endemic areas.[31,33,43,113]

The pathophysiology of primary DSS appears identical to that of secondary DSS.[43,112] It is not likely, however, that the sequence of events which has been hypothesized in "immune enhancement" occurs in primary infections since there is no preexisting antibody. One possibility is that nonantibody virus enhancement occurs. Thus, it has been demonstrated recently that dengue infection in mouse macrophage cultures is enhanced by substances of bacterial and parasitic origin.[114,115] These data are in support of another hypothesis that preexisting parasitic infections may alter the immune mechanisms of the host and enhance dengue virus infection of mononuclear phagocytes much in the same way that heterologous dengue antibody does.[116] It should be noted, however, that there is no good epidemiological evidence to support that latter hypothesis.

Most of the work on enhancement of dengue virus infection has been done in vitro. Furthermore, there is no good in vivo evidence that virus replication in humans is higher in secondary infections than in primary infections. In fact, available evidence from humans suggests the opposite.[32,117] Two studies cited as evidence for in vivo antibody-enhanced virus replication were done with monkeys.[118-120] In one study, peak DEN-2 viremias were shown to be higher in secondary than in primary infections.[118,119] It should be noted that these results were not repeatable in two subsequent studies on viremia in chimpanzees or monkeys.[30,180] In the second study cited as evidence for in vivo antibody-enhanced virus replication, 1 group of 5 monkeys was infused intravenously (i.v.) with pooled human sera that was known to have an in vitro enhancement titer of 1.2 million.[120] Another group of five monkeys was infused with a human serum that had no dengue antibody and both groups were then infected with DEN-2 virus. Higher viremias were observed in the group infused with dengue antibody. The experimental design of the latter study has been questioned, however, and it has been suggested that this may be just another in vitro experiment.[112] One possible explanation for the lower virus titers in secondary cases is that infectious viral antigen is tied up with the preexisting antibody, and therefore cannot be detected by currently used isolation systems. However, attempts to break up such antigen-antibody complexes with an oxidizing agent such as dithiotreitol have not succeeded in increasing isolation of dengue viruses.[183]

Another possible risk factor for severe and fatal dengue infection is the serotype and/or strain of the infecting virus. It is difficult to say if any of the four serotypes are more virulent than the others. In Thailand, DEN-2 has been the virus most frequently associated with DHF/DSS, although in recent years DEN-3 and DEN-4 appear to be more frequently associated with DSS.[69,100,108] All four dengue serotypes were isolated from severe and fatal DHF/DSS in Indonesia, but DEN-3 was the most important virus, with over 70% of fatal cases and 55% of all DHF/DSS cases associated with this serotype.[71,106] In addition, the 1966—1967 epidemic of DHF in the Philippines was also caused by DEN-3.[121] In Burma,

DEN-2 was the predominant virus isolated from DSS cases in 1976, although DEN-3 was isolated almost as frequently.[122] Finally, in Malaysia, the 1982 epidemic was caused by DEN-1 and DEN-3.[123] Also of interest was that in Colombo, Sri Lanka, where all four dengue serotypes are endemic and the majority of cases are secondary, increased DEN-2 transmission in 1982 was associated with mild flu-like illness, not with DHF/DSS.[124] DEN-4 has generally been less frequently isolated from DHF/DSS patients in Southeast Asia, although a small epidemic of DHF/DSS in Koh Samui, Thailand, in 1967—1968 was caused by this serotype.[125] Also, during 1985—1986, DEN-4 has been the predominant virus isolated from DSS cases in Rayong, Thailand.[188] Finally, a 1984 epidemic of DEN-4 in Yucatán, Mexico, resulted in eight confirmed cases of hemorrhagic disease, four of them fatal.[189] It must be concluded, therefore, that all four serotypes have the potential for causing severe hemorrhagic disease.

In recent years, there has been an accumulation of field and laboratory evidence that documents strain variation among dengue viruses (see Section II.C). Field evidence with three serotypes (DEN-1, -2, and -3) suggests that there are epidemic strains of dengue virus which are associated with more explosive transmission, higher viremia levels, and more severe disease.[31,44,106] The recent development of methods to study viral RNA has allowed more detailed study of virus strain variation. As noted in Section II.C, DEN-2 has been studied most extensively, but differences have not been noted between strains isolated from patients with mild vs. severe disease. Moreover, strains isolated from different epidemiological situations have not been examined fully so that conclusions cannot be drawn regarding endemic vs. epidemic strains of dengue virus. There are no good animal models that show virulence differences among dengue virus strains.

Age is another risk factor associated with severe and fatal dengue disease.[15] The modal age group of most confirmed cases of DHF/DSS in Southeast Asia is 5 to 9 years, with fewer cases occurring in younger and older children and in adults. The reasons for this age relationship are not known.

Other, less understood risk factors identified by some workers include sex, nutrition, and race or genetic background. In Thailand, the male:female ratio for all DHF cases was 1:1.4, and in unconfirmed, clinically diagnosed fatal cases, it was 1:1.2.[126,127] One possible explanation suggested for this difference was that females may produce more antibody than males.[69] In Indonesia, the male:female ratio for all confirmed DHF cases was 1:1.2, but for virologically confirmed fatal cases, it was 1.5:1, with 60% being males.[71] The fatality rate was also higher in males in a Philippine Island study.[128] Thus, there is still some question as to whether sex is an important risk factor for DHF/DSS.

In Bangkok, it was observed that DSS occurred only in those children who appeared healthy and well nourished, suggesting that good nutritional status increases the risk of DHF/DSS because well-nourished children are immunologically more competent.[69] In Indonesia, however, laboratory-confirmed DSS has been documented in malnourished children,[129] suggesting that nutrition may not be such an important risk factor for DHF.

Although not well studied, evidence is accumulating which suggests that the genetic makeup of the population or individual may be another important risk factor for severe and fatal dengue infection. Thus, in Thailand, Indonesia, and Niue Island, familial clustering of cases with severe and fatal outcome have been observed.[16,130,190] Furthermore, in Cuba, the incidence of hemorrhagic disease was significantly higher in whites than in blacks.[17] In Sri Lanka, where DHF occurs only sporadically, it is perhaps of significance that positive tourniquet tests, which indicate increased capillary fragility, are seldom observed in confirmed dengue patients.[191] Only one study has attempted to look at a genetic marker (histocompatibility antigen or HLA) in relation to DHF.[131] A positive correlation with DSS was observed for HLA-A2 and HLA-blank. A negative correlation was observed for HLA-B13. Finally, there is some evidence that those cases of dengue which present with neurological disorders may be patients with some underlying predisposition for neurological problems.[79]

In summary, until the pathogenesis of DHF is better understood, risk factors associated with this disease cannot be clearly defined. Current data suggest that age, immune status, virus strain, and genetics are the most important risk factors associated with severe and fatal dengue infection.

E. Serological Epidemiology

Seroepidemiological studies have been and are invaluable in obtaining background information on endemicity of dengue. Antibody surveys of the population, using the HI test, can determine age-stratified dengue infection. Because the HI test is not specific, however, some knowledge of the distribution of other flaviviruses in the area is necessary for correct interpretation.

Even in Asia where endemicity is highest, there are some countries that have endemic dengue but do not have epidemic DHF, while other countries have epidemics of DHF at regular intervals. Despite a great deal of research on DHF, however, the risk factors associated with epidemics are not well understood. It is not clear, e.g., whether all endemic countries are at risk, or whether there are epidemiological factors that can be used to predict epidemic DHF (see Section IV.D.1).

Prospective seroepidemiological studies are necessary to obtain background data on a study population and are essential to study the relative risk of primary and secondary dengue infections as a cause of DHF. Such studies are currently underway in Sri Lanka and Puerto Rico, areas where DHF does not regularly occur, and in Thailand, Burma, and Indonesia, areas where DHF is highly endemic. These studies should help define risk factors for DHF/DSS and determine whether immunological or virological factors are more important (see Section IV.D). The result will influence decisions regarding use of mono- or quadrivalent dengue vaccines.

V. TRANSMISSION CYCLES

Dengue viruses have three basic transmission cycles:[132] (1) a forest cycle involving lower primates and forest *Aedes* species, (2) a rural or semirural cycle involving humans and peridomestic *Aedes* species, and (3) an urban cycle involving humans and domesticated *Aedes* species (Figure 3). There may be some overlap between each of these cycles, depending on where they occur and the mosquito species involved. For example, in certain parts of Southeast Asia, evidence suggests that there is an overlap, with *Ae. albopictus* acting as a link between the three types of transmission cycles.

A. Evidence from Field Studies

1. Asia

Early work in the Philippines demonstrated that nonhuman primates were infected with dengue viruses in nature, leading to speculation that the natural maintenance cycle of dengue viruses involved nonhuman primates and forest mosquitoes in the jungles of Asia.[9,132] Subsequently, extensive field work was done in Malaysia over a 20-year period, eventually documenting that dengue viruses were maintained in the forests of Malaysia in a cycle involving canopy-dwelling *Ae. (Finlaya) niveus* mosquitoes and monkeys.[34,132,133] Evidence for this jungle cycle was as follows: (1) a high percentage (68%) of wild monkeys had flavivirus antibodies, primarily against dengue and Zika viruses; (2) natural dengue virus infection was demonstrated in sentinel monkeys in the forest by both isolation of virus (DEN-1, -2, and -4) and seroconversion (DEN-1, -2, and -3); and (3) DEN-4 virus was isolated from a pool of *Ae. (F.) niveus* collected from the canopy of the forest. It was concluded that dengue in Peninsular Malaysia exists in a silent enzootic jungle cycle involving canopy-

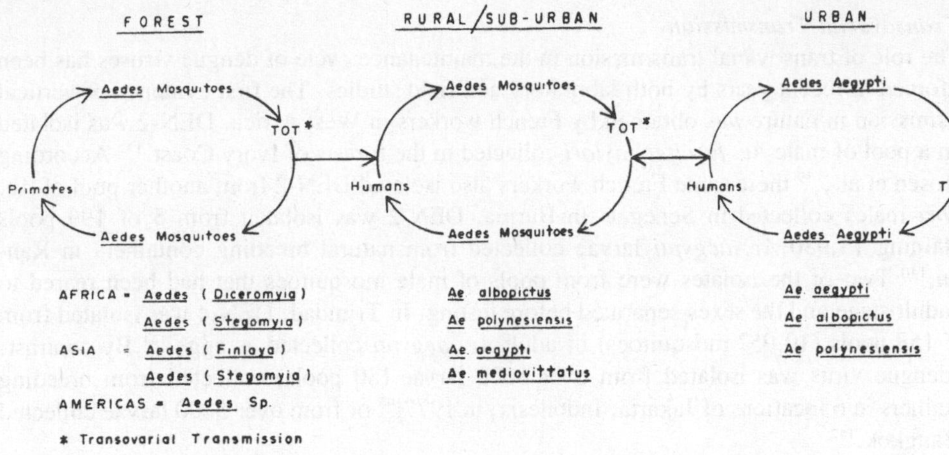

FIGURE 3. Transmission and maintenance cycles of dengue viruses.

dwelling mosquitoes of the *Ae. (F.) niveus* complex and monkeys, in a rural endemic cycle involving *Ae. (S.) albopictus* and humans, and in an urban cycle involving *Ae. (S.) aegypti* and humans. In Vietnam, dengue virus of an unknown serotype was isolated from *Ae. (F.) niveus* collected in a forest area.[134]

2. Africa

Studies in Nigeria in the early 1970s provided the first good evidence of a forest enzootic cycle of dengue in Africa.[135,136] Thus, both humans living in areas where *Ae. aegypti* was not prevalent and monkeys from both a lowland rain forest and a gallery forest had dengue N antibody. More recently, this forest cycle has been confirmed by isolation of over 300 DEN-2 viruses from wild-caught mosquitoes in West Africa during 1980 and 1981.[35,137] The isolates were from five species of mosquitoes: *Ae. (S.) africanus*, *Ae. (S.) leuteocephalus*, *Ae. (S.) opok*, *Ae. (D.) taylori*, and *Ae. (D.) furcifer*. Furthermore, two of the isolates were from pools of male mosquitoes, suggesting that transovarial transmission may play a role in the natural maintenance of dengue viruses in certain situations. This aspect of dengue ecology is discussed in more detail below.

3. Americas

In the Americas, evidence of a forest maintenance cycle for dengue viruses is more circumstantial because extensive field studies have not yet been carried out. Nevertheless, DEN-2 neutralizing antibody has been detected in Ayoreo Indians living in the remote Rincón del Tigre area of Bolivia.[138] These persons had not traveled outside the area, and *Ae. aegypti* was not present in that part of the country, suggesting that the infections may be part of a forest cycle similar to that observed in Asia and West Africa. Other more circumstantial evidence suggests that *Ae. (G.) mediovittatus* may be involved in a natural maintenance cycle for dengue on some Caribbean Islands.[36] This species is a forest mosquito that has moved into the peridomestic environment and shares many larval habitats with *Ae. aegypti*. *Ae. mediovittatus* is a very common mosquito in rural and suburban communities of Puerto Rico where dengue virus transmission has been maintained continuously for over 11 years.[192] Field studies have shown that *Ae. mediovittatus* feed avidly on humans, and their biting activity cycle is similar to that of *Ae. aegypti*.[36] It should be noted that there are no primates in the Caribbean Islands and, therefore, the role of *Ae. mediovittatus* in the maintenance of dengue viruses is probably analogous to that of *Ae. albopictus* in Asia.

4. Transovarial Transmission

The role of transovarial transmission in the maintenance cycle of dengue viruses has been reinforced in recent years by both laboratory and field studies. The first evidence of vertical transmission in nature was obtained by French workers in West Africa. DEN-2 was isolated from a pool of male *Ae. furcifer-taylori* collected in the forests of Ivory Coast.[137] According to Rosen et al.,[139] these same French workers also isolated DEN-2 from another pool of *Ae. taylori* males collected in Senegal. In Burma, DEN-2 was isolated from 5 of 199 pools containing 13,930 *Ae. aegypti* larvae collected from natural breeding containers in Rangoon.[140] Two of the isolates were from pools of male mosquitoes that had been reared to the adult stage and the sexes separated before testing. In Trinidad, DEN-4 was isolated from 1 of 158 pools (10,957 mosquitoes) of adult *Ae. aegypti* collected as eggs.[141] By contrast, no dengue virus was isolated from over 8000 larvae (80 pools) collected from breeding containers in 6 locations of Jakarta, Indonesia, in 1977[193] or from over 5000 larvae collected in Bangkok.[142]

B. Evidence from Experimental Infection Studies

1. Vectors

Both of the major mosquito vectors of dengue — *Ae. aegypti* and *Ae. albopictus* — have been studied extensively in the laboratory, and susceptibility to oral infection with dengue viruses has been shown to vary greatly among geographic strains of both species.[143,144] In general, however, it has been shown that *Ae. albopictus* has a higher susceptibility and is a more efficient host for dengue viruses than *Ae. aegypti*.[39,102,143,144] Other species such as *Ae. polynesiensis*, *Ae. mediovittatus*, and *Ae. triseriatus* have also been shown experimentally to have a higher susceptibility to oral infection with dengue viruses than *Ae. aegypti*.[36,37,39] Moreover, experimental studies on transovarial transmission have shown that the filial infection rate in *Ae. aegypti* is very low compared with other species of *Aedes*.[139] Collectively, the data suggest that *Ae. aegypti* is a less efficient host for dengue virus than certain other species of *Aedes*.[39,107]

There is little doubt that *Ae. aegypti* is the most important epidemic vector of dengue and DHF worldwide, primarily because of the highly domesticated habitats of this species and its close association with humans. Because of its generally low susceptibility to oral infection with dengue viruses, however, *Ae. aegypti* must feed on persons with high viremia to become infected, and it follows, therefore, that only those viruses associated with high human viremia would be transmitted by this species, while those viruses producing low viremia would probably not be transmitted.[39,107] This type of virus/vector relationship may be a major factor in selecting and propagating epidemic strains of dengue viruses in urban situations.[107] In contrast, viruses in semirural, rural, and forested areas could be maintained by more efficient vector species of *Aedes* in a cycle combining transovarial transmission with periodic amplification in humans or monkeys. This type of maintenance cycle also may occur in some urban areas where local *Ae. aegypti* populations are highly efficient hosts for dengue viruses.[140,141]

2. Vertebrate Hosts

There is considerable field evidence from both Malaysia and Africa that lower primates are involved in forest maintenance cycles of dengue viruses.[34,35,132,133,135-137] Moreover, experimental laboratory data show that chimpanzees, gibbons, and macaque monkeys are susceptible to infection with dengue viruses (see Section II.B).[27-30] All species develop detectable viremia in the absence of clinical illness. The experimental infection data suggest that dengue viruses have become well adapted to lower primates which, therefore, are not useful as laboratory animal models for the study of human disease.

VI. ECOLOGICAL DYNAMICS

As noted from the above discussion, the ecology of dengue viruses is influenced by many factors involving the vertebrate host, the mosquito vector, and the virus, as well as the environment which directly or indirectly affect all three populations involved and the interactions of each.

A. Macro- and Microenvironment

The environment is very important in determining the transmission dynamics of dengue viruses. The size and frequency of dengue epidemics are related directly to urbanization, which influences the number of susceptible human hosts available and the density of the mosquito vector population. Major epidemics are seldom observed in rural areas where human population densities are lower, and have never been described in forest areas where only lower primates and mosquitoes are involved in the transmission cycle. In the latter areas, survival of the virus probably depends more on the presence of a highly susceptible host mosquito that can transmit the virus transovarially to the next generation, with only periodic amplification in the vertebrate host.

The influence of the microenvironment on transmission dynamics is more difficult to define. Mosquito population densities are affected by the types of larval habitats available. The latter determines not only the number of mosquitoes, but also the nutritional status of the larvae, which in turn influences size, survival, and possibly vector competence. For example, *Ae. aegypti* that emerge from bromeliads will be smaller in number and body size than those emerging from cisterns or other larval habitats containing large volumes of water. Resting sites are another example of how microenvironment can influence transmission. Mosquitoes resting outdoors where they are exposed to the wind and rain as well as to numerous predators probably have a lower survival rate than those that seek resting sites inside houses where temperature, humidity, and air movement are more constant. Moreover, the generally warmer temperatures indoors would be more conducive to virus replication and thus a shorter EIP.

B. Climate and Weather

The importance of climate and weather is obvious since mosquitoes are cold-blooded animals and their activity cycles as well as virus replication in their tissues are influenced by temperature. Thus, dengue viruses are only endemic in tropical areas of the world where climate and weather allow continuous breeding populations of mosquitoes. In subtropical and temperate regions of the world, periodic epidemics of dengue may occur, but the viruses are not endemic and must be introduced before transmission is initiated.

Climate and weather also directly affect transmission dynamics of dengue viruses by acting as a selective pressure on the mosquito population. For example, long periods of drought and hot temperature may kill the majority of *Aedes* eggs, allowing only a few to hatch following flooding. This type of pressure could change the genetic makeup of the population considerably and, therefore, change the competence of the vector population for dengue viruses.[143-145] Obviously, these types of environmental pressures would be greater on those mosquito populations ovipositing outdoors where containers depend on rainwater for flooding than on those populations breeding indoors where containers are frequently refilled. Finally, as noted in Section IV.C, temperature has a direct effect on vector competence. High temperatures of around 30°C tend to speed up virus replication and shorten the EIP, whereas cooler temperatures prolong the EIP.[101]

C. Vector Oviposition

The oviposition behavior of the mosquito vectors of dengue viruses indirectly influences transmission dynamics by determining the proximity of adult mosquitoes to the human or primate hosts. Thus, even though *Ae. aegypti* is a less efficient host for dengue viruses,[39] it is the most efficient epidemic vector for dengue and DHF because it is highly domesticated. Since the adult female mosquito prefers to oviposit in containers with clean water located in and around human dwellings, emerging adult *Ae. aegypti* females do not usually have to fly very far to obtain a blood meal. Moreover, the engorged female does not have to migrate far to find a suitable resting place because she also prefers to rest indoors. The entire life cycle of *Ae. aegypti*, therefore, may take place in or very near human dwellings where there are numerous potential larval habitats in close proximity to human hosts.

By contrast, other mosquito species such as *Ae. albopictus, Ae. polynesiensis,* and *Ae. mediovittatus*, although generally more susceptible to infection with dengue viruses, are less efficient epidemic vectors of dengue because the gravid female mosquitoes prefer oviposition sites that are outdoors and frequently away from human habitations. Emerging female mosquitoes have less contact with humans, decreasing the probability that they will feed on a viremic human host, become infected, and subsequently transmit the infection to another susceptible human.

D. Vector Density, Fecundity, and Longevity

The influences of fecundity, longevity, and density of the mosquito vector population on dengue transmission are interrelated since fecundity and longevity influence density, and the probability of at least some infected mosquitoes surviving long enough to transmit the virus is directly affected by mosquito density and daily survival (longevity). The number of eggs a mosquito lays per gonotrophic cycle shows a positive correlation in *Ae. aegypti* with the size of the female and the amount and type of blood ingested.[146] Although there are exceptions, most *Ae. aegypti* females lay less than 100 eggs per batch.

Only limited data are available on longevity of dengue mosquito vectors in nature. In Bangkok, it was estimated that the average female *Ae. aegypti* survived approximately 8.5 days while in Calcutta, *Ae. albopictus* survived approximately 7 days.[147,148] Since the EIP for dengue viruses is 8 to 12 days, it is clear that high mosquito densities are necessary to maintain transmission. Daily survival of adult female mosquitoes is dependent upon many factors, including larval nutrition, availability of blood (protein) and carbohydrate, climate, presence of predators, resting, flight and feeding behavior, etc. Space does not allow a detailed discussion of all these factors, but many are interrelated and all influence mosquito density.

Assuming that the virus(es) is available in an area, there are two important factors that determine whether a pathogen is successfully maintained in nature: (1) adequate numbers of susceptible vertebrate hosts and (2) adequate numbers of susceptible adult mosquitoes. The rationale behind most disease control programs based on mosquito control is the reduction of mosquito densities to a point below the threshold where epidemic transmission can occur. Unfortunately, that threshold density has never been defined clearly for dengue viruses, possibly because there are still no good methods of measuring adult *Ae. aegypti* density. The usual approach is to do landing and/or biting counts, but this method is very time consuming, and at best, provides only a rough estimate of relative abundance. Three larval indexes are used most commonly to measure *Ae. aegypti* densities. The house or premise index simply reflects the percentage of premises that have *Ae. aegypti* breeding; the container index is the percentage of containers in an area with water that is positive for larvae; and the Breteau index is the number of positive containers per 100 houses.

Unfortunately, none of these indexes has shown a consistent correlation with dengue transmission which would allow definition of a threshold. Some workers have speculated that an *Ae. aegypti* house index of 5% is the threshold for dengue and urban yellow fever transmission, but major epidemics have occurred in areas with *Ae. aegypti* densities below that level.[149] Moreover, different geographic strains of mosquitoes vary in their vector competence for dengue (see Section VI.F), and daily survival rates also vary considerably. Thus, the threshold below which epidemic transmission will not occur probably varies greatly and must be determined for each location.

E. Biting Activity and Host Preference

The influences of biting activity and host preference on dengue transmission are also interrelated to each other and to mosquito density. Biting activity cycles determine when mosquitoes will have contact with the vertebrate hosts. In the case of dengue, both principal mosquito vectors *Ae. aegypti* and *Ae. albopictus* have similar cycles with two peaks of biting activity: one in the early morning right after daybreak and one in the late afternoon.[148,150,151] In most areas of the world where these mosquitoes occur, this biting activity cycle coincides with human activity and, therefore, increases the chances for human-mosquito contact and the probability of virus transmission.

An important factor that makes *Ae. aegypti* such an efficient vector of epidemic dengue is its habit of taking partial blood meals, or interrupted feeding. It is not uncommon for a single *Ae. aegypti* to bite several people in the same room or house before becoming satiated on blood. This is probably a survival mechanism, since even the slightest movement of the person being bitten or another person nearby will distract the mosquito and interrupt the blood feeding. Because of this habit, a single infected *Ae. aegypti* may transmit dengue virus to two or more persons within a short period of time. For example, in 1984, four members of an American family living in Honduras were probably infected with DEN-2 virus by a single infected female mosquito since all had onset of confirmed dengue within a 2-day period.[193]

Some mosquitoes have very strict host preferences, while others are less selective in choosing a host for a blood meal. The principal dengue vectors fall into the latter category. While domestic *Ae. aegypti* certainly prefer to feed on humans, they also readily engorge on dogs, cats, and other domestic animals.[150] *Ae. albopictus* is even less selective and randomly feeds on a number of domestic and peridomestic mammals and birds.[152] Moreover, being a peridomestic species, it has less contact with humans. Thus, if domestic animal populations are low, it is likely that the majority of the dengue mosquito vectors will seek out and feed on the human host, thereby increasing the probability of virus transmission. If domestic animal populations are high, however, the probability of feeding on human hosts and thus the probability of transmission are decreased.

On the other hand, there is no good evidence that increased densities of dengue mosquito vectors influence feeding patterns as has been described for other mosquitoes.[153] *Ae. aegypti* is an especially innocuous mosquito and persons who live in endemic areas seldom know that they are being bitten because they have become desensitized. It is possible, however, that increased mosquito densities in the domestic environment may stimulate people to implement some form of control or personal protection, which would then decrease adult mosquito density or contact and possibly transmission.

F. Vector Competence

An important aspect of dengue ecology involves genetic heterogeneity in the mosquito vector populations. Geographic strains of both *Ae. albopictus* and *Ae. aegypti* vary considerably in their susceptibility to oral infection with dengue viruses, and thus in their ability to transmit the infection.[143,144] These studies also showed that susceptibility to oral infection was dose related; therefore, a mosquito strain could become infected by increasing the amount of virus ingested, regardless of how resistant it was to dengue virus infection. In both mosquito species, the factors controlling susceptibility to oral infection were the same for all four dengue serotypes; susceptibility was controlled by a midgut or mesenteronal escape barrier, and there was no evidence of dissemination or salivary gland barriers to dengue infection.[143,144,193] Finally, susceptibility to oral infection with dengue viruses was genetically controlled in both species.[143,144] Thus, environmental or other selective pressures may cause changes in susceptibility of the mosquito population to dengue viruses, suggesting that vector competence may be an important risk factor for epidemic dengue transmission.[107]

Investigators in several laboratories are studying natural genetic variation among strains of *Ae. aegypti* by isoenzyme analysis using gel electrophoresis.[154-157] Attempts have been made to characterize strains of *Ae. aegypti* with high and low dengue susceptibility by this technique in an effort to identify genetic markers for susceptibility and refractoriness.[157] Preliminary results have failed to show any conclusive relationship between dengue susceptibility and enzyme banding patterns among either geographic strains or different morphotypes of *Ae. aegypti*.[192] The studies have shown, however, that this species can be separated into geographic groups based on genetic relationships, an observation that has obvious importance in determining the origin of mosquito strains that have reinfested an area after eradication.[155,156] Whether this type of genetic analysis of mosquito populations will increase our understanding of mosquito vector competence and its role in the distribution and spread of epidemic dengue must be the subject of further study.

G. Movement of Humans

Prior to the 1960s, major dengue epidemics occurred at infrequent intervals in most tropical areas of the world. Since World War II, there has been a constant increase in incidence of disease caused primarily by increased frequency of epidemics. The principal factor responsible for this increase has been the movement of people between and into areas that are permissive for dengue transmission.

Important changes in the ecology of dengue first occurred in Southeast Asia after World War II when millions of people seeking food, shelter, and work moved from rural areas to the urban centers of several countries. Environmental sanitation and water systems were not adequate to handle such a large influx of people, and as a result, mosquito densities increased followed by increased incidence of mosquito–borne diseases such as dengue. The emergence of epidemic DHF followed shortly, and within 20 years, this disease had become a major public health problem in most Southeast Asian countries.

The second important factor influencing the movement of humans and, consequently, the spread of dengue, was the advent of the jet airplane. Since the early 1960s, there has been a constant increase in air travel. The convenience and speed of jet air travel provides the ideal mechanism for the transport of dengue viruses between population centers of the tropics. For example, a traveler visiting a Southeast Asian city may become infected with dengue and travel around the world to permissive cities of the American or African tropics before becoming ill. Dengue viruses from Asia and tropical America are imported into the U.S. in this manner every year,[159] and it is not unlikely that similar introductions are made into other areas permissive for virus transmission. The movement of humans, plus the breakdown or absence of effective mosquito control in most cities of the tropics, are the most important reasons for the increased frequency of epidemic dengue and DHF in permissive areas of the world.

H. Human Behavior

Human behavior, directly or indirectly, has a significant impact on dengue transmission. It is difficult, however, to separate true cultural behavior from that dictated by economic conditions.

In most parts of the tropics, economic conditions are responsible for human behavioral patterns that influence dengue transmission, including the practice of water storage in the house and the intense human crowding that occurs in most urban centers of the tropics. These conditions allow large numbers of mosquitoes to live in close contact with a crowded human population, conditions which are highly favorable for transmission of mosquito–borne disease. Moreover, lack of air conditioning in many tropical areas influences people to sit outdoors in the late afternoon and early evening, a time of peak biting activity for dengue mosquito vectors.

In contrast, changes in human behavior in many areas of the southern U.S. constitute a deterrent to dengue transmission because people are generally inside air-conditioned offices and houses during peak biting activity periods of the mosquitoes. The probability of being bitten by a mosquito, therefore, is greatly decreased. Television and air conditioning have had a similar effect on transmission of arboviral encephalitis in California.[160]

Another human behavioral pattern influencing dengue transmission, especially in the Caribbean, is the accumulation of domestic trash. This is the primary source of mosquito breeding on some of the more affluent islands such as Puerto Rico where there is a piped water system to most communities. On the other hand, periodic clean-up campaigns carried out by some communities may influence selection of increased or decreased vector competence in the mosquito population by eliminating all but a few eggs (see Section VI.B).

VII. SURVEILLANCE

Surveillance for dengue and DHF can be of two basic types: reactive or active. Most surveillance for dengue has been of the reactive type, which depends upon monitoring a specific clinical illness such as dengue fever or DHF (case reports) by the medical community. This type of surveillance is insensitive because of the low index of suspicion on the part of physicians and the inherent difficulties in clinical differential diagnosis. Cases are often reported as influenza or nonspecific viral syndrome and, as a result, an epidemic may be near peak transmission before it is recognized as being caused by dengue. At that point, it is generally too late to have much impact on transmission, even though intensive mosquito control measures may be implemented. Although reactive surveillance has no predictive capability and is costly to the community, both in terms of human illness and the necessity for epidemic control measures, it is logistically and organizationally easy to set up. Moreover, once epidemic transmission is recognized, it is very easy to mobilize the government, medical, and lay communities to obtain support for epidemic control.

The alternative strategy, active surveillance, is based on the fact that during interepidemic periods and periods of silent transmission, most dengue infections are not recognized clinically. With the introduction of a new virus strain or serotype, there is usually a period of low-level transmission or a ''lag phase'' that may last from a few weeks to several months before epidemic transmission begins. The objective of active surveillance is to detect the new virus during this ''lag phase'' well in advance of increased transmission. If applied at this stage, mosquito control measures might effectively abort an incipient epidemic. To achieve predictive capability for epidemic dengue, the active surveillance system must utilize a laboratory that is capable of carrying out relatively rapid and sensitive diagnostic methods.

A. Virological Surveillance

The most important type of active surveillance for providing an early warning, predictive capability for epidemic dengue is virologic. A sensitive, relatively rapid and inexpensive virus isolation system is necessary. Such a system using mosquito cell cultures for isolation and monoclonal antibodies for virus identification is now available,[62,66] and dengue viruses can be isolated and identified in as short a period as 2 to 3 days.[183] New methods designed to detect antigen in viremic sera also show promise for even more rapid specific diagnosis (see Section III). Examples include a multisite monoclonal RIA,[91] an immunofiltration assay,[92] and the use of cDNA hybridization probes that can be constructed to detect specific parts of the viral genome.[51]

The principal objective of an active virological surveillance system is to monitor the dengue viruses being transmitted in an area during interepidemic periods and to determine their movement in a specified area. For example, in Puerto Rico, dengue activity is monitored during interepidemic periods by obtaining five to ten blood samples per week from viral syndrome patients who visit selected health centers and private physicians on the island. These samples are processed immediately for virus isolation without regard for serology. While not as sensitive as it should be, the system has allowed detection of newly introduced viruses well in advance of increased transmission.[192] With the ability to rapidly identify the introduction of a new virus serotype, it should be possible to prevent major epidemics by implementing mosquito control measures immediately after detection of the virus.

An important additional benefit from having a good virological surveillance system is that unpassaged viruses are available for study. The recent application of the RNA oligonucleotide fingerprint technique has provided a method for determining the geographic origin of dengue viruses.[48-51] Moreover, techniques for geographic classification employing cDNA probes and antigen signature analysis have been developed recently and may be more useful for rapid characterization of a newly introduced virus.[51,52]

The disadvantage of active virological surveillance is that it is difficult to motivate the medical community to cooperate and take blood samples from patients who do not appear to have dengue. Probably the most difficult part of this type of surveillance is to change the way health authorities and physicians think, and convince them to place emphasis on the interepidemic period and prevention rather than the epidemic period and control. This change in thinking is critical, however, to developing an early warning surveillance system that is necessary for epidemic prevention.

B. Epidemiological Surveillance

The objective of epidemiological surveillance is to monitor disease activity to learn how much dengue-like illness is occurring in the community. The first step is to make dengue a reportable disease and have physicians make weekly reports of dengue-like illness to central health authorities. This type of epidemiological surveillance is reactive and, therefore, not very sensitive. However, it serves a useful purpose in providing statistical data for reporting purposes.

More important for developing a predictive capability for epidemic dengue is the reporting of increased fever of unknown origin (FUO) activity, something that is seldom done in endemic dengue areas. Increased febrile illness in a community may be due to any number of etiological agents, including dengue. When an increase in FUO of any kind is noted by the medical community, health authorities should be notified immediately and the cases investigated by epidemiology and laboratory personnel. Blood samples can be taken and processed for virological and/or serological diagnosis without delay to determine whether dengue is the etiological agent responsible.

C. Clinical Surveillance

Clinical surveillance for classical dengue and DHF/DSS is relatively insensitive, depending upon the awareness and interest of the medical community. Surveillance for fatal hemorrhagic disease, however, may be more effective as an early warning of epidemic activity. An example of this was observed in Indonesia where virologically confirmed fatal dengue infections were monitored from 1975 to 1978.[106] During the first 5 months of the study, a single DEN-1 virus was isolated. In March and April of 1976, there was an increase in the number of cases with a fatal outcome, and most of those were associated with DEN-3 infection. Six months later, it was this virus that caused a series of epidemics throughout Indonesia, suggesting that the DEN-3 strain isolated in Jakarta in the spring of 1976 was a new epidemic strain of virus. Thus, it may be feasible to detect more virulent or epidemic strains of virus well in advance of actual epidemic transmission by monitoring a viral illness with a fatal outcome.

In areas where DHF/DSS is not endemic, the disease should be defined through educational programs for the medical community. The clinical surveillance system in these areas should include reporting all cases of hemorrhagic disease in the community. In addition to allowing early detection of DHF/DSS, this program will help define the disease on the severe end of the clinical spectrum in nonendemic areas.

D. Serological Surveillance

In the past, serological surveillance has been of limited use for predicting epidemic activity because it was necessary to have paired acute and convalescent sera collected 14 days apart for testing, and laboratory reporting was slow. When used in conjunction with epidemiological case reporting, however, it is very useful to determine what proportion of reported cases is actually dengue. With the development of new methods for measuring dengue IgM antibody, serological diagnosis is somewhat faster. The MAC-ELISA has been found to be extremely useful for dengue surveillance because persons with specific IgM antibody are known to have had a dengue infection at some time in the previous 2 months.[192] IgM is detectable in most patients by day 5 of illness and persists for as long as 90 days after infection. Most patients, however, have lost IgM antibody by 60 days postinfection.

E. Entomological Surveillance

Entomological surveillance deals with the mosquito vectors and requires knowledge of the species present, species associations, distribution, types and productivity of larval habitats, seasonal changes in population density, and something about the behavior of the principal vector. Once this type of information is available about a permissive area (*Ae. aegypti* is present), constant surveillance for the mosquitoes is not necessary unless an active control program is in operation. However, periodic larval surveys and tests for insecticide susceptibility should be carried out to determine whether changes are occurring in distribution, larval ecology, and susceptibility of the principal vector species. Unless mosquito population densities are exceptionally low (house index of less than 5%), entomological surveillance has little or no predictive value for epidemic transmission.

In areas where *Ae. aegypti* has been eradicated, however, entomological surveillance should be routine using ovitraps and regular periodic larval surveys for detection of introduced mosquitoes. Emphasis should be placed in those areas where introductions are most likely to occur such as ports, airports, and used tire depots.

VIII. INVESTIGATION OF EPIDEMICS

Epidemic dengue and DHF generally occur at regular intervals in endemic areas and at irregular intervals in nonendemic areas. Unfortunately, the factors responsible for the distribution and spread of epidemic dengue are not well understood. The roles that virus strain/serotype and vector competence play are not well defined, and there may be other intrinsic or extrinsic host-, vector-, and virus-related factors that influence transmission. Active investigation of these parameters during epidemic transmission will provide more information on risk factors in a shorter period of time than can be obtained during inter-epidemic periods or by retrospective investigation of epidemics. Moreover, epidemic transmission provides the ideal opportunity to clinically define the disease in nonendemic areas because large numbers of cases are available for study.

In Chapter 12 of Volume I, Gregg outlines the basic information that should be sought by epidemiological investigation of an arbovirus disease outbreak. The same principles apply to dengue. In addition, however, the investigation should attempt to identify host-,vector-, virus-, and environmentally related factors that influence transmission and thus, the distribution and spread of dengue viruses. The clinical disease associated with dengue virus infection in epidemic areas should be carefully defined in relation to risk factors associated with severe and fatal hemorrhagic disease.

With the first suggestion of epidemic transmission, a population-based, age-stratified randomized serological survey should be carried out in the epidemic area. It is important to obtain an adequate sample from age groups below 15 years because this is the population at highest risk for DHF. After epidemic transmission has subsided, another blood sample should be taken, preferably from the same persons. The paired samples are then tested by HI, PRNT, or ELISA to determine the rate of transmission and to provide age-stratified baseline denominator data on background immunity in the study population, data which are essential for study of risk factors associated with severe and fatal disease. At the same time as the second serological survey, a random household clinical questionnaire survey should be conducted in the community. This survey, combined with data from the serosurvey, will provide information on the apparent to inapparent rate of illness, and combined with entomological and environmental data, will provide information on transmission risk factors. In addition, case control studies can be very useful to study host-related factors that require other types of testing such as HLA and the affect of dengue infection on pregnancy outcome.

A clinical-virological surveillance program should be set up in the epidemic area to provide specimens for virus isolation and identification, and to provide clinical information on disease associated with dengue infection. Emphasis should be placed on severe disease, but this type of study is best set up in collaboration with selected local physicians who can screen patients and collect the specimens and clinical information required. Alternatively, or in addition, an epidemiologist-physician can be assigned to a clinic or hospital in the area where most of the cases of severe disease would be referred.

Entomological studies that should be carried out during the epidemic are determined by the local epidemiological situation. In highly urban areas where *Ae. aegypti* is the only mosquito vector, larval surveys are adequate to provide data on mosquito densities, although landing, resting, and ovitrap collections may also be made.

In areas where other potential *Aedes* mosquito vectors occur, larval surveys should be carried out to determine associations and relative abundance of the species involved. In addition, adult mosquito collections should be made, the species separated and tested for virus infection to determine the relative importance of each species in actual transmission.

In both situations, larvae and/or eggs should be collected from the epidemic area, reared to the fourth larval instar, and processed for virus isolation to determine whether transovarial transmission is involved. Additionally, the mosquito species from the epidemic area may be colonized and tested for their vector competence for the epidemic and other strains of dengue viruses.

IX. PREVENTION AND CONTROL

The options available to us for prevention and control of epidemic dengue and DHF are somewhat limited. Obviously, the most effective method of prevention would be to eradicate the principal vector mosquito *Ae. aegypti*. Unfortunately, experience in the American region as well as in other parts of the world indicates that eradication is not a very realistic option, even though it may be technically feasible. Thus, to be truly effective, all countries in a particular region must achieve eradication and then maintain constant surveillance for reintroduction of the species. A second option is to regulate air travel since we know that dengue viruses are repeatedly and frequently transported between population centers of the tropics by travelers.[159] Although implementing more strict quarantine regulations might help, it is unlikely that the international cooperation necessary to make quarantine really effective could be organized. Additionally, this would not prevent epidemic transmission in those areas where multiple serotypes are endemic. A third option, immunization, is discussed in more detail below. That leaves only surveillance (see Section VII) and mosquito control as the main lines of defense against epidemic dengue and DHF.

A. Mosquito Control

1. Background

Historically, control of *Ae. aegypti*, the principal vector of epidemic dengue, has been source reduction. The rationale was based on the fact that this highly domesticated species thrived in situations where water was stored or accumulated in artificial containers in the domestic environment. Thus, elimination of those artificial larval habitats would eliminate, or at least greatly reduce, the density of the species. Early programs using this labor-intensive approach were very effective.[162] More recently, larvicides such as temephos (Abate) formulated on sand granules have been very effective.

In the late 1960s, ultralow volume (ULV) application of insecticides was developed as an adulticide and tested against a number of mosquito species,[163,164] and since about 1969, ULV applied malathion has been the standard method for control of *Ae. aegypti* in most parts of the world. Unfortunately, little or no emphasis was given to development of new control measures and there were no new breakthroughs in the control of this species for over 15 years. Moreover, little or no emphasis was placed on source reduction using larvicides or environmental sanitation and this, plus development of resistance to insecticides by the mosquitoes and economic problems in many tropical countries of the world, led to a near breakdown in routine mosquito control for dengue. The result was reinfestation of many countries that had achieved eradication.[162]

2. Current Approaches

A major problem with using routine mosquito control for prevention of epidemic dengue is that to be effective, it must be continuous, especially during periods of low, silent, or no dengue transmission. During these interepidemic periods, however, most people, including control agencies, lose interest in mosquito control and as a result, mosquito population densities are allowed to increase. Equally unfortunate is that surveillance methods are usually too insensitive to detect increased dengue activity until the epidemic has reached near-peak transmission (see Section VII). By that time, it is too late to effectively intervene and control the epidemic by controlling mosquitoes.

A solution to this problem is to transfer the responsibility for *Ae. aegypti* control from government agencies to those who are responsible for the problem in the first place — the people themselves. A program for prevention and control of epidemic dengue and DHF which is currently being developed in Puerto Rico has five basic components. The first is to improve surveillance to the point where it becomes useful as an early warning system that has a predictive capability for epidemic dengue. If epidemics can be predicted, they can be prevented. Closely tied in with surveillance is the second component of the program: a rapid-response emergency vector control program. If the surveillance program is sensitive enough to detect new virus strains or serotypes shortly after introduction and well in advance of peak epidemic transmission, it is necessary to have an effective emergency vector control capability that can be implemented immediately. The objective of this component, therefore, is to prevent the spread of the epidemic to other areas. Emergency vector control should be fully integrated and combine appropriate adulticiding methods, targeted source reduction using both cleanup and insecticide approaches, community education on how the people can help control the mosquitoes, and on personal protective measures.

A third component of the program is education of the medical community to increase awareness of dengue disease and to train physicians in diagnosis and treatment of DHF.

Closely tied to the medical education program is the fourth component: an emergency hospitalization program designed to provide the most effective use of hospital and treatment facilities. This program should be adapted to each geopolitical region and should be population based, with local and central coordinating committees to oversee activities.

Ultimately, prevention of epidemic dengue and DHF will depend on the last component: effective, long-term mosquito control. To be cost effective, the program must be community based and integrated. Community involvement will depend on effective communication to the people that epidemics can be prevented and that it is their responsibility to help implement preventive measures. Disease prevention and control have historically been, and continue to be, responsibilities of government. To prevent epidemic dengue, however, all segments of society, including government, business, and community organizations, must be involved. Ultimately, the objective of the program should be to develop enforceable legislative controls patterned after the program that was so effective in Singapore.[149] In most democratic societies, however, such programs can be successful only after the public has been educated and accepts its responsibility for playing the principal role in prevention and control.

B. Dengue Vaccines

Considerable effort has been put into dengue vaccine development in the past several years. Unfortunately, little progress has been made. Most attention has been directed toward DEN-2 virus because that serotype is considered by some to be the most important in causing severe and fatal disease.[72,108] A live DEN-2 candidate vaccine (PR-159/S-1) was developed from a Puerto Rican strain of virus by attenuation in fetal rhesus monkey lung cells (DBS-FRhL-2).[166-168] It is temperature sensitive, produces uniform small plaques in cell culture, and has decreased mouse neurovirulence. Human trials have shown that only about 61% of recipients without previous flavivirus infection develop antidengue antibody compared with 90% in persons with previous yellow fever vaccination.[169] Clinically, recipients have presented with a variety of symptoms including low-grade fever, chills, abdominal pain, headache, night sweats, nausea, and anorexia.[169,170] The PR-159/S-1 vaccine strain of virus has been shown to produce viremia in human recipients, a potentially beneficial trait since the viruses isolated from vaccine recipients have shown no evidence of reversion to wild-type growth characteristics.[171,172]

More recently, a DEN-4-candidate vaccine, attenuated by passage in primary dog kidney cells, was prepared in fetal rhesus lung cells and tested in five human volunteers.[173-177] Only two of the recipients developed N antibody, and both of these persons had detectable viremia and symptoms compatible with mild dengue, including a rash. The virus recovered from the two recipients showed characteristics of the parental virus, suggesting reversion to the wild type.

Attenuation of DEN-1 and DEN-3 viruses has also been attempted, but results with monkeys and humans indicated lack of immunogenicity and/or attenuation. Current work on development of dengue vaccines is ongoing in at least three laboratories in the U.S., Thailand, and the People's Republic of China. Unfortunately, progress is slow and it is unlikely that a safe, immunogenic dengue vaccine will be available for general use in the near future.

X. FUTURE RESEARCH

A. Field Studies

Well-designed, long-term, prospective seroepidemiological studies are necessary to determine the risk factors associated with epidemic transmission and severe and fatal disease and to determine the relative importance of primary vs. secondary infection as a cause of DHF/DSS. Companion clinical-virological studies set up in the same areas can provide information on another important risk factor: strain variation among dengue viruses.

More detailed studies are needed to define the spectrum of illness associated with dengue infection. The disease should be characterized in different parts of the world, in association with different human populations and after infection with different strains and serotypes of dengue viruses. This is especially important in the American and African regions where hemorrhagic disease has not been common, but may become more frequent in the future.

More extensive research must be done to improve surveillance methods that will provide an early warning predictive capability for epidemic transmission of dengue and DHF. This includes virological, serological, epidemiological, clinical, and entomological surveillance as discussed in Section VII.

Field studies are required to determine whether rural or forest maintenance cycles for dengue viruses actually occur in endemic areas. Mosquito species, transovarial transmission, and subhuman primate involvement, in areas where they occur, should be studied. Equally important, the significance of such cycles in the distribution and spread of epidemic dengue must be determined.

B. Experimental Studies

The recent development of molecular technology for dengue viruses must be expanded and used to develop new, more rapid, sensitive, and specific diagnostic tests for dengue. Molecular characterization of the dengue virus genome should also lead to genetically engineered, quadrivalent vaccines that are economical and immunogenic.

New molecular technology should be used to study dengue virus strain differences and to characterize viruses transmitted in different geographic areas, in different epidemics, and in association with mild and severe disease. Studies should be done to determine whether there are epidemic and endemic strains of virus circulating in nature and how frequently and what kind of changes they undergo. Since it has been documented that dual infections occur in humans in nature,[178] the possibility of recombination should be studied. Also important are the types of pressures involved in natural selection of dengue viruses.

Finally, studies are necessary to determine the role of vector competence in dengue epidemiology and the role that vector/virus relationships of different mosquito species play in natural maintenance cycles.

REFERENCES

1. **Rush, B.**, An account of the bilious remitting fever, as it appeared in Philadelphia in the summer and autumn of the year 1780, in *Medical Inquiries and Observations*, Prichard & Hall, Philadelphia, 1789, 104.
2. **Sabin, A. B.**, Research on dengue during World War II, *Am. J. Trop. Med. Hyg.*, 1, 30, 1952.
3. **Hammon, W. McD., Rudnick, A., and Sather, G. E.**, Viruses associated with epidemic hemorrhagic fevers of the Philippines and Thailand, *Science*, 131, 1102, 1960.
4. **Westaway, E. G., Brinton, M. A., Gaidamovich, S. Ya., Horzinek, M. C., Igarashi, A., Kääriäinen, L., Lvov, D. K., Porterfield, J. S., Russell, P. K., and Trent, D. W.**, Flaviviridae, *Intervirology*, 24, 183, 1985.
5. **Theiler, M. and Downs, W. G.**, *The Arthropod-Borne Viruses of Vertebrates. An Account of the Rockefeller Foundation Virus Program, 1951 to 1970*, Yale University Press, New Haven, 1973.
6. **Graham, H.**, The dengue. A study of its pathology and mode of propagation, *J. Trop. Med. (London)*, 6, 209, 1903.
7. **Bancroft, T. L.**, On the etiology of dengue fever, *Aust. Med. Gaz. (Sydney)*, 25, 17, 1906.
8. **Siler, J. F., Hall, M. W., and Hitchens, A. P.**, Dengue, its history, epidemiology, mechanism of transmission, etiology, clinical manifestations, immunity and prevention, *Philipp. J. Sci.*, 29, 1, 1926.
9. **Simmons, J. S., St. John, J. H., and Reynolds, F. H. K.**, Experimental studies of dengue, *Philipp. J. Sci.*, 44, 1, 1931.

10. **Snijders, E. P., Dinger, E. J., and Schuffner, W. A. P.,** On the transmission of dengue in Sumatra, *Am. J. Trop. Med.*, 11, 171, 1931.
11. **Rosen, L., Rozeboom, L. E., Sweet, B. H., and Sabin, A. B.,** The transmission of dengue by *Aedes polynesiensis* Marks, *Am. J. Trop. Med. Hyg.*, 3, 878, 1954.
12. **Carey, D. E.,** Chikungunya and dengue: a case of mistaken identity?, *J. Hist. Med. Allied Sci.*, 26, 243, 1971.
13. **Ehrenkranz, N. J., Ventura, A. K., Cuadrado, R. R., and Pond, W. L., and Porter, J. E.,** Pandemic dengue in Caribbean countries and the southern United States — past, present and potential problems, *N. Engl. J. Med.*, 285, 1460, 1971.
14. **Halstead, S. B.,** Selective primary health care: strategies for control of disease in the developing world. XI. Dengue, *Rev. Infect. Dis.*, 6, 251, 1984.
15. **Anon.,** *Guide for Diagnosis, Treatment and Control of Dengue Hemorrhagic Fever*, 2nd ed., Technical Advisory Committee on DHF for the Southeast Asian and Western Pacific Regions, World Health Organization, Geneva, 1980.
16. **Barnes, W. J. S. and Rosen, L.,** Fatal hemorrhagic disease and shock associated with primary dengue infection on a Pacific Island, *Am. J. Trop. Med. Hyg.*, 23, 495, 1974.
17. **Guzman, M. G., Kouri, G. P., Bravo, J., Soler, M., Vázquez, S., Santos, M., Villaescusa, R., Basanta, P., Indan, G., and Ballester, J. M.,** Dengue haemorrhagic fever in Cuba. II. Clinical investigations, *Trans. R. Soc. Trop. Med. Hyg.*, 78, 239, 1984.
18. **Von Allmen, S. D., López-Correa, R. H., Woodall, J. P., Morens, D. M., Chiriboga, J., and Casta-Vélez, A.,** Epidemic dengue fever in Puerto Rico, 1977: a cost analysis, *Am. J. Trop. Med. Hyg.*, 28, 1040, 1979.
19. **Gubler, D. J.,** Dengue and dengue hemorrhagic fever in the Americas, in *Dengue Hemorrhagic Fever*, Thongeharoen, P., Ed., World Health Organization, New Delhi, in press.
20. **Monath, T. P.,** Socioeconomic impact of arbovirus diseases — a world view, in *Proc. 6th Arbovirus Symp.*, Brisbane, Australia, QIMR, 1986, 3.
21. **Armada Gessa, J. A. and Figueredo Gonzalez, R.,** Application of environmental management principles in the program for eradication of *Aedes (Stegomyia) aegypti* (Linneus, 1792) in the Republic of Cuba, 1984, *PAHO Bull.*, 20, 186, 1986.
22. **Clarke, D. H. and Casals, J.,** Techniques for hemagglutination and hemagglutination-inhibition with arthropod-borne viruses, *Am. J. Trop. Med. Hyg.*, 7, 561, 1958.
23. **Russell, P. K. and Nisalak, A.,** Dengue virus identification by the plaque reduction neutralization test, *J. Immunol.*, 99, 291, 1967.
24. **Henchal, E. A., Gentry, M. K., McCown, J. M., and Brandt, W. E.,** Dengue virus-specific and flavivirus group determinants identified with monoclonal antibodies by indirect immunofluorescence, *Am. J. Trop. Med. Hyg.*, 31, 830, 1982.
25. **Blok, J., Henchal, E. A., and Gorman, B. M.,** Comparison of dengue viruses and some other flaviviruses by cDNA-RNA hybridization analysis and detection of a close relationship between dengue virus serotype 2 and Edge Hill virus, *J. Gen. Virol.*, 65, 2173, 1984.
26. **Blok, J.,** Genetic relationships of the dengue virus serotypes, *J. Gen. Virol.*, 66, 1323, 1985.
27. **Rosen, L.,** Experimental infection of new world monkeys with dengue and yellow fever viruses, *Am. J. Trop. Med. Hyg.*, 7, 406, 1958.
28. **Whitehead, R. H., Yuill, T. M., Gould, D. J., and Simasathien, P.,** Experimental infection of *Aedes aegypti* and *Aedes albopictus* with dengue viruses, *Trans. R. Soc. Trop. Med. Hyg.*, 65, 661, 1971.
29. **Halstead, S. B., Shotwell, H., and Casals, J.,** Studies on the pathogenesis of dengue infection in monkeys. I. Clinical laboratory responses to primary infection, *J. Infect. Dis.*, 128, 7, 1973.
30. **Scherer, W. F., Russell, P. K., Rosen, L., Casals, J., and Dickerman, R. W.,** Experimental infection of chimpanzees with dengue viruses, *Am. J. Trop. Med. Hyg.*, 27, 590, 1978.
31. **Gubler, D. J., Reed, D., Rosen, L., and Hitchcock, J. C., Jr.,** Epidemiologic, clinical, and virologic observations on dengue in the Kingdom of Tonga, *Am. J. Trop. Med. Hyg.*, 27, 581, 1978.
32. **Gubler, D. J., Suharyono, W., Tan, R., Abidin, M., and Sie, A.,** Viraemia in patients with naturally acquired dengue infection, *Bull. WHO*, 59, 623, 1981.
33. **Gubler, D. J., Suharyono, W., Lubis, I., Eram, S., and Sulianti Saroso, J.,** Epidemic dengue hemorrhagic fever in rural Indonesia. I. Virological and epidemiological studies, *Am. J. Trop. Med. Hyg.*, 28, 701, 1979.
34. **Rudnick, A.,** The ecology of the dengue virus complex in Peninsular Malaysia, in *Proc. Int. Conf. Dengue/DHF*, Pang, T. and Pathmanathan, R., Eds. University of Malaysia Press, Kuala Lumpur, 1984, 7.
35. **Cordellier, R., Bouchité, B., Roche, J. C., Monteny, N., Diaco, B., and Akoliba, P.,** Circulation silvatique du virus Dengue 2 en 1980, dans les savannes sub-soudaniennes du Côte d'Ivoire, *Cah. ORSTOM Ser. Entomol. Méd. Parasitol.*, 21, 165, 1983.

36. **Gubler, D. J., Novak, R. J., Vergne, E., Colón, N. A., Vélez, M., and Fowler, J.,** *Aedes (Gymnometopa) mediovittatus* (Diptera:Culicidae), a potential maintenance vector of dengue viruses in Puerto Rico, *J. Med. Entomol.*, 22, 469, 1985.
37. **Freier, J. E. and Grimstad, P. R.,** Transmission of dengue virus by orally infected *Aedes triseriatus, Am J. Trop. Med. Hyg.*, 32, 1429, 1983.
38. **Xue-dong, L. and Gui-fang, Z.,** Studies on the experimental transmission of dengue virus by *Culex fatigens, Chin. J. Microbiol. Immunol.*, 5, 247, 1985.
39. **Rosen, L., Rozeboom, L. E., Gubler, D. J., Lien, J. C., and Chaniotis, B. N.,** Comparative susceptibility of various species and strains of mosquitoes to oral and parenteral infection with dengue and Japanese encephalitis viruses, *Am. J. Trop. Med. Hyg.*, 34, 603, 1985.
40. **Moreau, J., Rosen, L., Saugrain, J., and Lagraulet, J.,** An epidemic of dengue on Tahiti associated with hemorrhagic manifestations, *Am. J. Trop. Med. Hyg.*, 22, 237, 1973.
41. **Maguire, T., Miles, J. A. R., MacNamara, F. N., Wilkinson, P. J., Austin, F. J., and Mataika, J. U.,** Mosquito-borne infections in Fiji. V. The 1971-72 dengue epidemic, *J. Hyg. (Cambridge)*, 73, 263, 1974.
42. **Loison, G., Rosen, L., Papillaud, J., Tomasini, J., Vaujany, J., and Chanalet, G.,** La dengue en Nouvelle-Calédonie (1971-1972), *Bull. Soc. Pathol. Exot.*, 66, 511, 1973.
43. **Rosen, L.,** The emperor's new clothes revisited, or reflections on the pathogenesis of dengue hemorrhagic fever, *Am. J. Trop. Med. Hyg.*, 26, 337, 1977.
44. **Gubler, D. J., Suharyono, W., Lubis, I., Eram, S., and Gunarso, S.,** Epidemic dengue 3 in Central Java, associated with low viremia in man, *Am. J. Trop. Med. Hyg.*, 30, 1094, 1981.
45. **Russell, P. K. and McCown, J. M.,** Comparison of dengue-2 and dengue-3 virus strains by neutralization tests and identification of a subtype of dengue-3, *Am. J. Trop. Med. Hyg.*, 21, 97, 1972.
46. **Henchal, E. A., Repik, P. M., and McCown, J. M., and Brandt, W. E.,** Identification of an antigenic and genetic variant of dengue 4 virus from the Caribbean, *Am. J. Trop. Med. Hyg.*, 35, 393, 1986.
47. **Gubler, D. J. and Rosen, L.,** Quantitative aspects of replication of dengue viruses in *Aedes albopictus* (Diptera:Culicidae) after oral and parenteral infection, *J. Med. Entomol.*, 13, 469, 1977.
48. **Vezza, A. C., Rosen, L., Repik, P., Dalrymple, J., and Bishop, D. H. L.,** Characterization of the viral RNA species of prototype dengue viruses, *Am. J. Trop. Med. Hyg.*, 29, 643, 1980.
49. **Trent, D. W., Grant, J. A., Rosen, L., and Monath, T. P.,** Genetic variation among dengue 2 viruses of different geographic origin, *Virology*, 128, 271, 1983.
50. **Repik, P. M., Dalrymple, J. M., Brandt, W. E., McCown, J. M., and Russell, P. K.,** RNA fingerprinting as a method for distinguishing dengue 1 virus strains, *Am. J. Trop. Med. Hyg.*, 32, 577, 1983.
51. **Kerschner, J. H., Vorndam, A. V., Monath, T. P., and Trent, D. W.,** Genetic and epidemiologic studies of dengue type 2 viruses by hibridization using synthetic deoxyoligonucleotides as probes, *J. Gen. Virol.*, 67, 2645, 1986.
52. **Monath, T. P., Wands, J. R., Hill, L. J., Brown, N. V., Marciniak, R. A., Wong, M. A., Gentry, M. K., Burke, D. S., Grant, J. A., and Trent, D. W.,** Geographic classification of dengue-2 virus strains by antigen signature analysis, *Virology*, 154, 313, 1986.
53. **Yuill, T. M., Sukhavachana, P., Nisalak, A., and Russell, P. K.,** Dengue-virus recovery by direct and delayed plaques in LLC-MK_2 cells, *Am. J. Trop. Med. Hyg.*, 17, 441, 1968.
54. **Rosen, L. and Gubler, D. J.,** The use of mosquitoes to detect and propagate dengue viruses, *Am. J. Trop. Med. Hyg.*, 23, 1153, 1974.
55. **Kuberski, T. T. and Rosen, L.,** A simple technique for the detection of dengue antigen in mosquitoes by immunofluorescence, *Am. J. Trop. Med. Hyg.*, 26, 533, 1977.
56. **Thet-Win,** Detection of dengue virus by immunofluorescence after intracerebral inoculation of mosquitoes, *Lancet*, 1(8262), 53, 1982.
57. **Lam, S. K., Pang, T., Chew, C. B., Poon, G. K., and Ramalingam, S.,** The use of *Toxorhynchites* larvae in the detection of dengue viruses, in *Proc. Int. Conf. Dengue/DHF*, Pang, T. and Pathmanathan, R., Eds., University of Malaysia Press, Kuala Lumpur, 1984, 446.
58. **Igarashi, A.,** Isolation of Singh's *Aedes albopictus* cell clone sensitive to dengue and chikungunya viruses, *J. Gen. Virol.*, 40, 530, 1978.
59. **Varma, M. G. R., Pudney, M., and Leake, C. J.,** Cell lines from larvae of *Aedes (Stegomyia) malayensis* Colless, and *Aedes (S.) pseudoscutellaris* (Theobald) and their infection with some arboviruses, *Trans. R. Soc. Trop. Med. Hyg.*, 68, 374, 1974.
60. **Kuno, G.,** Dengue virus replication in a polyploid mosquito cell culture grown in serum-free medium, *J. Clin. Microbiol.*, 16, 851, 1982.
61. **Tesh, R. B.,** A method for the isolation and identification of dengue viruses using mosquito cell cultures, *Am. J. Trop. Med. Hyg.*, 28, 1053, 1979.
62. **Kuno, G., Gubler, D. J., Vélez, M., and Oliver, A.,** Comparative sensitivity of three mosquito cell lines for isolation of dengue viruses, *Bull. WHO*, 63, 279, 1985.

63. **Leake, C. J., Nisalak, A., and Burke, D. S.,** Comparative isolation of dengue viruses from DHF patients by mosquito inoculation and on three mosquito cell lines, in *Proc. Int. Conf. Dengue/DHF*, Pang, T. and Pathmanathan, R., Eds., University of Malaysia Press, Kuala Lumpur, 1984, 437.
64. **Kuberski, T. T. and Rosen, L.,** Identification of dengue viruses using complement fixing antigen produced in mosquitoes, *Am. J. Trop. Med. Hyg.*, 23, 1153, 1977.
65. **Gentry, M. K., Henchal, E. A., McCown, J. M., Brandt, W. E., and Dalrymple, J. M.,** Identification of distinct determinants on dengue 2 virus using monoclonal antibodies, *Am. J. Trop. Med. Hyg.*, 31, 548, 1982.
66. **Gubler, D. J., Kuno, G., Sather, G. E., Vélez, and Oliver, A.,** Use of mosquito cell cultures and specific monoclonal antibodies for routine surveillance of dengue viruses, *Am. J. Trop. Med. Hyg.*, 33, 158, 1984.
67. **Gubler, D. J.,** Application of serotype specific monoclonal antibodies for identification of dengue viruses, in *Arbovirus Cultivation in Arthropod Cells in Culture*, Yunker, C., Ed., CRC Press, Boca Raton, Fla., 1987.
68. **Kuno, G., Gubler, D. J., and Santiago de Weil, N.,** Antigen capture ELISA for the identification of dengue viruses, *J. Virol. Methods*, 12, 93, 1985.
69. **Halstead, S. B.,** Observations related to pathogenesis of dengue hemorrhagic fever. VI. Hypotheses and discussion, *Yale J. Biol. Med.*, 42, 350, 1970.
70. **Nimmanitya, S.,** Clinical spectrum and management of dengue hemorrhagic fever, in *Proc. Int. Conf. Dengue/DHF*, Pang, T. and Pathmanathan, R., Eds., University of Malaysia Press, Kuala Lumpur, 1984, 34.
71. **Sumarmo, S. P. S., Wulur, H., Jahja, E., Gubler, D. J., Suharyono, W., and Sorensen, K.,** Clinical observations on virologically confirmed fatal dengue infections in Jakarta, Indonesia, *Bull. WHO*, 61, 693, 1983.
72. **Halstead, S. B.,** Different dengue syndromes — the perspective from a pathogenetic point of view, *Asian J. Infect. Dis.*, 29, 59, 1978.
73. **Sumarmo, S. P. S.,** Demam Berdarah Dengue Pada Anak di Jakarta, Ph.D. thesis, University of Indonesia, Djakarta, 1983.
74. **Eram, S., Setyabudi, Y., Sadono, T. I., Sutrisno, D. S., Gubler, D. J., and Sulianti Saroso, J.,** Epidemic dengue hemorrhagic fever in rural Indonesia. II. Clinical studies, *Am. J. Trop. Med. Hyg.*, 28, 711, 1979.
75. **Venzon, E. L., Rudnick, A., Marchette, N. J., Fabie, A. E., and Dukellis, E.,** The greater Manila hemorrhagic fever epidemic of 1966, *Philipp. J. Sci.*, 48, 297, 1972.
76. **Chan, Y. C., Lim, K. A., and Ho, B. C.,** Recent epidemics of hemorrhagic fever in Singapore, *Jpn. J. Med. Sci. Biol.*, 20, 81, 1967.
77. **Halstead, S. B., Voulgaropoulos, E., Thien, N. H., and Udomsakdi, S.,** Dengue hemorrhagic fever in South Vietnam: report of the 1963 outbreak, *Am. J. Trop. Med. Hyg.*, 14, 819, 1965.
78. **Thaung, U., Ming, C. K., and Thein, M.,** Dengue hemorrhagic fever in Burma, *Southeast Asian J. Trop. Med. Public Health*, 6, 580, 1975.
79. **Gubler, D. J., Kuno, G., and Waterman, S. H.,** Neurologic disorders associated with dengue infection, in *Proc. Int. Conf. Dengue/DHF*, Pang, T. and Pathmanathan, R., Eds., University of Malaysia Press, Kuala Lumpur, 1984, 290.
80. **Sumarmo, S. P. S., Wulur, H., Jahja, E., Gubler, D. J., Sutomeggolo, T. S., and Sulianti Saroso, J.,** Encephalopathy associated with dengue infection, *Lancet*, 1, 449, 1978.
81. **George, R. and Duraisamy, G.,** Bleeding manifestations of dengue hemorrhagic fever in Malaysia, *Acta Trop.*, 38, 71, 1978.
82. **Thaung, U.,** Panel and general discussion in "Conference on DHF: New Developments and Future Research, Singapore, 1977", *Asian J. Infect. Dis.*, 2, 78, 1978.
83. **Nimmannitya, S. and Thisyakorn, U.,** Dengue hemorrhagic fever with unusual manifestations, in *Proc. Int. Conf. Dengue/DHF*, Pang, T. and Pathmanathan, R., Eds., University of Malaysia Press, Kuala Lumpur, 1984, 268.
84. **Mendoza, H., Jimenez, J. R., Gubler, D. J., and Feris, J.,** Infecciones virales asociadas a encefalitis en niños, *Arch. Dom. Pediatr.*, 20, 83, 1984.
85. **Bhamarapravati, N., Tuchinda, P., and Boonyapaknavik, V.,** Pathology of Thailand haemorrhagic fever: a study of 100 autopsy cases, *Ann. Trop. Med. Parasitol.*, 61, 500, 1967.
86. **Casey, H. L.,** Standardized Diagnostic Complement Fixation Method and Adaptation to Micro-test, Public Health Monogr. No. 74, U.S. Government Printing Office, Washington, D.C., 1965.
87. **Dulbecco, R., Vogt, M., and Strickland, A. G. R.,** A study of the basic aspects of neutralization of two animal viruses, western equine encephalitis virus and poliomyelitis virus, *Virology*, 2, 162, 1956.
88. **Burke, D. S., Nisalak, A., and Ussery, M. A.,** Antibody capture immunoassay detection of Japanese encephalitis virus immunoglobulin M and G antibodies in cerebrospinal fluid, *J. Clin. Microbiol.*, 15, 1034, 1982.

89. **Halstead, S. B.,** Etiologies of the experimental dengues of Siler and Simmons, *Am. J. Trop. Med. Hyg.*, 23, 974, 1974.
90. **Halstead, S. B., Rojanasuphot, S., and Sangkawibha, N.,** Original antigenic sin in dengue, *Am. J. Trop. Med. Hyg.*, 32, 154, 1983.
91. **Monath, T. P., Wands, J. R., Hill, L. J., Gentry, M. K., and Gubler, D. J.,** Multisite monoclonal immunoassay for dengue viruses: detection of viraemic human sera and interference by heterologous antibody, *J. Gen. Virol.*, 67, 639, 1986.
92. **Richman, D. D., Cleveland, P. H., Redfield, D. C., Oxman, M. N., and Wahl, G. M.,** Rapid viral diagnosis, *J. Infect. Dis.*, 149, 298, 1984.
93. **Rosen, L.,** Dengue — an overview, in *Viral Diseases in Southeast Asia and West Pacific*, Mackenzie, J. S., Eds., Academic Press, Sydney, 1982, 484.
94. **Carey, D. E., Causey, O. R., Reddy, S., and Cooke, A. R.,** Dengue viruses from febrile patients in Nigeria, 1964 to 1968, *Lancet*, 1, 105, 1971.
95. **Fagbami, A. H. and Fabiyi, A.,** Epidemiology of dengue infections in Nigeria: virus isolations and clinical observations, 1972 to 1975, *J. Trop. Med. Hyg.*, 79, 226, 1976.
96. **Johnson, B. K., Ocheng, D., Gichogo, A., Okiro, M., Libondo, D., Kinyanjui, P., and Tukey, P. M.,** Epidemic dengue fever caused by dengue type 2 virus in Kenya: preliminary results of human virological and serological studies, *East Afr. Med. J.*, 59, 781, 1982.
97. **Saleh, A. S., Hassan, A., Scott, R. McN., Mellick, P. W., Oldfield, E. C., III, and Podgore, J. K.,** Dengue in Northeast Africa, *Lancet*, 2, 211, 1985.
98. **Saluzzo, J. F., Cornet, M., Castagnet, P., Rey, C., and Digoutte, J. P.,** Isolation of dengue 2 and dengue 4 viruses from patients in Senegal, *Trans. Soc. Trop. Med. Hyg.*, 80, 5, 1986.
99. **Gubler, D. J., Sather, G. E., Kuno, G., and Cabral, J. R.,** Dengue 3 virus transmission in Africa, *Am. J. Trop. Med. Hyg.*, 35, 1280, 1986.
100. **Halstead, S. B.,** Dengue haemorrhagic fever — a public health problem and a field for research, *Bull. WHO*, 58, 1, 1980.
101. **Watts, D. M., Burke, D. S., Harrison, B. A., Whitmire, R. E., and Nisalak, A.,** Effect of temperature on the vector efficiency of *Aedes aegypti* for dengue 2 virus, *Am. J. Trop. Med. Hyg.*, 36, 143, 1987.
102. **Jumali, S., Gubler, D. J., Nalim, S., Eram, S., and Sulianti Saroso, J.,** Epidemic dengue hemorrhagic fever in rural Indonesia. III. Entomological studies, *Am. J. Trop. Med. Hyg.*, 28, 717, 1979.
103. **Rice, L.,** Dengue fever: preliminary report of an epidemic at Galveston, *Tex. State J. Med.*, 18, 217, 1922.
104. **Hayes, G. R., Jr., Scheppf, P. P., and Johnson, E. B.,** An historical review of the last continental U.S. epidemic of dengue, *Mosq. News*, 31, 422, 1971.
105. **Hafkin, B., Kaplan, J. E., Reed, C., Elliott, L. B., Fontaine, R., Sather, G. E., and Kappus, K.,** Reintroduction of dengue fever into the Continental United States, *Am. J. Trop. Med. Hyg.*, 31, 1222, 1982.
106. **Gubler, D. J., Suharyono, W., Sumarmo, S. P. S., Wulur, H., Jahja, E., and Sulianti Saroso, J.,** Virological surveillance for dengue haemorrhagic fever in Indonesia using the mosquito inoculation technique, *Bull. WHO*, 57, 931, 1979.
107. **Gubler, D. J.,** Current research on dengue, in *Current Topics in Vector Research*, Vol. 3, Harris, K. F., Ed., Springer-Verlag, New York, 1987, 37.
108. **Sangkawibha, N., Rojanasuphot, S., Abandrik, S., Viryaponse, S., Jatanasen, S., Salitul, V., Phanthumchindas, B., and Halstead, S. B.,** Risk factors in dengue shock syndrome. A prospective epidemiological study in Rayong, Thailand. I. The 1980 outbreak, *Am. J. Epidemiol.*, 120, 653, 1984.
109. **Guzmán, M. G., Kouri, G. P., Bravo, J., Calunga, M., Soler, M., Vázquez, S., and Venereo, C.,** Dengue hemorrhagic fever in Cuba. I. Serological confirmation of clinical diagnosis, *Trans. R. Soc. Trop. Med. Hyg.*, 78, 235, 1984.
110. **Sumarmo, S. P. S., Hoffman, S. L., Burke, D. S., Converse, J. D., and Punjabi, N. H.,** Clinical and virological observations in nine infants with dengue haemorrhagic fever, Jakarta, 1981, in *Proc. Int. Conf. Dengue/DHF*, Pang, T. and Pathmanathan, R., Eds., University of Malaysia Press, Kuala Lumpur, 1984, 281.
111. **Halstead, S. B.,** The pathogenesis of dengue. Molecular epidemiology in infectious disease (the Alexander D. Langmuir Lecture), *Am. J. Epidemiol.*, 114, 632, 1981.
112. **Rosen, L.,** The pathogenesis of dengue hemorrhagic fever — critical appraisal of current hypotheses, *SAMJ Suppl.*, 11 October, 40, 1986. *Proc. 1st Int. Semin. DHF Am.*, San Juan, Puerto Rico, in press.
113. **Scott, R. McN., Nimmannitya, S., Bancroft, W. H., and Mansuwan, P.,** Shock syndrome in primary dengue infections, *Am. J. Trop. Med. Hyg.*, 25, 866, 1976.
114. **Hotta, H., Hotta, S., Takada, H., Kotani, S., Tanaka, S., and Ohki, M.,** Enhancement of dengue virus type 2 replication in mouse macrophage cultures by bacterial cell walls, peptidoglycans and a polymer of peptidoglycan subunits, *Infect. Immunol.*, 41, 462, 1983.

115. **Wiharta, A. S., Hotta, H., Hotta, S., Sato, Y., and Sato, H.,** Enhanced production of dengue virus in mouse peritoneal macrophage cultures treated with highly purified *Bordetella pertussis* toxin, *ICMR Ann.*, 3, 61, 1983.
116. **Pavri, K. M. and Prasad, S. R.,** T suppressor cells: role in dengue hemorrhagic fever and dengue shock syndrome, *Rev. Infect. Dis.*, 2, 142, 1980.
117. **Kuberski, T., Rosen, L., Reed, D., and Mataika, J.,** Clinical and laboratory observations on patients with primary and secondary dengue type 1 infections with hemorrhagic manifestations in Fiji, *Am. J. Trop. Med. Hyg.*, 26, 775, 1977.
118. **Halstead, S. B., Shotwell, H., and Casals, J.,** Studies on the pathogenesis of dengue infection in monkeys. II. Clinical laboratory responses to heterologous infection, *J. Infect. Dis.*, 128, 15, 1973.
119. **Marchette, N. J., Halstead, S. B., Falkler, W. A., Jr., Stenhouse, A., and Nash, D.,** Studies on the pathogenesis of dengue infection in monkeys. III. Sequential distribution of virus in primary and heterologous infections, *J. Infect. Dis.*, 128, 23, 1973.
120. **Halstead, S. B.,** *In vivo* enhancement of dengue virus infection in rhesus monkeys by passively transferred antibody, *J. Infect. Dis.*, 140, 527, 1979.
121. **Venzon, E. L., Rudnick, A., Marchette, N. J., et al.,** The greater Manila dengue hemorrhagic fever epidemic of 1966, *Philipp. J. Sci.*, 48, 297, 1972.
122. **Thaung, U.,** Dengue haemorrhagic fever in Burma, *Asian J. Infect. Dis.*, 2, 23, 1978.
123. **Fang, R., Lo, E., and Lim, T. W.,** The 1982 dengue epidemic in Malaysia: epidemiological, serological and virological aspects, *Southeast Asian J. Trop. Med. Public Health*, 15, 51, 1984.
124. **Vitarana, T. and Jayasekera, N.,** A study of dengue in a low DHF area — Sri Lanka, in *Proc. Int. Conf. Dengue/DHF*, Pang, T. and Pathmanathan, R., Eds., University of Malaysia Press, Kuala Lumpur, 1984, 103.
125. **Smith, T. J., Winter, P. E., Nisalak, A., and Udomsakdi, S.,** Dengue control on an island in the Gulf of Thailand, *Am. J. Trop. Med. Hyg.*, 20, 715, 1971.
126. **Nimmannitya, S., Halstead, S. B., Cohen, S. N., and Margiotta, M. R.,** Dengue and chikungunya virus infection in man in Thailand, 1962 to 1964. I. Observations on hospitalized patients with hemorrhagic fever, *Am. J. Trop. Med. Hyg.*, 17, 954, 1969.
127. **Nisalak, A., Halstead, S. B., Singharaj, P., Udomsakdi, S., Nye, S. W., and Vinijchaikul, K.,** Observations related to pathogenesis of dengue hemorrhagic fever. III. Virologic studies of fatal disease, *Yale J. Biol. Med.*, 42, 293, 1970.
128. **Dizon, J. J.,** The occurrence of dengue hemorrhagic fever in the Philippines, *Asian J. Infect. Dis.*, 2, 15, 1978.
129. **Sumarmo, S. P. S., Nassar, S. S., and Suparman, A. K.,** Dengue shock syndrome in protein calorie malnutrition (PCM) children, in *Proc. Conf. DHF, Current Knowledge*, Bangkok, 1976, 18.
130. **Halstead, S. B., Scanlon, J. E., Umpaivit, P., and Udomsakdi, S.,** Dengue and chikungunya virus infection in man in Thailand, 1962-1964. IV. Epidemiologic studies in the Bangkok metropolitan area, *Am. J. Trop. Med. Hyg.*, 18, 997, 1969.
131. **Chiewsilp, P., Scott, R. McN., and Bhamarapravati, N.,** Histocompatibility antigens and dengue hemorrhagic fever, *Am. J. Trop. Med. Hyg.*, 30, 1100, 1981.
132. **Rudnick, A.,** Studies of the ecology of dengue in Malaysia: a preliminary report, *J. Med. Entomol.*, 2, 203, 1965.
133. **Rudnick, A.,** Ecology of dengue virus, *Asian J. Infect. Dis.*, 2, 156, 1978.
134. **Anon.,** Dengue hemorrhagic fever in the Democratic Republic of Viet Nam, in *Dengue Newsletter, Southeast Asian and Western Pacific Regions*, Vol. 2, WHO, Geneva, 1976, 1.
135. **Monath, T. P., Lee, V. H., Wilson, D. C., Fagbami, A., and Tomori, O.,** Arbovirus studies in Nupeko Forest, a possible natural focus of yellow fever virus in Nigeria. I. Description of the area and serological survey of humans and other vertebrate hosts, *Trans. R. Soc. Trop. Med. Hyg.*, 68, 30, 1974.
136. **Fagbami, A. H., Monath, T. P., and Fabiyi, A.,** Dengue virus infections in Nigeria: a survey for antibodies in monkeys and humans, *Trans. R. Soc. Trop. Med. Hyg.*, 71, 60, 1977.
137. **Roche, J. C., Cordellier, R., Hervy, J. P., Digoutte, J. P., and Monteny, N.,** Isolement de 96 souches de virus dengue 2 a partir de moustiques capturés en Côte d'Ivoire et Haute-Volta, *Ann. Virol.*, 134E, 233, 1983.
138. **Roberts, D. R., Peyton, E. L., Pinheiro, F. P., Balderrama, F., and Vargas, R.,** Associations of arbovirus vectors with gallery forests and domestic environments in southeastern Bolivia, *Bull. PAHO*, 18, 337, 1984.
139. **Rosen, L., Shroyer, D. A., Tesh, R. B., Freier, J. E., and Lien, J. C.,** Transovarial transmission of dengue viruses by mosquitoes: *Aedes albopictus* and *Aedes aegypti*, *Am. J. Trop. Med. Hyg.*, 32, 1108, 1983.
140. **Khin, M. M. and Than, K. A.,** Transovarial transmission of dengue 2 virus by *Aedes aegypti* in nature, *Am. J. Trop. Med. Hyg.*, 32, 590, 1983.

141. **Hull, B., Tikasingh, E., de Souza, M., and Martínez, R.,** Natural transovarial transmission of dengue 4 virus in *Aedes aegypti* in Trinidad, *Am. J. Trop. Med. Hyg.*, 33, 1248, 1984.
142. **Watts, D. M., Harrison, B. A., Pantuwatana, S., Klein, T. A., and Burke, D. S.,** Failure to detect natural transovarial transmission of dengue viruses by *Aedes aegypti* and *Aedes albopictus* (Diptera:*Culicidae), J. Med. Entomol.*, 22, 261, 1985.
143. **Gubler, D. J. and Rosen, L.,** Variation among geographic strains of *Aedes albopictus* in susceptibility to infection with dengue viruses, *Am. J. Trop. Med. Hyg.*, 25, 318, 1976.
144. **Gubler, D. J., Nalim, S., Tan, R., Saipan, H., and Sulianti Saroso, J.,** Variation in susceptibility to oral infection with dengue viruses among geographic strains of *Aedes aegypti, Am. J. Trop. Med. Hyg.*, 28, 1045, 1979.
145. **Nalim, S., Gubler, D. J., Basuno, E., Suwasono, H., Masran, M., and Djuarti, W.,** Studies on the susceptibility of a large urban population of *Aedes aegypti* to infection with dengue viruses, *Southeast Asian J. Trop. Med. Public Health*, 9, 494, 1978.
146. **Clements, A. N.,** *The Physiology of Mosquitoes*, Pergamon Press, New York, 1963, 185.
147. **Sheppard, P. M., MacDonald, W. W., Tonn, R. J., and Grab, B.,** The dynamics of an adult population of *Aedes aegypti* in relation to dengue hemorrhagic fever in Bangkok, *J. Anim. Ecol.*, 38, 661, 1969.
148. **Gubler, D. J.,** The ecology of *Aedes albopictus*, in *The Johns Hopkins University CMRT Annual Report*, JHU, Baltimore, 1970, 74.
149. **Chan, K. L.,** Singapore's dengue haemorrhagic fever control programme: a case study on the successful control of *Aedes aegypti* and *Aedes albopictus* using mainly environmental measures as a part of integrated vector control, SEAMIC Publ. No. 45, Tokyo, 1985.
150. **Christophers, S. R.,** *Aedes aegypti (L.), The Yellow Fever Mosquito: Its Life History, Bionomics and Structure*, Cambridge University Press, London, 1960.
151. **Ho, B. C., Chan, Y. C., and Chan, K. L.,** Field and laboratory observations on landing and biting periodicities of *Aedes albopictus* (Skuse), *Southeast Asian J. Trop. Med. Public Health*, 4, 238, 1973.
152. **Tempelis, C. H., Hayes, R. O., Hess, A. D., and Reeves, W. C.,** Blood-feeding habits of four species of mosquito found in Hawaii, *Am. J. Trop. Med. Hyg.*, 19, 335, 1970.
153. **Edman, J. D.,** Host-feeding patterns of Florida mosquitoes. I. *Aedes, Anopheles, Coquillettidia, Mansonia* and *Psorophora, J. Med. Entomol.*, 8, 687, 1971.
154. **Tabachnick, W. J. and Powell, J. R.,** A worldwide survey of genetic variation in the yellow fever mosquito, *Aedes aegypti, Genet. Res.*, 34, 215, 1979.
155. **Powell, J. R., Tabachnick, W. J., and Arnold, J.,** Genetics and the origin of a vector population: *Aedes aegypti*, a case study, *Science*, 208, 1385, 1980.
156. **Tabachnick, W. J., Aitken, T. H. G., Beaty, B. J., Miller, B. R., Powell, J. R., and Wallis, G. P.,** Genetic approaches to the study of vector competency in *Aedes aegypti*, in *Recent Developments in the Genetics of Insect Disease Vectors*, Steiner, W. W. M., Tabachnick, W. J., Rai, K. S., and Narang, S., Eds., Stipes, Champaign, Ill., 1982, 413.
157. **Gubler, D. J., Novak, R., and Mitchell, C. J.,** Arthropod vector competence — epidemiological, genetic, and biological considerations, in *Recent Developments in the Genetics of Insect Disease Vectors*, Steiner, W. W. M., Tabachnick, W. J., Rai, K. S., and Narang, S., Eds., Stipes, Champaign, Ill., 1982, 343.
158. **Wallis, G. P., Tabachnick, W. J., and Powell, J. R.,** Genetic heterogeneity among Caribbean populations of *Aedes aegypti, Am. J. Trop. Med. Hyg.*, 33, 492, 1984.
159. **Gubler, D. J.,** Dengue in the United States, 1983-1984. *Morbid. Mortal. Wkly Rep.*, 34, 5SS, 1985.
160. **Gahlinger, P. M., Reeves, W. C., and Milby, M. M.,** Air conditioning and television as protective factors in arboviral encephalitis risk, *Am. J. Trop. Med. Hyg.*, 35, 601, 1986.
162. **Schliessman, D. J. and Calheiros, L. B.,** A review of the status of yellow fever and *Aedes aegypti* eradication programs in the Americans, *Mosq. News*, 34, 1, 1974.
163. **Mount, G. A., Lofgren, C. S., Pierce, N. W., and Husman, C. N.,** Ultra-low volume nonthermal aerosols of malathion and naled for adult mosquito control, *Mosq. News*, 28, 99, 1968.
164. **Lofgren, C. S.,** Ultra-low volume applications of concentrated insecticides in medical and veterinary entomology, *Ann. Rev. Entomol.*, p. 321, 1970.
166. **Eckels, K. H., Brandt, W. E., Harrison, V. R., McCown, J. M., and Russell, P. K.,** Isolation of a temperature-sensitive dengue-2 virus under conditions suitable for vaccine development, *Infect. Immun.*, 14, 1221, 1976.
167. **Harrison, V. R., Eckels, K. H., Sagartz, J. W., and Russell, P. K.,** Virulence and immunogenicity of a temperature-sensitive dengue-2 virus in lower primates, *Infect. Immun.*, 18, 151, 1977.
168. **Eckels, K. H., Harrison, V. R., Summers, P. L., and Russell, P. K.,** Dengue-2 vaccine: preparation from a small-plaque virus clone, *Infect. Immun.*, 27, 175, 1980.
169. **Bancroft, W. H., Top, F. H., Jr., Eckels, K. H., Anderson, J. H., Jr., McCown, J. M., and Russell, P. K.,** Dengue-2 vaccine: virological, immunological, and clinical responses of six yellow fever-immune recipients, *Infect. Immun.*, 31, 698, 1981.

170. **Bancroft, W. H., Scott, R. McN., Eckels, K. H., Hoke, C. H., Jr., Simms, T. E., Jesrani, K. D. T., Summers, P. L., Dubois, D. R., Tsoulos, D., and Russell, P. K.,** Dengue virus type 2 vaccine: reactogenicity and immunogenicity in soldiers, *J. Infect. Dis.*, 149, 1005, 1984.
171. **Bancroft, W. H., Scott, R. McN., Brandt, W. E., McCown, J. M., Eckels, K. H., Hayes, D. E., Gould, D. J., and Russell, P. K.,** Dengue-2 vaccine: infection of *Aedes aegypti* mosquitoes by feeding on viremic recipients, *A. J. Trop. Med. Hyg.*, 31, 1229, 1982.
172. **Miller, B. R., Beaty, B. J., Aitken, T. H. G., Eckels, K. H., and Russell, P. K.,** Dengue-2 vaccine: oral infection, transmission, and lack of evidence for reversion in the mosquito, *Aedes aegypti*, *Am. J. Trop. Med. Hyg.*, 31, 1232, 1982.
173. **Halstead, S. B., Diwan, A. R., Marchette, N. J., Palumbo, N. E., and Srisukonth, L.,** Selection of attenuated dengue 4 viruses by serial passage in primary kidney cells. I. Attributes of uncloned virus at different passage levels, *Am. J. Trop. Med. Hyg.*, 33, 654, 1984.
174. **Halstead, S. B., Marchette, N. J., Diwan, A. R., Palumbo, N. E., and Putvatana, R.,** Selection of attenuated dengue 4 virus by serial passage in primary kidney cells. II. Attributes of virus cloned at different dog kidney passage levels, *Am. J. Trop. Med. Hyg.*, 33, 666, 1984.
175. **Halstead, S. B., Marchette, N. J., Diwan, A. R., Palumbo, N. E., Putvatana, R., and Larsen, L. K.,** Selection of attenuated dengue 4 viruses by serial passage in primary kidney cells. III. Reversion to virulence by passage of cloned virus in fetal rhesus lung cells, *Am. J. Trop. Med. Hyg.*, 33, 672, 1984.
176. **Halstead, S. B., Eckels, K. H., Putvatana, R., Larsen, L. K., and Marchette, N. J.,** Selection of attenuated dengue 4 viruses by serial passage in primary kidney cells. IV. Characterization of a vaccine candidate in fetal rhesus lung cells, *Am. J. Trop. Med. Hyg.*, 33, 679, 1984.
177. **Eckels, K. H., Scott, R. McN., Brancroft, W. H., Brown, J., Dubois, D. R., Summers, P. L., Russell, P. K., and Halstead, S. B.,** Selection of attenuated dengue 4 viruses by serial passage in primary kidney cells. V. Human response to immunization with a candidate vaccine prepared in fetal rhesus lung cells, *Am. J. Trop. Med. Hyg.*, 33, 684, 1984.
178. **Gubler, D. J., Kuno, G., Sather, G. E., and Waterman, S. H.,** A case of natural concurrent human infection with two dengue viruses, *Am. J. Trop. Med. Hyg.*, 34, 170, 1985.
179. **Gubler, D. J., and Rosen, L.,** unpublished data.
180. **Rosen, L. and Gubler, D. J.,** unpublished data.
181. **Trent, D. W. and Gubler, D. J.,** unpublished data.
182. **Trent, D. W.,** personal communication.
183. **Gubler, D. J. and Kuno, G.,** unpublished data.
184. **Gubler, D. J. and Bourdony, C.,** unpublished data.
185. **Sather, G. E., Gómez, I., and Gubler, D. J.,** unpublished data.
187. **Waterman, S. B.,** personal communication.
188. **Sangkawibha, N.,** personal communication.
189. **Loroño, M. A. and Gubler, D. J.,** unpublished data.
190. **Sumarmo, S. P. S. and Gubler, D. J.,** unpublished data.
191. **Vitarana, T. and Gubler, D. J.,** unpublished data.
192. San Juan Laboratories, Center for Disease Control, unpublished data.
193. **Gubler, D. J.,** unpublished data.

INDEX

A

B

C

I

J

K

L

P

Q

R

S

T

U

V

W

Y

Z